CW01034173

Praise for *Zero-Sum Victory*

"*Zero-Sum Victory,* the product of a rare combination of incisive scholarship and years of hard-won and occasionally bitter firsthand experience, is an essential guide to twenty-first-century western military strategy and its many dysfunctions. It is critical reading for anyone seeking to understand what went so wrong in Iraq and Afghanistan, and indeed, what might still go wrong now."—Simon Anglim, author of *Orde Wingate: Unconventional Warrior*

"War termination is among the most important responsibilities of the state. Yet the United States has systematically failed at it since 2001. In *Zero-Sum Victory,* Christopher Kolenda makes a landmark contribution toward improving the country's baleful record by highlighting the crucial importance of negotiation in ending wars. The book is essential reading for both policymakers and scholars."—Stephen Biddle, author of *Nonstate Warfare: The Military Methods of Guerillas, Warlords, and Militias*

"Few others have Christopher Kolenda's diversity of perspectives on war, and the framework he provides makes this book essential reading for understanding how bureaucratic silos and strategic narcissism led to our failures in Afghanistan and Iraq. He expertly lays the groundwork for a much-needed assessment of where we went wrong—and how to avoid those mistakes in future conflicts."—Jason K. Dempsey, author of *Our Army: Soldiers, Politics, and American Civil-Military Relations*

"*Zero-Sum Victory* is mandatory reading for any student or observer of the global war on terror. In this compelling and highly readable text, Christopher Kolenda draws on all aspects of his impressive career as a combat commander, strategic policy advisor, and an academic. His extensive hands-on experience provides authentic insights into what we are getting wrong about war. He highlights the absence of strategic consideration on a war termination framework, the mistaken conflation of military plans with national strategies, and the prevalence of confirmation bias masking emerging risks. As Kolenda argues, these errors were further compounded by the conduct of 'the bureaucratic way of war' that confused authorities, responsibilities, and accountabilities to such an extent that 'the whole became less than the sum of its parts.' Kolenda's coverage of the failure of Afghan peace negotiations is the most authoritative firsthand account available. *Zero-Sum Victory* concludes with a superb analysis of the implications for US foreign policy and scholarship."—Maj. Gen. Adam Findlay, AM, former Special Operations Commander, Australian Defence Force

"Christopher Kolenda, one of the great soldier-scholars of our time, has produced a pathbreaking new study on a critical policy question that has yet to be sufficiently answered: Why have American wars in the post-9/11 era experienced so much failure? Kolenda offers a rich and thoughtful analysis that will make a lasting contribution to the war studies field. *Zero-Sum Victory* is essential reading for subject specialists and the broader public alike."—Michael Kugelman, senior associate for South Asia, Woodrow Wilson Center

"Christopher Kolenda writes from a position of strength, for he was a man in the arena. War in all its forms has always been a very human affair. It is these failures of human enterprise that he shines a searchlight upon, and in so doing helps us understand the 'why' of how our best endeavors in Afghanistan and Iraq fell short." —Lt. Gen. Sir Graeme Lamb, KBE, CMG, DSO

"Christopher Kolenda has been a commander in combat, a diplomat, and a humanitarian trying to protect civilians from harm. Few are as qualified to write about America at war. In *Zero-Sum Victory,* he explains why we have had such trouble winning wars since 1945 and how we can do better. It is a groundbreaking work that is both analytically rigorous and informed by firsthand experience."—Carter Malkasian, author of *The American War in Afghanistan: A History*

"Christopher Kolenda's mesmerizing account of how US interventions in Afghanistan and Iraq became quagmires highlights a singular truth we ignore at our peril: the purpose of war is not to keep the game going in multiple overtimes until one side prevails, but rather to achieve an acceptable political outcome. He concludes with a set of practical but challenging recommendations for the structural changes needed in the bureaucracy to ensure more objective and better-integrated approaches to ending the inevitable irregular wars of the future. Policymakers, take note!"—Robin L. Raphel, US ambassador to Tunisia (1997–2000)

"In *Zero-Sum Victory,* Christopher Kolenda provides insightful and compelling answers to the question of why the United States wasn't able to 'win' in Vietnam, Iraq, and Afghanistan. His conclusion that the United States systematically defaulted to a goal of zero-sum victory in its unconventional wars is a powerful one for policymakers confronting the coming decades of challenges to US military power."—Jonathan Schroden, director of CNA's Countering Threats and Challenges Program

"Christopher Kolenda is a rarity: a US Army commander who fought in Afghanistan and then spent years jump-starting peace talks with the Taliban. His rigorous, unsparing assessment of what went wrong during the longest war in American history deserves widespread attention."—Craig Whitlock, author of *The Afghanistan Papers: A Secret History of the War*

Zero-Sum Victory

Zero-Sum Victory

What We're Getting Wrong about War

Christopher D. Kolenda

Scholarly publisher for the Commonwealth,
serving Bellarmine University, Berea College, Centre
College of Kentucky, Eastern Kentucky University,
The Filson Historical Society, Georgetown College,
Kentucky Historical Society, Kentucky State University,
Morehead State University, Murray State University,
Northern Kentucky University, Spalding University,
Transylvania University, University of Kentucky,
University of Louisville, and Western
Kentucky University.

Unless otherwise noted, illustrations are by Christopher D. Kolenda.

Editorial and Sales Offices: The University Press of Kentucky
663 South Limestone Street, Lexington, Kentucky 40508-4008
www.kentuckypress.com

Library of Congress Cataloging-in-Publication Data

Names: Kolenda, Christopher D., author.
Title: Zero-sum victory : what we're getting wrong about war / Christopher D. Kolenda.
Other titles: What we're getting wrong about war
Description: Lexington, Kentucky : The University Press of Kentucky, [2021] | Includes index.
Identifiers: LCCN 2021023493 | ISBN 9780813152769 (hardcover) | ISBN 9780813152837 (pdf) | ISBN 9780813152899 (epub)
Subjects: LCSH: United States—Military policy—History—21st century—Case studies. | Afghan War, 2001—Case studies. | Iraq War, 2003–2011—Case studies. | Prolonged war—United States—Case studies. | Irregular warfare—Case studies.
Classification: LCC UA23 .K65 2021 | DDC 355/.35573—dc23

This book is printed on acid-free paper meeting
the requirements of the American National Standard
for Permanence in Paper for Printed Library Materials.

Manufactured in the United States of America.

Member of the Association of
University Presses

To those who served.

Contents

List of Illustrations x
Glossary of Key Actors xi

Introduction 1
The Past as Prologue: The Vietnam War 18

Part I. Toward a War Termination Framework

1. Further Defining War Termination 25
2. The Decisive Victory Paradigm Undermines Strategy for Irregular War 31

Part II. The Pursuit of Decisive Victory in Afghanistan

3. Light Footprints to a Long War 43
4. Plans Hit Reality: A Recent History of Bad Neighbors and Worse Governance 49
5. The Fall of the Taliban and the Bonn Conference 57
6. America's Bureaucratic Way of War 67
   Conclusion to Part II 78

Part III. Persisting in a Failing Approach

7. Accelerating Success, 2003–2007 83
8. Failing to Keep Pace with the Insurgency, 2007–2009 92
9. The Good War Going Badly 98
10. Surging into the Good War 103
11. More Shovels in the Quicksand 108
12. Misapplying the Iraq Formula 111

13. Assessments and Risks 114
Conclusion to Part III 122

Part IV. Ending the War in Afghanistan

14. Reconciliation versus Transition 127
15. Reconciling Reconciliation 133
16. Competing Visions: Karzai, Taliban, and Pakistan 141
17. Exploratory Talks: Building and Damaging Confidence 148
18. Coming Off the Rails 155
19. Fallout: BSA, Bergdahl, and the 2014 Elections 160
Conclusion to Part IV 165

Part V. Pursuit of Decisive Victory in Iraq

20. Operation Iraqi Freedom: Plans without a Strategy 171
21. A Complicated Approach to a Complex Situation 180
22. From Decisive Victory to Transition 187
Conclusion to Part V 192

Part VI. Staying the Course in Iraq

23. Achieving Milestones While Losing the War 197
24. Trapped by Partners in a Losing Strategy 202
25. Mirror Imaging Civil-Military Relations 207
26. To Surge or Not to Surge: A Possible Win Beats a Certain Loss 210
27. A New Plan on Shaky Foundations 215
Conclusion to Part VI 218

Part VII. Ending the War in Iraq

28. The Surge Misunderstood 223
29. The Absence of a Political Strategy Erodes US Leverage 225
30. New Administration, Similar Challenges 231
Conclusion to Part VII 242

Part VIII. Implications

31. Iraq and Afghanistan Compared 249
32. Implications for US Foreign Policy 254

33. Implications for Scholarship 264

Abbreviations 273
Key Events in the Afghanistan Conflict 275
Key Events in the Iraq Conflict 281
Notes 285
Acknowledgments 369
Index 371

Illustrations

1. Critical Factors Framework 34
2. Perception of Corruption as a Problem in Afghanistan 87
3. Perception of Corruption as a Problem in Daily Life 88
4. US Troop Numbers in Afghanistan 104
5. US Public Optimism on the Afghanistan War 106
6. Enemy-Initiated Attacks, Afghanistan, 2009–2013 118
7. National Mood in Afghanistan 163
8. Perceived Fairness of 2005 Elections 189
9. Future Government Legitimacy 190
10. Annotated Enemy-Initiated Attacks, Iraq, 2004–2010 198
11. Iraqis' Satisfaction with Their Lives 232
12. Iraqis' Satisfaction with Their Country 233
13. Troop Numbers in Afghanistan and Iraq Compared 252

Glossary of Key Actors

Abdullah, Abdullah. Shura-e-Nazar; close companion of Ahmad Shah Massoud; former minister of foreign affairs under Hamid Karzai; runner-off in 2009, 2014, and 2019 presidential elections.

Agha, Tayyab. Mullah Omar's secretary; chief of the Taliban political commission, 2009–2015; led talks with the United States.

Allawi, Ayad. Shi'a and former Ba'athist; conspired against Saddam Hussein; named head of Iraqi National Accord Party; president of the Governing Council of Iraq in 2003, interim prime minister, 2004–2005; vice president, 2014–2015.

al Qaeda (AQ). A Sunni terrorist organization; founded by Osama bin Laden; based in Sudan, then fled to Afghanistan in 1996; orchestrated the 9/11 attacks; led by Ayman al-Zawahiri after bin Laden's death.

al Qaeda in Iraq (AQI). Founded in 1999 as "Tawhid wal-Jihad"; renamed after pledging allegiance to al Qaeda in 2004; grew into ISIS.

Ba'ath Party. Sunni Arab party founded by Saddam Hussein.

bin Laden, Osama. Founder of al Qaeda; fought during Soviet-Afghan War; orchestrated 9/11 attacks; killed by US special forces on May 1, 2011.

Blair, Tony. UK prime minister, 1997–2007.

Brahimi, Lakhdar. UN special envoy for Afghanistan, 2001–2004; chairperson of the 2001 Bonn Conference.

Bremer, L. Paul. US diplomat; led the Coalition Provisional Authority in Iraq, 2003–2004.

Bush, George W. 43rd US president, January 2001–January 2009.

Casey, George. US Army general; commander of MNF-I from 2004 to 2007; succeeded by David Petraeus.

Clinton, Hillary. US secretary of state, 2009–2013.

Crocker, Ryan. US ambassador to Syria, 1999–2001; acting US ambassador to Afghanistan, 2002; US ambassador to Pakistan, 2004–2007; US ambassador to Iraq, 2007–2009; US ambassador to Afghanistan, 2011–2012.

Dobbins, James. US ambassador to Afghanistan, 2001–2002; special representative for Afghanistan and Pakistan, 2013–2014.

Fahim, Mohammad Quassim. Formerly Ahmad Shah Massoud's deputy, led Northern Alliance after Massoud's death in 2001; vice president of Afghanistan from 2009 until his death in 2014.

Franks, Tommy. US Army general; CENTCOM commander, 2000–2003.

Gates, Robert. US secretary of defense, 2006–2011, covering both the Bush and Obama administrations.

Grossman, Marc. US special representative for Afghanistan and Pakistan, 2011–2012.

Haqqani network. Afghan insurgent group within the Taliban; designated by United States as a terrorist organization in 2012 for alleged ties to al Qaeda.

Hekmatyar, Gulbuddin. Pashtun; founded Hizb-i-Islami during Soviet-Afghan War; fought Afghan civil war, 1993–1996; as prime minister in 1996 he bombed Kabul; led HiG insurgency, 2004–2014.

Hill, Chris. US ambassador to Iraq, 2009–2010.

Hizb-i-Islami Gulbuddin (HiG). Founded in 1975 by Gulbuddin Hekmatyar; one of the Peshawar Seven during the Soviet-Afghan War; fought against ISA in Afghan civil war, 1993–1996; fought against United States and Afghan government, 2004–2014.

Holbrooke, Richard. US diplomat who served as ambassador to the United Nations, 1999–2001; special representative for Afghanistan and Pakistan from 2009 until he died in 2010.

Hussein, Saddam. Sunni; Iraqi president, 1979–2003; executed in 2006.

Islamic Emirate of Afghanistan. The official name of the Taliban.

Islamic State of Afghanistan. The so-called warlord government, 1992–1996.

Islamic State of Iraq and Syria (ISIS). Salafi terrorist organization that emerged in 2014 led by Abu Bakr al-Baghdadi; declared a caliphate in parts of Iraq and Syria, 2014–2017.

Jamiat-e-Islami. Founded by Burhanuddin Rabbani; supported by India.

Karzai, Hamid. Pashtun tribal leader; returned to Afghanistan during OEF to fight the Taliban; president of Afghanistan, 2004–2014.

Khalilzad, Zalmay. US special envoy for Bonn Conference; US ambassador to Afghanistan, 2003–2005; US Ambassador to Iraq, 2005–2007; appointed special representative for Afghanistan Reconciliation in 2018; negotiated February 2020 US-Taliban agreement.

al-Maliki, Nouri Kamal. Shi'a; Iraqi prime minister, 2006–2014.

Massoud, Ahmad Shah. Founded Shura-e-Nazar during the Soviet-Afghan War; led Northern Alliance against Taliban; killed by al Qaeda on September 9, 2001.

McChrystal, Stanley. US Army general; led the Joint Special Operations Command from 2003–2008; commanded ISAF, 2009–2010; succeeded by David Petraeus.

Northern Alliance. A military front against the Taliban government led by Ahmad Shah Massoud; partnered with the United States in 2001 during OEF.

Obama, Barack. 44th US president, January 2009–January 2017.

Odierno, Raymond. US Army general; succeeded Petraeus as MNF-I commander in 2008–2010.

Omar, Mullah Mohammad. Insurgent during Soviet-Afghan War; founded the Taliban in 1994; died in 2013.

Peshawar Seven. Alliance of seven mujahideen parties who fought against the Soviet-backed government in Afghanistan during the Soviet-Afghan War; six parties formed the Islamic State of Afghanistan.

Petraeus, David. US Army general; commander of MNF-I, 2007–2008; commander of the US Central Command, 2008–2010; commander of ISAF, 2010–2011.

Rabbani, Burhanuddin. Tajik; founded Jamiat-e-Islami during Soviet-Afghan War; president of Afghanistan, 1992–1996; killed September 2011.

Rice, Condoleezza. US national security advisor, 2001–2005; US secretary of state, 2005–2009.

Richards, Sir David. UK Army general; commander of ISAF, 2006–2007.

Rumsfeld, Donald. US secretary of defense, 2001–2006.

Sadrists. Followers of Muqtada al-Sadr.

al-Sadr, Muqtada. Shi'a. Founded Jaish al-Mahdi (JAM), opposed US presence in Iraq.

Shura-e-Nazar. "Supreme Council of the North"; created by Ahmad Shah Massoud in 1984; aligned with Jamiat during Soviet-Afghan War; de facto leader of the Northern Alliance.

Taliban. "Religious students" in Pashtu; founded by Mullah Mohammad Omar in 1994; ousted ISA in 1996; ruled Afghanistan, 1996–2001; harbored al Qaeda.

Talabani, Jalal. Sunni; founded Patriotic Union of Kurdistan to promote Kurdish rights and democracy in Iraq; Iraqi president, 2006–2014; died in 2017.

Tawhid wal-Jihad. See al Qaeda in Iraq.

al-Zarqawi, Abu Musab. Jordanian; founded AQI; killed by US special operations forces in 2006.

Introduction

Why have the major post-9/11 US military interventions turned into quagmires, and what can we learn from these conflicts about war termination and its role in policy and strategy? Despite huge power imbalances, major capacity-building efforts, and repeated tactical victories during the wars in Afghanistan (2001–present) and Iraq (2003–2011; 2014–present), the US military and its coalition partners have been unable to defeat, quickly and decisively, poorly trained and equipped insurgencies.

Former Joint Special Operations Command (JSOC) chief General Stanley A. McChrystal, reflecting on his fight against al Qaeda in Iraq, observed that the world's most elite counterterrorist force struggled initially "against a seemingly ragtag band of radical fighters." Despite the massive advantages in technology, resources, education, and training, he recalls, "things were slipping away from us."[1]

This study is important because the war termination expectations for major American military interventions (quick, zero-sum decisive victory) are very different than the results (quagmire), and the US national security community does not know the cause. Part of the challenge is that the outcomes in these wars are not solely in the hands of the United States. Adversaries, partners, allies, and voters, for instance, all make choices that affect the outcomes of the conflict—none of which the US National Security Council can fully control or influence.

I have undertaken this study to understand and isolate what appear to be systemic US policy and strategy errors in recent large-scale interventions against insurgencies. This volume focuses on the problems that are largely under the control or influence of the US National Security Council and that have played an important role in the significant deviation between expectations and results. Understanding these issues is an important step in undertaking reforms to address them.

A very common explanation is that the American military is not suited for such messy wars waged "amongst the people" within states.[2] The United States should thus avoid irregular wars against insurgents and nonstate actors, argues Lieutenant General (Ret.) Daniel Bolger, and concentrate on conventional state-on-state wars against uniformed militaries. Bolger blames the quagmires on the generals for their failure to develop a winning strategy and for misusing the armed forces in counterinsurgency campaigns for which they were ill-suited.[3] Bolger, to his credit, accepts his own responsibility for these problems. This sentiment is understandable but raises serious civil-military questions. The president might feel compelled to intervene militarily in what is or likely to become an irregular war—such as President Bush did after 9/11. Should the military be allowed to veto such a decision because *it doesn't do counterinsurgencies*?

A related argument is made by Brendan Gallagher, currently a serving US Army officer, in *The Day After: Why America Wins the War but Loses the Peace.* He blames American political leaders and the foreign policy establishment for setting unrealistic goals and for failing to plan for what happens the "day after" the military wins the war.[4] The military, however, was engaged every step of the way in the policy and strategy making for these conflicts and signed up for each one of them.

These points of view, moreover, reflect the conventional wisdom of the national security establishment that the military is responsible solely for winning wars. This assumption is well-expressed by political scientist William C. Martel in his work on victory in war. "Policymakers," he notes, "have the primary responsibility for determining what victory means . . . and how precisely the use of military force will meet those goals."[5] Then it is up to the generals to develop a strategy to deliver those objectives.[6] Most irregular wars, however, seem to elude military solutions.

A different argument is posited by Lieutenant General (Ret.) H. R. McMaster, President Trump's former national security adviser, who suggests that so-called forever wars are the new normal and that Americans need to show greater patience. Burden sharing, he argues, makes these wars manageable.[7] Instead of trying to bring wars to a successful conclusion, this argument suggests the United States should simply keep them going, as long as the costs are tolerable, in hopes that the country can wear down the adversaries until they are no longer a threat.

In this absence of ideas, the war in Afghanistan alone has racked up an estimated $1 trillion cost to the American taxpayer.[8] As of 2018, the war, with its smaller military footprint of about 14,000 American troops, is costing about $45 billion per year. Having troops tied down in the far corners of the world reduces America's ability to address arguably larger priorities, such as Social Security solvency, infrastructure, heath care, and education. It could also reduce the capacity to deal with larger threats.

There are some merits to the arguments made by Bolger, Martel, and McMaster, as well as some obvious challenges. However, there is a more central problem, which originates in a basic misunderstanding of war's purpose: that the military is to achieve a zero-sum decisive victory so that the United States clearly wins and gets all it wants, while the enemy clearly loses and gets nothing. The 2015 US *National Military Strategy,* for instance, states that "[t]he U.S. military's purpose is to protect our Nation and win our wars. We do this through military operations to defend the homeland, build security globally, and project power and *win decisively.*"[9] According to the US Army's *Doctrine Publication 1* (ADP 1), "The Army of 2028 will be ready to deploy, fight, and *win decisively* against any adversary, anytime and anywhere, in a joint, multi-domain, high-intensity conflict, while simultaneously deterring others and maintaining its ability to conduct irregular warfare."[10]

American college football is a zero-sum game: the game continues into multiple overtimes until one side wins the match. War is not college football. Its purpose is not to score points by crossing the goal line, as it were, in capturing an enemy capital or destroying an enemy conventional army. Nor is it to simply keep the game going into multiple overtimes, continuing the killing and human suffering, until one side wins and one side loses.

The national security establishment's focus on what is sometimes called "decisive victory" thus misses the point. The 19th-century Prussian military philosopher Carl von Clausewitz famously wrote that war is a means to political ends, hence the purpose is to achieve a political outcome, not to achieve military ends for their own sake. Real, durable, political outcomes from the use of force require a primary focus *not* on some clear-cut military victory, as desirable as that might be, but on stable political outcomes that address the political issues that gave rise to the fighting in the first place.

This means that real victory in war is usually a matter of harnessing military violence to a political strategy that will normally require some kind of negotiated settlement to end the fighting. It thus follows that war termination decisions should hinge on gaining durable policy outcomes, not on whether clear-cut military victory is achieved. The American obsession with decisive victory too often gets in the way of this kind of politico-military strategizing by substituting destruction for negotiation, and violence for politics. For wars that hinge on political legitimacy, as many irregular wars do, military-centric strategies may have low probabilities of success. The zero-sum decisive victory presumption thus establishes an impossible standard, and one that is inducing major policy and strategy errors.

In short, *the American national security establishment's assumptions about war and war termination seem dangerously misguided.* The belief in the military's centrality to waging war until a zero-sum decisive victory is achieved has limited the presidents' policy and strategy options, damaged America's reputation and strategic position, given the generals and admirals an inappropriately large voice in national security affairs, and heightened the risk of quagmires.

What Is War Termination?

This mindset has clear implications for what political scientists call war termination—the set of activities associated with the end of a conflict. Historian Lawrence Freedman describes strategy as "the best word we have for expressing attempts to think about our actions in advance, in light of our goals and capacities."[11] The quality of a strategy may be revealed in how well an actor finishes the war, not simply in how well it is fought. "If war is simply a 'grammar' in which the logic of political purpose is expressed in violent terms," explains historian Bradford Lee, "then it is in war termination that the purpose finally achieves its ultimate definition or refinement in political demands."[12]

The US Department of Defense, however, has no definition or doctrine for this seemingly critical aspect of war. Options other than decisive victory do not exist in the national security lexicon.[13] Scholars, too, have struggled to provide comprehensive guidance on how to approach war termination in irregular wars. They discuss war termination as a phase of activity nestled between the end of organized hostilities and the beginning of peace.[14] "*In the time between war and peace,*" writes historian Matthew Moten, "it is easy to lose sight of the objectives for which one embarked upon war in the first place, and to forfeit the grasp on accomplishments bought at great expense to the treasury and the lives and health of the nation's soldiery."[15] This view of war termination as a sort of transition from the military to the diplomats seems too narrow, even for conventional wars. A strategy that stops at the military objectives is short-sighted, Bradford Lee notes. The United States "has failed again and again to translate military success . . . into the most favorable and durable political results."[16]

Decisive victory, if achieved, makes the transition from military to diplomatic activity easy to comprehend. The fighting stops and the talking starts. This, however, is not the only way wars end. There are four broad outcomes for wars against insurgencies: decisive victory, negotiated settlement, transition-withdrawal, and decisive loss. As a consideration for policy and strategy, this work defines an intended war termination outcome as *how the United States seeks to achieve a favorable and durable result that meets policy aims at acceptable cost.*

Bradford Lee found that throughout the wars of the 20th century US political and military leaders rarely, if ever, properly considered war termination during the policy and strategy process.[17] Lack of forethought about war termination can lead to decisions that undermine the ability to achieve a favorable and durable outcome. During the 1991 Gulf War, Chairman of the Joint Chiefs of Staff General Colin L. Powell waited until the Iraqi Army was in headlong retreat before discussing war termination. When he offered to bring a recommendation to President George H. W. Bush the next day, the latter responded, "If that is the case . . . why not end it today?"[18] Michael Gordon and Bernard Trainor discuss how this shortfall undermined America's ability to capitalize on such a stunning battlefield victory.[19] This lack of thinking through the endgame is part of the reason the United States, in political scientist Gideon Rose's words, tends to "trip across the finish line."[20]

Gaps in War Termination Scholarship for US Interventions

The existing scholarship focuses mainly on the termination of conventional wars. The United States is consistently surprised by the complexity of endgame challenges, Gideon Rose argues, and often must improvise through uncertain and ambiguous strategic terrain.[21] Interventions in irregular wars may present very different challenges. The lack of attention to war termination in such conflicts creates three significant gaps in scholarship.

First, the conventional war paradigm may overestimate the military factors and underplay the political and diplomatic issues that tend to be more salient for wars against insurgencies. This could lead scholars to ignore implicit assumptions in intervention strategies that sow the seeds for war termination problems.[22] For instance, conventional wars tend to be fought between military forces until one side wins or a stalemate ensues. Because the clash between fielded forces is so central to success, political and military leaders work to "define a military objective that, if achieved, can deliver the political objective."[23] Once hostilities conclude, diplomats negotiate agreements that seek to maximize the advantage of battlefield results. Interventions against insurgencies tend not to hinge upon the clash of opposing forces. There might not be a realistic military objective that could deliver the political aims. Ivan Arreguín-Toft, in fact, argues that stronger powers have been losing to weaker ones increasingly often since the 19th century.[24] Insurgencies tend to avoid rather than bring about decisive battles. Several longitudinal studies show that the ability of the insurgency to sustain tangible support and of the government to win the battle of political legitimacy play more decisive roles in the war's outcome than battlefield engagements.[25] Thus, intervention strategies that assume or fixate upon decisive military victory for war termination may have a low probability of success against insurgencies.

A second major gap in the existing literature is that it does not help us understand why the United States has been slow to abandon losing or ineffective strategies. Realists such as Kenneth Waltz and John J. Mearsheimer argue that a state at war will adopt strategies with the highest likelihood of success and take advantage of conflict situations when the benefits outweigh the costs.[26] If this is the case, then states should abandon a losing strategy quickly. As will be discussed in this book, the United States did not change strategies in Vietnam until 1969, in Iraq until 2007, and in Afghanistan until 2009—five, four, and eight years into the conflicts, respectively. By each point the American public tired of the war and demanded withdrawal. What accounts for the strategic paralysis? Scholars differ on the factors that may impede decisions about war termination. Elizabeth Stanley and Fred Iklé describe important political and bureaucratic obstacles that can affect decision-making, such as uncertain battlefield outcomes, poisonous "patriots versus traitors" domestic politics, and bureaucratic politics.[27] In examining cases from the US Civil War to the Korean War, political scientist Dan Reiter asserts that domestic politics has been of "curious insignificance . . . in war-termination decision-making."[28] Stanley, however, marshals significant amounts of evidence that suggest a change in domestic coalitions was a key driver in bringing about an end to the Korean War.[29] Distinguishing between conventional wars and interventions against insurgencies might yield some answers to this puzzle.

The amount of ground taken or lost is normally a reasonable indicator of progress in conventional war. This is not necessarily the case for irregular war. Valid strategic metrics may be more difficult to develop, which could increase the risk of poor decision-making. The presence of bureaucratic silos may shape how progress is measured and could lead to in-silo positive indicators even as the war is going poorly overall. The interface between interagency silos that are forced to work together may leave important but undetected vulnerabilities. The nature of the host nation government likely plays a heightened role in counterinsurgency. Governments most likely to need assistance from an external power might also be the ones most vulnerable to insurgency, and more resistant to changes in strategy that require them to make reforms or compromises. The increasingly rich fields of decision theory, organizational management, behavioral economics, and agency theory may be of assistance in understanding potential causes of strategic paralysis.[30]

A third issue with current scholarship is that it tends to ignore that bargaining behavior could be very different for interventions against insurgencies. Conventional wars that settle into stalemates could be more likely to produce opportunities in which both sides are motivated to end the conflict. Thomas Schelling suggests that "most conflict situations are essentially bargaining situations."[31] Historian Roger Spiller argues that combatants always "converge toward an agreement to stop fighting."[32] This symmetry makes the fallback to negotiations possible. "Uncertainty causes war, combat provides information and reduces uncertainty," Reiter argues, "and war ends when enough information has been provided."[33] I. William Zartman refers to this as ripeness: a "mutually hurting stalemate" in which both actors are in "an uncomfortable and costly predicament," which makes parties ripe for negotiation.[34] Thus, if decisive victory does not work out, combatants can fall back to negotiations. Are those same opportunities present for irregular wars? Once an intervening power decides the war is unwinnable and wants to withdraw, bargaining leverage may shift to the insurgency. The former may seek negotiations to end the conflict, but the latter might have large incentives to play for time. The insurgency might calculate that its negotiating leverage will be much higher after the intervening power withdraws. Why settle for half the loaf now when a little patience might give you three-fourths of it? Alternatively, the intervening power, as it withdraws, might seek to transition security responsibility to the host government as it builds their capacity. Capacity-building efforts, however, might create crippling dependencies that impede battlefield performance or give the host government false confidence that it can maintain the predatory and exclusionary practices that gave rise to the insurgency in the first place.

Overlaying these gaps in scholarship is a presumption in much of the literature that war termination boils down to a political decision to end a war when the enemy offers to surrender, or when decisive victory is deemed no longer possible or cost effective. Is it possible that restricting the range of successful outcomes to decisive victory is heightening the risk of quagmires? Martel offers a useful typology of victory, all of which are variations on the decisive victory theme.[35] Wars, however, can end successfully with negotiated settlements, which do not necessarily have to come about after bloody stalemates have settled in. For interventions against insurgencies, a successful conclusion can also result, in theory, by building host nation capacity to carry on the war successfully as the intervening power withdraws—an outcome described as transition. Exploring these different outcomes during the policy and strategy processes could open wider possibilities for success.

Getting War and War Termination Wrong

There are many reasons that major post 9/11 US military interventions have turned into quagmires. Adversaries, host nation actors, and regional powers all played a role. So did unpreparedness for fighting large-scale irregular wars. This book explores the systemic problems in US policy and strategy that emerge from flawed presumptions about war termination.

The United States has no organized way to think about war termination or its place in policy and strategy. The US government has a military doctrine that covers how to *fight* wars but no shared interagency framework for how to *wage* wars, which should include war termination.[36] As Colin S. Gray argues, there is more to war than warfare and thus more to strategy than military strategy.[37] Fighting war is a competency of the military. Waging war includes the integration of all elements of national power into a coherent policy and strategy that define and outline how to achieve the political purpose of the war. This difference between waging and fighting gets lost in practice. One result is what might be called America's *bureaucratic way of war.* The US government tends to view war in agency-centric stages: diplomacy to prevent war or build a coalition (State), armed conflict (Defense), diplomacy to negotiate peace (State), reconstruction (USAID), which may be useful for conventional war but badly limits strategic options for waging wars that rely less on military, force-on-force outcomes. The failure to distinguish between waging and fighting has reinforced the military's exaggerated influence in national security decision-making. The recent habit of presidents to turn to the military in times of national security challenges for "military options" creates a situation in which one of the means (military) is controlling the ends and limiting the exploration of alternatives.

An important part of this problem includes the absence of an agreed terminology for successful war outcomes. The United States has no authoritative language for describing results that could meet the political purpose of the war. The default position is *decisive victory*—a vague and dangerous belief that the expert application of military power will force an adversary to capitulate, thereby giving up on its own interests and agreeing to give the United States all it wants. American political and military leaders thus work to "define a military objective that, if achieved, can deliver the political objective."[38]

Certainly, a clear battlefield victory that obliterates the opposing conventional force can put the United States into a strong, perhaps dominant, position to impose a settlement, but this option is often unavailable in irregular wars. There normally is no realistic military objective that could deliver the political objective. Without a shared understanding of alternative outcomes, such as a negotiated settlement or transition, the United States struggles to define plausible outcomes that could meet the political purpose of the war.

Consequently, there is no authoritative body of knowledge for how to organize instruments of national power for outcomes such as *negotiated settlements* or for what has become known as *transition*—the handover of security responsibilities to the host nation and withdrawal of American forces. These outcomes tend to rely more on diplomatic and political rather than military instruments, so the absence of a shared framework undermines policy and strategy development, often creates civil-military tensions, and impedes interagency communication and coordination, which can lead to major problems in execution.

These broader problems have led to specific, systemic policy and strategy errors in Afghanistan and Iraq that have undermined the prospects of success.[39]

First, the United States entered both wars with military-centric strategies that had low probabilities of success. Once the spectacular military campaigns overthrew the sitting governments, the US government was at a loss for how to achieve a favorable and durable outcome. In both cases, the United

States, intent on *decisive victory,* ignored or dismissed early opportunities to negotiate with the militarily defeated parties and unwittingly allowed predatory and exclusionary governments to form. The combination led to insurgencies that became sustainable. The stage was set for quagmire.

Second, policymakers had difficulty changing these failing or inadequate strategies due to cognitive bias, political and bureaucratic frictions, and relationship problems with the host nation. In both cases, US officials believed their strategies were working even as the situations deteriorated. Confirmation bias was evident as officials emphasized positive indicators of progress and rationalized or, in some cases, even used negative indicators as evidence of success. Political frictions combined with confirmation bias and loss aversion led the Bush administration to dig in its heels on Iraq, and similar factors led the Obama administration to resist changes to drawdown timelines in both conflicts.

The tendency in both administrations to operate in bureaucratic silos reinforced the narrative of progress. Many in-silo milestones and indicators were positive. The problems were happening at the interface of the silos. Seams or gaps between them were ably exploited by the host governments and the insurgencies. In Iraq and Afghanistan, political elites in host governments took advantage of these gaps to manipulate unwitting US officials into advancing their predatory and exclusionary agendas. When silos interacted with one another, fault lines could result in which efforts in one silo damaged progress in others. Civilian casualties from military operations, for instance, undermined the legitimacy of both governments and international missions.

Patron-client problems diluted American capacity-building efforts, created frictions between the United States and the host governments, and, in some cases, between Washington and American military and diplomats in Baghdad or Kabul. These toxic factors intensified the conflicts, reduced public support in both countries, and impeded US decision-making. Although the United States made important strategic changes in both conflicts, it paid penalties in public support along the way as the wars dragged on, inconclusively. The so-called crossover point, when the host nation government could assume the security lead and defeat the insurgency after the US troop withdrawal, never occurred.

Third, when Americans tire of the war and the president decides to withdraw, the United States has lost critical leverage for successful negotiations or transition. This results in worse outcomes than those that were available earlier in the war. In Iraq, Bush was unable to gain Iraqi prime minister Nouri Maliki's agreement for a "conditions-based" withdrawal and had to settle for a fixed timeline. The very limited amount of support Obama was willing to commit for an enduring presence led Maliki to calculate that the benefits were not worth the political cost. Just over two years after the last American troops withdrew, the Islamic State of Iraq and Syria (ISIS) arose and humiliated the Iraqi Army, which America had spent so much blood and treasure to build.

In Afghanistan, initial efforts from 2010 to 2013 to bring about negotiations with the Taliban ended in disaster. Although the Taliban were willing to discuss the conditions of the US withdrawal, they were uninterested in negotiating an end to the conflict. They could play for time until international forces left. Deep problems in the US-Afghan relationship led Afghanistan's President Karzai to refuse to sign a bilateral security agreement to extend American troop presence beyond 2014. This was later signed by the incoming National Unity Government of President Ashraf Ghani and Chief Executive Abdullah Abdullah, but not by one of these elected leaders. Obama's drawdown timeline, meanwhile, had to be modified because of larger than expected gains in territory by the Taliban and major problems within the Afghan security forces. Peace talks would not start again until late 2018.

America's bureaucratic way of war creates an absence of in-theater authority and accountability, which impedes sound execution, undermines the United States' ability to keep pace with a dynamic environment, and thus heightens the risk of quagmire. The US government deploys to conflict zones in bureaucratic silos. As such, there is no organizational entity on the ground that is responsible and accountable for achieving US aims. The upshot is that no one below the president of the United States has the authority to direct and manage US efforts in the combat theater. The result on the ground is chaos. Efforts among agencies are unprioritized, tend to be poorly coordinated, and often conflict with one another. Host nation actors and adversaries exploit these silos to their own advantage. The whole of US efforts in these wars is less than the sum of their parts.

How This Study Began

This book originates, in part, from my own frustration at the inability of the US government to develop and execute a sensible strategy in Afghanistan. A fellowship at King's College, London, under the direction of Professor Theo

Farrell, then head of the War Studies Department, gave me the opportunity to investigate why the US government could not seem to get its act together to defeat a "rag-tag" militant group like the Taliban. It also provided a way to reflect upon my experiences—both successes and failures—and contribute my insights to help improve policy and strategy decisions over war and peace.

I offer a unique perspective on this issue as one of the very few Americans to have commanded in combat at the unit level in Afghanistan and served as a senior adviser to three four-star commanders in Afghanistan as well as to cabinet-level officials in the Pentagon. My role in the grand scheme of things has been relatively minor, but I've been able to support and observe the key actors and decision-makers.

The paratroopers I was privileged to command in Kunar and Nuristan are the only ones, to my knowledge, in the now 20-year history of this war to have motivated a large insurgent group (in this case, a branch of Hizb-i-Islami Gulbuddin [HiG]) to stop fighting and join the government. I later helped develop the Obama administration's strategies for Afghanistan and the military's implementation strategy in Kabul, served as the secretary of defense's personal representative during the 2011–2013 exploratory talks with the Taliban, and later, through Track 2 efforts in 2017–2018, helped convince the Trump administration to resume talks with the Taliban.

In short, I have witnessed some of the most extraordinary aspects of the war from different positions and occupations. Everyone experiences war in their own way. Others have had equally unique and interesting experiences in this war, and I hope they, too, will enrich the study of war and peace with their insights.

As the detailed work on reconciliation began in 2010, I realized very quickly that, despite the talent and experience of those engaged in the process, the US government had no idea what it was doing. We had no useful doctrine, concepts, or body of knowledge for bringing about a negotiated outcome with an insurgent group like the Taliban. We often wasted time negotiating with ourselves while alienating our partners in the Afghan government, leading the Taliban to conclude that the United States was operating in bad faith. None of this was due to malfeasance or lack of effort by American officials.

I fully accept my own errors in this endeavor, which include, but are not restricted to, failing to frame issues with sufficient clarity to promote better policy decisions and unintentionally upsetting my colleagues in Defense, State, and the White House when we talked past each other when debating issues and approaches. Conducting research for this book has helped me to understand these failings even further. Throughout these strategy-making endeavors, I combed the existing literature to determine whether some useful models or analogies existed that could guide decision-making. What I found was very interesting, but ultimately disappointing. This book seeks to address this gap in the literature.

To address the concern that America's challenges in Afghanistan are entirely unique, I examined the US intervention in Iraq. While the Iraq conflict developed its own trajectory and unique characteristics, I found the three major policy and strategy errors were present there, too.

Methodology

This book develops a new understanding of the systemic policy and strategy errors that exacerbate the war termination challenges facing the United States during large-scale interventions against insurgencies. The universe of possible cases to analyze for this study is naturally limited. Since the United States emerged from the Second World War as a global power with a new national security architecture and powerful bureaucracies, it has not successfully prosecuted a large-scale military intervention against an insurgency. This work defines large-scale military intervention as one in which the United States deploys military forces of division-size or greater (over 10,000 soldiers) to play a leading role in ground combat operations to fight an insurgency. Only three conflicts qualify: Vietnam, Iraq, and Afghanistan. They have all been quagmires.

Comparing the Afghanistan and Iraq wars to one another to understand strategy failures very broadly follows John Stuart Mill's method of agreement; when similar systems produce similar outcomes, their similarities, not their differences, best explain the outcome.[40] Both wars were started in the wake of the 9/11 attacks as part of the Global War on Terrorism, with the aim of overthrowing an existing government; the people involved on the US side were similar and often the same in both wars; so were US bureaucratic structures and assumptions about war termination and the use of force. Eventually, both wars turned into counterinsurgencies. These similar systems produced similar outcomes—quagmires.

War is complex, so it might be possible that even within similar systems different paths can lead to the same outcome. Hence, in addition to cross-case comparison, a detailed within-case analysis has been used to test existing theories and derive new hypotheses. Within-case variation also helped to eliminate certain variables. The fact that challenges within each war continued after the administration changed, for example, rules out party affiliation or any other personal characteristics of the president and his team as explanations on the US side.

The Vietnam War, the only other post–World War II large-scale US counterinsurgency, has been used as a plausibility probe to see whether recent war termination challenges are unique to their post 9/11 context. The different time and context of the Vietnam War helped eliminate some further possible explanatory variables and supported the outcome of the Iraq-Afghanistan comparison. The fact that all three systemic errors were present in all three wars supports their explanatory value.

The gaps in theory and scholarship noted in the introduction meant that an abductive research approach was likely to generate the richest insights. Abduction combines "deductively derived hypotheses" and "inductively derived insights." It involves "moving back and forth between the two to produce an account that will be 'verisimilar and believable to others looking over the same events.'"[41] Rather than proceeding linearly from theory to case research, this approach intertwines theory and empirical observation to enhance the understanding of both. The result is a pragmatic framework that enables theory to confront the real world and empirical observation to challenge theory.[42] Hypotheses have been deduced from existing theories of war termination, strategy theory, and theories of rational decision-making. When reality did not correspond to what these theories suggest, recourse to other fields, especially economics and organizational behavior, has been used to understand these deviations.

Abductive inference is more appropriate for this effort than a primarily deductive approach for several reasons. First, existing war termination theory focuses mostly on state-on-state conventional war. New concepts are needed to understand the complexity of war termination challenges for intervening powers in irregular conflicts. Second, the universe of cases is limited. There have been no instances of a successful US large-scale military intervention against an insurgency since World War II. Comparing smaller-scale interventions and advisory missions, for instance, with large-scale interventions can yield useful insights that will nevertheless have to remain limited due to their different character. Likewise, comparing these cases to the successful US suppression of the Philippine Insurrection (1899–1902) would surely be an interesting exercise, but might be limited in the conclusions one can draw for contemporary decision-makers, due to the vastly different national security decision-making architecture and the powerful bureaucracies that emerged after the Second World War. Comparing American interventions with those by other powers, such as the Soviet intervention in Afghanistan or the British interventions in Sierra Leone and Malaysia, also presents methodological challenges due to vast amount of dissimilarities in doctrine, national security architecture, and decision-making, to say nothing of domestic politics, strategic culture, and span of global responsibilities and interests. Studies of the above are beyond the scope of this book but would yield rich insights nonetheless.

Given the methodological challenges, the claims in this book are rather modest: the United States' lack of an organized way of thinking about war termination and thus its implicit fixation on zero-sum decisive victory are inducing three major, systemic errors that are increasing the likelihood of quagmire: first, the United States has tended to adopt military-centric intervention strategies that have a low likelihood of attaining a favorable and durable outcome. Second, the United States has had difficulty modifying a losing or ineffective strategy due to cognitive biases, internal bureaucratic and political frictions, and patron-client challenges with the host nation, thus extending the length of the war and increasing penalties in public support. Third, when the United States tires of the war and announces its intention to withdraw, bargaining asymmetries with the host nation and the adversary occur, impeding the ability to use transition or negotiations to secure a favorable and durable outcome.

This volume does not claim, for instance, that the US policy and strategy problems in Afghanistan and Iraq are the *only* explanation for the quagmires. To be sure, the resilience of the insurgencies, the existence of external sanctuary and indigenous support for the insurgencies, and deep legitimacy problems in the host nations, among other issues, contributed substantially. Local actors in Afghanistan and Iraq made decisions to advance their particular interests, not US objectives. Key decisions by those actors that played important roles in each war's trajectory, such as President Karzai's 2005 decision to allow warlords into his cabinet, the Sunni-Shi'a civil war in Iraq, Prime Minister Maliki's Charge of the Knights operation into Basra, and Pakistan's arrest of Taliban deputy Mullah Abdul Ghani Baradar in January 2010 had little to do with the United States. The policy and strategy errors on the

American side include its inability or reluctance to recognize the strategic salience of these factors and adjust the war termination policy accordingly.

Likewise, there is no claim to a universal theory of war termination challenges for intervening powers in irregular conflicts. The extent to which these three systemic errors can be generalized beyond the cases addressed in this book is an important subject for future research. It would be interesting, for instance, to examine the extent to which the Soviet Union's quagmire in Afghanistan (1979–1987) exhibited similar policy and strategy errors. Likewise, an analysis of the United Kingdom's rather successful interventions in Sierra Leone and Malaysia could yield insights on the extent to which these errors were avoided or had less impact. Of course, conclusions from such case comparisons would be tenuous as well, given substantial differences in matters such as strategic culture and decision-making architecture.

In sum, the methodology I have chosen is imperfect but reasonably sound to yield insights on the war termination-related policy and strategy errors that heightened the likelihood of turning America's major post-9/11 interventions into quagmires. Uncovering the systemic errors common to the interventions addressed in this book provides a highly useful starting point for further scholarship as well as for important national security reforms. These suggestions are addressed in the concluding chapters of this book.

A Note on Sources

Data for the case study analysis has come from a variety of sources. This research has benefitted from personal access to key decision-makers in both the Bush and Obama administrations. I have interviewed over 30 former senior officials who were instrumental in developing policy and strategy for Iraq and Afghanistan and were present for discussions in the White House Situation Room, in Baghdad, and/or in Kabul. Many of these former senior officials agreed to speak "on background" due to the sensitive nature of the issues.[43]

The interviews are complemented by a significant number of published memoirs, which include those of a president, a national security adviser, three former secretaries of defense, and two theater-level commanders. This book has also made use of a wide range of US government documents that are publicly available. Among these, congressionally mandated Department of Defense semiannual reports beginning in 2006 have provided exceptional detail on policy, strategy, operations, and assessments in Iraq and Afghanistan.

Reporting cables have offered insights into how officials viewed and interpreted issues at the time they occurred. These sources plus a wide array of journalistic reporting and academic studies have helped to control for first-person biases or incomplete recollections. The book has also benefitted from my personal experience as a combat commander in Afghanistan, on strategy in headquarters in Kabul, on policy in Washington, D.C., and in secret talks with the Taliban.

Structure of the Book

I have organized the book into eight major parts. After the prologue gives an overview of key war termination challenges during the Vietnam War, Part I works toward a taxonomy on war termination and potential integration into the policy and strategy processes. It also explores the focus on decisive victory in developing a strategy for Afghanistan. Parts II to VI further explore the Afghanistan and Iraq case studies. Each country case study is organized according to the major problems that spring from the fixation on zero-sum decisive victory: military-centric intervention strategies that had low probabilities of success; cognitive bias, political and bureaucratic frictions, and emerging problems in relations with the host nations that created strategic paralysis within the US government; and the loss of leverage as the United States grew exhausted, decided to withdraw, and tried in vain to transition the war to the host nation or negotiate. Key theoretical concepts that are relevant to the argument are introduced at the beginning of chapters (mainly in the parts on Afghanistan), and then imbedded into the narrative. Part VII discusses implications for US foreign policy and scholarship.

America's policy and strategy failures have received little scrutiny. Congress fails to ask hard questions. Senior civilian and military officials work countless hours in their silos. Most soldiers and civilians on the ground perform their jobs, as they understand them, to expected standards. That performance, however, is not producing sustainable success in any post-9/11 conflicts. In fact, the 1991 Gulf War stands alone since World War II as an outright military success in a large-scale intervention, but even there the United States fumbled the negotiations. Key competitors, meanwhile, have figured out how to take advantage of America's bureaucratic way of war. Whether the United States can overcome its fixations and bad habits could become one of the defining global security issues in the twenty-first century.

The Past as Prologue
The Vietnam War

The war termination problems that the United States encountered during the major post-9/11 interventions may not be an aberration. The Vietnam conflict suffered from the same factors, albeit in subtly different ways. Explanations of the poor outcome in Vietnam include three schools of thought. The counterinsurgency school argues that General Westmoreland and the US Army were fixated on conventional war and unable to adapt their tactics to meet the demands of fighting an insurgency.[1] An alternative, suggested by Westmoreland in his memoirs, is that the nature of the conflict required the US military to focus on fighting the primary threat from North Vietnamese and Viet Cong main force units, while the South Vietnamese military not committed to the conventional fight would need to take on the guerrillas.[2] This school suggests that the campaign design and tactics were right but the war was largely unwinnable due to factors beyond Westmoreland's control. A third school examines some of these factors from a civil-military lens, including civilian micromanagement of military operations, the failure of military officials to provide candid advice, and how the political decision not to mobilize the country may have undercut public support for the war.[3]

Examination of war termination challenges enriches these perspectives. An important difference from the Iraq and Afghanistan case studies is that a negotiated outcome—in the form of North Vietnamese capitulation—was discussed from 1964 to 1966 during deliberations over whether to escalate the Vietnam War.[4] President Johnson wanted to limit the costs of supporting South Vietnam and had no intention of conducting a large-scale ground invasion of North Vietnam or using nuclear weapons to force them to sue for peace. The successful use of graduated pressure during the Cuban Missile Crisis became an important reference point for a lower-cost alternative in Vietnam.[5] Defense Secretary Robert S. McNamara and other civilian leaders believed the United States could use a similar model to carefully raise the pressure on the Democratic Republic of Vietnam (DRV) to compel them to stop supporting the National Liberation Front (Viet Cong) insurgency against the South Vietnam government.[6] This approach, they believed, would limit the costs to the United States and avoid the risk of Chinese or Russian intervention. The uniformed military, however, wanted to either sharply escalate the conflict to compel the DRV to give up its proxies (which meant widespread bombing of North Vietnam and perhaps a ground force invasion) or get out.[7]

The odds against success in Vietnam may have been greater than in post-9/11 Afghanistan or Iraq. In the latter two, the United States overthrew existing regimes. These were replaced by new governments before the insurgencies fomented (although resistance began immediately in Iraq, and within a year in Afghanistan). In Vietnam, the United States needed to rescue a deeply troubled client.[8] By 1964 the NLF had significant internal support as well as external support and sanctuary from North Vietnam, and it controlled roughly 40 percent of the country.[9] The South Vietnamese government was deeply kleptocratic and losing popular legitimacy.[10] Either situation, unless reversed, normally results in a loss for the government. South Vietnam had both from the start of the US intervention.[11]

The United States was in a difficult position. As the discussions over strategy continued into 1965, the civilian leadership understood the prospects of success to be low.[12] The Johnson administration began to test the graduated pressure concept by escalating the conflict while attempting to start negotiations.[13] They discussed unilateral suspension of military actions, which prompted significant resistance from the military.[14] Johnson believed he needed to show the American people and the world that he was as serious about peace as he was about fighting.[15] "The weakest chink in our armor," he surmised, "is public opinion."[16] When Johnson said that he was very reluctant to go against the views of the Joint Chiefs, McNamara advised, "We decide what we want and impose it on them. They see this as a total military problem—nothing will change their views."[17]

Johnson eventually approved a 37-day bombing pause beginning on December 24 and a 30-hour Christmas cease-fire.[18] Washington wanted to make clear that the halt was a serious move toward peace that required a suitable concession from North Vietnam to keep the process moving forward. The United States engaged in efforts with 34 countries to communicate to the North Vietnamese and the world America's desire for a peaceful resolution.[19] As Rusk put it, "We have put everything into the basket of peace except the surrender of South Viet-Nam."[20]

The graduated pressure concept overestimated the effects of strategic bombing and underestimated the strength of the DRV's resolve and the consequences of South Vietnam's political dysfunction.[21] The Americans expected the bombing pause to signal a willingness to bargain. The DRV dangled the faint hope of negotiations to reduce military pressure. By linking bombing to peace talks, the Johnson administration unwittingly fell into this trap. "Hanoi used negotiations as a tactic of warfare to buy time to strengthen its military capabilities in South Vietnam and weaken the will of those on the side of Saigon," Goodman summarizes. "Rather than serving as an alternative to warfare, consequently, the Vietnam negotiations were an extension of it."[22] This pattern, he argues, protracted the war and played to the DRV's advantage.[23]

The result of the civil-military frictions over strategy was a thinly camouflaged bureaucratic struggle. Unable to convince the uniformed military of their logic, the Johnson administration micromanaged military operations and authorities in the hope of using graduated pressure to bring about peace talks. The uniformed military, on other hand, played bureaucratic games to prod McNamara and Johnson into escalating the war toward the troop levels and authorities they believed were necessary to force DRV capitulation.[24] US troop levels surged from approximately 200,000 in 1965 to over 500,000 by 1968.[25] The United States adopted several different operational approaches to defeating the insurgency, to include taking over the war effort in 1965, but it was never able to reduce the insurgency's sustainable support or pressure the South Vietnamese government to govern effectively enough to win the battle of legitimacy in the insurgent heartlands.[26] As a South Vietnamese official explained to journalist Stanley Karnow in late 1964, "Our big advantage over the Americans is that they want to win the war more than we do."[27]

This losing strategy became intractable. Assessments were made within bureaucratic silos and then aggregated to convey an overall picture. Doing so painted a misleading image. Officials remained upbeat despite the worsening security situation.[28] Such assurances impeded strategic adaptation, as debates between advocates and skeptics grew poisonous.[29] The Johnson administration and the military command began losing credibility. Even though Defense Secretary Robert McNamara in May 1967 counseled President Johnson to "negotiate an unfavorable peace," he could not overcome the status quo bias.[30] The 1968 Tet Offensive was a psychological shock to the United States, irrevocably damaged public support for the war, and was a key factor in Johnson's declining to run for reelection.[31] Although the insurgency suffered heavy losses and never fully recovered, their residual strength combined with that of the North Vietnamese Regular Army sustained the conflict. The Soviet Union reportedly agreed in 1968 to facilitate talks between the Johnson administration and Hanoi. Biographer John A. Farrell argues that Nixon sabotaged the effort by convincing South Vietnam's President Thieu to object.[32]

A change in the administration was necessary to alter the strategy. Richard Nixon won the 1968 election in part by promising to end the Vietnam War. By 1969 he began troop withdrawals and the process of "Vietnamization" to turn the war back over to the South Vietnamese.[33] Transition efforts, however, were undermined by severe patron-client problems. The South Vietnamese government remained unable to win the battle of legitimacy in insurgent-controlled areas. Meanwhile, Secretary of State Henry Kissinger began secret talks with North Vietnam in August 1969.[34] The clear US intention to withdraw from the war limited American bargaining leverage.[35] The United States and the DRV concluded an initial agreement in October 1972, but South Vietnam rejected the accord and talks deadlocked.[36] To break the impasse, Nixon authorized Operation Linebacker II, which unleashed a massive bombing campaign against the DRV from December 18 to December 29, while pressuring South Vietnam's President Thieu to accept the agreement. The Paris Peace Accords were signed a month later, on January 27, 1973.[37] The accords called for national elections and allowed the North Vietnamese Army to remain in the South, but they were only to receive reinforcements sufficient to replace losses. The United States had 60 days to withdraw all forces from South Vietnam. That last article, explained historian Peter Church, "proved . . . to be the only one of the Paris Agreements which was fully carried out."[38]

This short examination of the Vietnam conflict suggests that the war termination problems the United States has experienced in post-9/11 interventions are probably not a new phenomenon. The Johnson administration assumed a decisive victory outcome was possible at low-cost by using graduated pressure to compel North Vietnam to cut off support to the NLF. The costs, however, were insufficiently compelling to force the DRV to capitulate to American demands. The Joint Chiefs were never persuaded of the logic. Instead of recommending the exploration of alternative strategies that met Johnson's intentions to limit the costs of the war, the uniformed military went all-in for decisive victory. They manipulated McNamara and Johnson into escalating troop levels to the amount they felt was necessary to win, even though they were unable to gain approval for greater actions against North Vietnam. American public opinion turned against the war. The Nixon administration attempted negotiations while withdrawing American forces. Predictably, the North Vietnamese were willing to agree to some concessions to ease the US withdrawal, but not to end the conflict. By 1975, the US Congress slashed funding on military aid to South Vietnam from $2.8 billion in 1973 to $300 million. North Vietnamese and NLF forces took Saigon on April 30, 1975.

Part I

Toward a War Termination Framework

1

Further Defining War Termination

Three Successful War Termination Outcomes in Irregular War

A country goes to war to achieve certain aims. These aims could include vanquishing an existential threat, territorial conquest, regaining lost territory, regime change, retribution, coercing the adversary to change certain policies, and the like. Success, quite logically, means the *durable* attainment of those aims. This simple concept is at the heart of many of America's troubles with irregular wars.

Decisive victory, gaining the enemy's capitulation or annihilation, is the most easily understood way to succeed. Decisive military victory is central to Martel's argument about winning. Policymakers, he argues, "should understand that using the correct combination of military instruments of power will permit them to achieve the levels of victory to which they aspire."[1] Sometimes, decisive victory is the best path to a favorable and durable outcome. In World War II, for instance, vanquishing the Nazi regime—permanently ending its threat—was necessary. In the American Civil War, defeating the Confederate armies was the most logical way to preserve the Union and end slavery.

Successful warfighting, even to the point of defeating an opposing army, is normally not sufficient for a durable political outcome. War, in political scientist Thomas Schelling's formulation, is violent bargaining.[2] Even wars that have involved the surrender of the adversary's armed forces usually involve some form of negotiation. US general Ulysses S. Grant, for instance, feared that Confederate general Robert E. Lee's Army of Northern Virginia would continue fighting, perhaps conducting a guerrilla campaign, if not offered parole and the promise of no post-war trials. The Allies in World War II feared the Germans would do the same if not offered acceptable terms. In the Pacific theater, despite being pummeled by two atomic bombs, the Japanese offered to surrender only if they could retain the emperor.

Similarly, the American Revolution aimed to coerce Great Britain into recognizing the newly formed United States as an independent country. The British fought from 1776 to 1783 in an effort to retain colonial rule. They could have carried on the struggle, even after the surrender of Cornwallis at Yorktown in October 1781, but they calculated that the probability of success at acceptable cost was too low. The British accepted American independence in the Treaty of Paris, but they extracted important concessions, too.

Winning, Schelling points out, should not have a competitive meaning. Adversaries tend to have a combination of conflicting and compatible interests. Winning in war means gaining relative to one's aims, not in relation to the adversary. Success may be realized through bargaining and mutual accommodation and by the avoidance of mutually damaging behavior. There are, he argues, a range of variable-sum outcomes available.[3] Viewing success in zero-sum terms closes off a range of possibilities for winning.

Decisive victory, while highly desirable, is thus not the only possible war termination outcome and might not be realistic or cost-effective. Even Prussian military theorist Carl von Clausewitz, who many mistakenly believe is an advocate of total war, recognizes the simple fact that there can be other ways to win. In a rarely quoted passage, he notes, "it is possible to *increase the likelihood of success without defeating the enemy's forces* . . . many roads lead to success [and] they do not all involve the opponent's outright defeat."[4] In other words, even the old Prussian figured out that winning does not always mean zero-sum decisive victory.

Some wars, for example, end in a *negotiated outcome.* In this case, neither party surrenders. The combatants negotiate an agreement that ends the conflict. The Good Friday Agreement ending the "Troubles" in Northern Ireland is an example. Neither the British government nor the allied and insurgent Northern Irish parties vanquished the other or got everything they wanted in the negotiations. The British government retained control of Northern Ireland, making the negotiated outcome a win for them even if the insurgency did not surrender. Sinn Fein, the political wing of the Irish Republican Army, was granted important concessions, including recognition as a political party and a role in governance. The two sides each achieved some but not all of their aims.

The Korean War (1950–1953) is another example of a negotiated settlement. The United States had positioned troops in South Korea after World War II to defend the latter's territorial integrity. When North Korea invaded in 1950 to unify the peninsula under its leadership, the United States and its allies fought back to save South Korea. After a stunning victory at Inchon in late 1950, the United States expanded its war aims to include unification of the Korean peninsula under South Korean leadership. This led to Chinese intervention to save its North Korean ally. The war settled into a grinding stalemate near the original borders. By 1953, a negotiated settlement was reached. The United States and South Korea achieved their original aim to preserve the existence of South Korea, but not the expanded aim of unifying the peninsula under South Korean leadership. The Chinese succeeded in preserving North Korea. North Korea failed in its aim to unify the peninsula under its leadership. Thus, a war that ends in a negotiated settlement can have multiple winners even if no one capitulates.

A third option, available to an intervening power, is to end its direct combat role before the war ends. This concept is called *transition*. In theory, once the intervening power deems a host nation's capability to be sufficient—or determines that further efforts are no longer needed—it can withdraw its troops. To achieve this outcome, the intervening power aims to build the capacity of the host nation until it overmatches the capability of its adversary. Once the estimated "crossover" point is reached, the intervening power can scale down its commitment and eventually withdraw altogether. President George W. Bush expressed this concept during the Iraq War: "as the Iraqis stand up, we will stand down."[5]

In short, a combatant may have multiple options to achieve a favorable and durable outcome in a given conflict. Decisive victory is neither the only option nor necessarily the best one in certain circumstances. In sum, favorable war termination outcomes can include:

Decisive victory, in which the adversary capitulates and ceases military resistance. This often results in an imposed settlement to avoid further bloodshed. The win is lopsided in favor of one side, but not necessarily zero-sum.

Negotiated settlement or mixed outcome, in which neither side vanquishes the other. Parties compromise to end the war and settle remaining differences through peaceful politics. These are normally variable-sum outcomes. Success occurs if the negotiated settlement enables the combatant to achieve its main war aims.

Transition, in which the intervening power degrades the adversary while building the capacity of the host nation government and security forces. As these forces become superior to those of the enemy (thus reaching the so-called crossover point), the intervening power transfers security responsibility to the host nation and withdraws without concluding a peace agreement. An intervening power wins if its interests remain secured after withdrawal.

Clausewitz's insight that a combatant can *increase* the likelihood of success without defeating the enemy's forces is critically important. After all, seeking an unrealistic outcome could prolong a war and heighten its costs. A combatant fixated on decisive victory might become blind to opportunities to achieve its aims through other means. Alternatively, an intervening power that seeks a favorable and durable outcome through transition could meet with unexpected success that opens an opportunity for a decisive victory. Knowing when to adopt an alternative is thus a significant decision and requires a major change in strategy.

Each of these broadly defined successful outcomes in irregular war normally involves some sort of bargaining. The surrender of enemy forces tends to be predicated upon the attainment of an agreement sufficient to avoid further bloodshed. Transition relies on bargaining with the host nation, and perhaps the insurgency as well, for the protection of the intervening power's interests. Rejecting negotiations in the mistaken belief that success may only be attained in zero-sum fashion heightens the probability of quagmires.

Likewise, it is important to note that the various outcomes may entail different levels of durability. A decisive victory that is perceived to impose unnecessarily harsh terms may sow the seeds of revanchism. A negotiated settlement approved by one governing administration could be overturned by the next administration and risk a resumption of conflict. Negotiated settlements or transitions that place the protection of a country's interests into the hands of a host nation government could find those interests marginalized or abandoned by that actor.

These war termination considerations seem to be important enough to be considered at the start of the policy and strategy process rather than after the default (*decisive victory*) has been exhausted. Do not take the first step, cautions Clausewitz, before considering the last.[6]

Strategy and War Termination

Historian Lawrence Freedman describes strategy as "the best word we have for expressing attempts to think about our actions in advance, in light of our goals and capacities."[7] Strategy is the process of determining how to employ one's capabilities to achieve a favorable and durable outcome in the face of cooperative and competitive actors and factors. Strategy, as Schelling notes above, exists in a dynamic environment because these actors and factors, which are beyond one's control, affect the outcome. Historian Michael Handel observes that strategy development ought to be a rational process, evidenced by identifying political goals, analyzing the character of the conflict, developing a strategy to achieve stated aims, and then making peace when the goals have been attained or when the costs and risks have come to outweigh the value of the political object.[8] The process is intentional—the calculated and purposeful use of force, or the potential use of force, to achieve desired political outcomes.

This is not to imply that statesmen are perfectly rational or that they do not make bad decisions. Nobel Prize-winning behavioral scientist Daniel Kahneman defines rationality as "logical coherence—reasonable or not." Logical coherence and reasonableness together is a high standard, and, as Kahneman admits, "impossibly restrictive."[9] That wars can escape rationality is a central point of Clausewitz's theory of war.[10] Passions, fears, personal and bureaucratic interests, entrapment, and biases, among other challenges, can undermine decision-making.[11] Factors such as these lead military historians Braford Lee and Karl Walling to argue that it is the responsibility of the state leadership to make strategy "as rational an instrument of policy as the circumstances of a particular war admit."[12]

How Do Outcomes Fit into Strategy?

Ideally, the strategy-making process should examine several paths to success. Strategy is not a crystal ball or blueprint for the future; rather, it is a choice about the most likely way to succeed in a competitive environment. As noted earlier, there are normally many roads that can lead to success. Returning to our taxonomy of war termination outcomes—*decisive victory, negotiated settlement,* and *transition*—is it possible that each option might require a different strategy?

A strategy designed to achieve a *decisive victory,* for instance, could focus on defeating the enemy's military forces. This is likely to require a very high level of commitment and perhaps the prioritization of military efforts. A strategy that seeks a win through a *negotiated settlement,* on the other hand, might prioritize diplomacy and require lower levels of military effort that are tightly linked to negotiating leverage. A strategy seeking *transition* might prioritize host nation political legitimacy and capability, with security efforts focused on host nation military professionalism over enemy attrition. In each case, the use and prioritization of the military instrument can vary significantly. The same is true for other elements of national power.

Even so, for the chosen strategy to be sound the assumptions underpinning it must remain valid. The environment is dynamic, so the assumptions require constant monitoring. Actors at war are engaged in continuous and competitive action-reaction-counteraction cycles. As these interactions unfold and create new situations, paths that were open at one point in time could close. Other, better paths to a successful outcome may emerge. Examining several plausible paths to success keeps decision-makers alert to these alternatives and sensitive to changes in the environment that may necessitate modifications to the strategy. In this way, strategy becomes a continuous process of diagnosis, decision, implementation, and assessment, rather than a fixed formula. Without an authoritative language and set of concepts, the United States has repeatedly experienced problems integrating its elements of power into a coherent whole.

2

The Decisive Victory Paradigm Undermines Strategy for Irregular War

Among the many different concepts that are used to classify war and warfare, the distinction between conventional and irregular wars is one of the most basic. Conventional wars are clashes between uniformed militaries, fighting largely under the same rules, until one side wins or stalemate ensues. Of the major wars fought by the United States before September 11, 2001, nearly all of them pitted uniformed American soldiers against uniformed adversaries—notable exceptions being the American Revolution (many irregular forces fought for independence outside the control of the Continental Army), the Philippine Insurrection (1899–1902), and the Vietnam War (1964–1975).

Irregular war, by contrast, is a violent struggle among state and nonstate actors to win political legitimacy.[1] The latter can be an insurgent or militant group that seeks to win the support of the population to its cause. As a nonstate actor, it often operates in and among the population; many of its members simply operate from their homes and wear everyday civilian attire, making them difficult to distinguish from civilians. The group relies on the population for recruitment, supplies, and protection. Such groups often employ guerrilla tactics to strike targets at times and places of their choosing, after which they disappear into the local population. Their goal is to wear down the conventional force.

These wars are often fought in the developing world, where local populations have major grievances against the ruling elite. Historically, irregular wars occur more often than conventional wars. These are so-called "away games" for the United States—wars in other countries—in which no existential threat exists. The comparatively small stakes involved in irregular wars make it natural for the United States to focus on conventional wars. This bias is part of what can be termed the conventional war paradigm.

The US national security establishment is organized intellectually and bureaucratically around the ability to wage conventional war. This has had major implications on how the United States develops strategy. Critically, it has shaped a mindset within the US national security establishment that sets zero-sum decisive victory as the default option for success. Once decisive victory was deemed no longer possible in the wars in Vietnam, Afghanistan, and Iraq, the United States struggled to develop a coherent alternative.

How Realistic Is Decisive Victory Against Insurgencies?

Can a strategy that aims for or implicitly assumes zero-sum decisive victory place an intervening power at higher risk of quagmire? Clausewitz's statement above suggests this is possible.[2] Competitive interaction creates complexity and uncertainty. These factors make war unpredictable and increase the risks that the intervening power will make strategy errors that undermine the likelihood of success.

A huge imbalance in the correlation of military forces, for instance, could lead an intervening power to place excessive faith in military force to achieve decisive victory. Insurgents recognize these imbalances and seek to minimize opponent strengths while exploiting inherent vulnerabilities. Thus, insurgent forces tend to wage wars of exhaustion that avoid decisive battles. They focus on population control and support. They use violence both to wear down stronger opponents and to communicate their own relevance and staying power.[3]

The military contest tends to gain the most attention, but the political, diplomatic, and economic dimensions of the conflict tend to be more decisive.[4] An intervening power that prioritizes the military effort heightens the risk of making self-induced errors. The mixed record of counterinsurgencies, even those with external support from sophisticated western powers, suggests that strategies based primarily upon military factors can increase the likelihood of failure.[5]

Political scientist Chris Paul and his colleagues at RAND were fascinated by this paradox and sought to determine the key to successful counterinsurgency. They analyzed 71 insurgencies since 1944 for common themes, approaches, and practices that led to the counterinsurgents' success. They classify 42 as insurgent wins (a 59 percent success rate), and tally 29 for the counterinsurgent (a 41 percent success rate).[6] Their study led them to examine the salience of 24 counterinsurgency approaches, and then to arrive at a scorecard of 15 good and 11 bad practices.[7]

Each counterinsurgent win, they point out, had two critical factors. First, successful counterinsurgents were able to reduce significantly the tangible support for insurgency. The keys were external sanctuary and indigenous support. Second, the host nation had to be committed to winning the battle of legitimacy in contested and insurgent-controlled areas. The absence of one factor or the other consistently led to a counterinsurgent loss.[8]

Tangible support is the ability of the insurgency to recruit manpower, obtain matériel, sustain financing, gather critical intelligence, and have access to sanctuary.[9] The ability of the insurgents to sustain such tangible support almost perfectly correlates to the outcome of the war. Tangible support is not necessarily the same as popular support. A counterinsurgent can have the support of the majority but still not win if the insurgency sustains enough tangible support to pose a material threat to the government.[10] These findings are reinforced in Jason Lyall's analysis of 286 insurgencies: material support was one of the top three determinants of the war's length and outcome.[11] If the counterinsurgents failed to reduce the tangible support available to the insurgency significantly, the result was a loss in every case. In only two cases did the counterinsurgent disrupt tangible support and still lose.[12]

The second critical factor is the commitment of the host nation to win the battle of legitimacy. Perverse practices that demonstrate failure of resolve include maximizing personal wealth and power at the expense of the state and citizenry, protecting unfair divisions of power and support, extending the conflict to bilk external supporters, and/or avoiding combat.[13] Governments motivated by such perverse incentives become predatory and exclusionary; in so being, they alienate significant portions of the population and create grievances that feed into the tangible support of the insurgency. Put succinctly, the problems articulated above undermine the legitimacy of the government.

Host nation governments experiencing such problems become unable to regain and retain support in contested and insurgent-controlled areas. Paul and his colleagues found 17 cases in which the host nation lacked commitment to winning the battle of legitimacy.[14] The host nation lost each time. In 26 of 28 wins, the host nation government met the critical test of commitment in the early phases of the conflict.[15] This suggests that highly damaging predatory and exclusionary practices become entrenched over time, making reform more difficult. At some point, reversing such problems may become impractical.

External intervention seems unable to overcome unfavorable critical factors. In 28 cases, an external actor intervened to support a host nation government

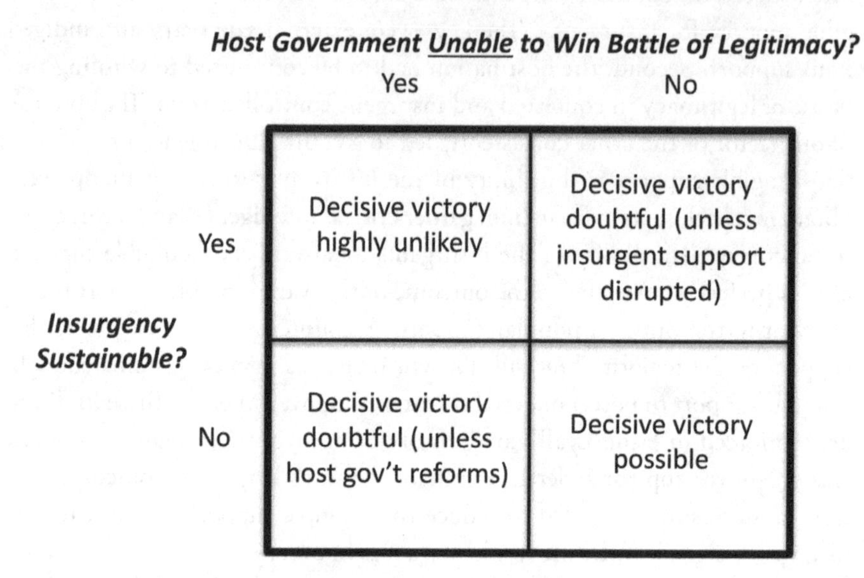

Figure 1. Critical Factors Framework

battling an insurgency. Externally supported counterinsurgents won no more often than wholly indigenous counterinsurgents.[16] Not a single intervention by a great power prevailed if one or both critical factors remained unfavorable, as seen in the cases of Vietnam (1960–1975), Afghanistan (1978–1992; 2001–2016), and Iraq (2003–2016).[17]

The presence of an unfavorable critical factor did not exclude the possibility of decisive victory. In some cases, the counterinsurgent managed to change the unfavorable variable. The insurgent's tangible support was sufficiently disrupted in Sierra Leone (1991–2002), Uganda (1986–2000), and Turkey against the PKK (1984–1999).[18] The governments of Senegal (1982–2002), Peru (1980–1992), and Angola (1975–2002) enacted a sufficient number of reforms to succeed.[19]

In sum, two critical factors appear to be salient: is the insurgency sustainable and is the host government unable to win the battle of legitimacy in contested and insurgent-controlled areas? Figure 1 above illustrates the likely outcomes against those factors. If both answers are *yes,* then the probability of the host nation achieving a decisive victory—even with external intervention—is highly unlikely.

For an external power, this table can be highly useful in determining whether or not to intervene on behalf of a host nation. If both factors are negative, it might be wiser for the external power to remain uninvolved, or to devise a strategy that aims for a negotiated outcome or transition.

On the other hand, if the insurgency is not sustainable (i.e., tangible support is effectively disrupted or nonexistent) and the government can take and retain contested and insurgent-controlled areas, the chances for a decisive victory improve significantly. External support could expedite the host nation's victory.

Decisive victory, of course, is not the only way to succeed. Insurgencies can end in a negotiated settlement—that is, a "mixed" outcome in which neither side forces the other to surrender but one does relatively better in attaining its aims. Paul and his colleagues code that actor as the winner. The reality does not need to be so binary; multiple actors can be considered successful if their aims are attained durably. This is particularly true for an intervening power that probably has less at stake than the local actors. Theoretically, a negotiated settlement in which the host nation government remains intact, the insurgency becomes part of the polity and is no longer under military threat from the intervening power, and the government and insurgency agree to measures that protect the intervening power's interests could be viewed as a win-win-win outcome.

An intervening power could also transition security responsibility for an ongoing conflict to the host nation, and then proceed to withdraw. This is premised on the "crossover" point: as the foreign counterinsurgent degrades the insurgency and builds the capacity of the host nation's security forces, the latter will become capable of defeating the insurgency. Once this threshold is reached, the foreign force can withdraw knowing that the partnering host nation is capable of succeeding on its own, or with minimal support.

Paul et al. identify six cases in which an intervening power adopted the transition method; out of these six, the government won only twice.[20] While this is too small a sample size to draw inference, as the authors themselves recognize, it is nevertheless possible to articulate some hypotheses.

The importance of the host government's ability to win the battle of legitimacy suggests that a transition strategy should place greater weight on factors that improve or prevent damage to legitimacy rather than on fighting the insurgency. Otherwise, the theoretical crossover point might never occur. As discussed in more detail later, a strategy that initially seeks a decisive victory but then changes to transition may require that the intervening force shifts its strategic and operational priorities.

The critical factors above are important characteristics of a conflict. Policymakers and strategists should consider them when determining which war termination outcomes are feasible. If both critical factors are unfavorable, the external power should avoid seeking decisive victory unless there are compelling reasons to believe the factors can be reversed. Otherwise, aiming for a different outcome or declining intervention altogether are the better choices.

While weighing the likelihood of success, the external power should also consider the element of time. The average (mean) duration of the 71 insurgencies since 1944 was 128 months (10.7 years), while the median duration was 118 months, or 9.8 years. Counterinsurgent wins tend to take longer than losses (132 months versus 72 months).[21] The potential costs in blood, treasure, and time should factor in strategic decision-making.

These calculations should also inform interventions that seek to replace an existing regime with a new, friendlier one. In such cases, the intervening power must ensure that the two factors do not become unfavorable in the aftermath. If a sustainable insurgency develops or the host nation becomes predatory and exclusionary, there is a low likelihood that regime change will result in a decisive victory.

Applying War Termination Concept to Cases

The introductory chapters have outlined a taxonomy of successful war termination outcomes for irregular war interventions, how these could factor in the policy and strategy processes, and the conditions under which decisive victory is possible.

The US government, however, has no organized way of differentiating successful war termination outcomes and their role in policy and strategy considerations. This is creating three systemic errors that have heightened the risk of major irregular war interventions becoming quagmires. First, the presumption that success requires decisive victory is restricting policy options, placing undue emphasis on the use of military force, and inducing military-centric strategies that have low probabilities of success. Second, the US government has difficulty modifying losing or ineffective strategies due to cognitive obstacles, political and bureaucratic frictions, and entrapment by the host nation. Third, as the war drags on, public support wanes, and the administration gives up on decisive victory, the United States forfeits critical leverage in announcing a withdrawal and a desire to seek negotiations or transition. The insurgency calculates that its leverage will be higher after the US military leaves and thus resists negotiating an end to the conflict. The host nation government, meanwhile, has become so dependent upon US financial and military support that transition becomes unrealistic. The United States thus becomes trapped in conflicts longer than may be necessary. The following case studies will illustrate the consequences of these errors.

Part II

The Pursuit of Decisive Victory in Afghanistan

Late November 2001. Taliban leader Mullah Mohammad Omar was in hiding. His forces were being pummeled by the Americans and their Northern Alliance allies. The Taliban's northern front had disintegrated.

As Kabul fell to the Northern Alliance in early November, Taliban formations were collapsing across the country. In the south, Hamid Karzai narrowly escaped an American air strike as he was leading local resistance forces against the Taliban.

Soon thereafter, Afghan leaders (with US urging) chose Karzai to head the new Afghan Transitional Administration—a caretaker government until Afghans could establish a new constitution and hold elections. A leader of the Popalzai tribe from Kandahar, Karzai came from a distinguished Pashtun family. International figures, such as US special envoy Ambassador Zalmay Khalilzad, recognized the appeal of having a southern Pashtun as the interim head. They hoped this might rob the Taliban of support and lower the risk of southern Pashtun resistance.

The Taliban continued fighting in the Loya Kandahar region—an area encompassing the provinces of Kandahar, Helmand, Oruzgan, and Zabul—the Taliban's traditional heartland. It is here that, almost 25 years ago, the Taliban first arose. The group's original purpose was to end the sickening butchery of innocent Afghans. The people had welcomed the "religious students" movement (Taliban means religious students in Pashtu) because they wanted the violence to end.

The bloodshed that catalyzed the birth of the Taliban was, at the time, perpetuated by the warlords. These latter were the former mujahideen (holy warrior) leaders who led the Afghan insurgency against the Soviets from 1979 to 1987. By 1992, these fighters had overthrown the Afghan communist government.

Unable to agree on a leader, the warlords created the Islamic State of Afghanistan and distributed government ministries among themselves. Like riders in the Afghan sport bushkazi, the warlords were fighting over the country as if it were a goat carcass. A civil war ensued. The warlords and their followers inflicted horrific atrocities on the Afghan people.

Overthrowing the Islamic State of Afghanistan was thus the most popular thing the Taliban ever did. Their policies and practices while in government, including human rights abuses, destroying the historic Buddha statues in Bamiyan province, and harboring al Qaeda, made the Taliban deeply unpopular among the Afghan people and the international community.

With the United States and seemingly the entire world backing Karzai, Taliban leader Mullah Mohammad Omar reportedly weighed the risks and opportunities of ending the fighting and throwing his support behind Karzai.

The one-eyed leader of the Taliban decided to give it a shot.

He sent word to Karzai that he was willing to negotiate an end to the fighting. Representatives of the two leaders agreed to meet on a hill in the Kandahar district of Shah Wali Kot. The Taliban delegation was led by Mullah Obaidullah Akhund. Mohammad Tayyeb Agha, the Taliban leader's secretary and trusted confidante, and others accompanied him.[1]

Haji Ibrahim joined the Karzai delegation. A Popalzai kinsman of Karzai, he explained the meeting to me years later. He wore the traditional southern Pashtun black turban over his bald head and a white shalwar kameez (loose fitting clothing worn by Afghans) with a black pinstripe vest. Deep grooves wrinkled his leathery face and framed a set of piercing eyes.

Weeks before the fateful meeting in Shah Wali Kot, when Karzai was trapped by Taliban forces, Haji Ibrahim had sped Karzai out to safety on his motorcycle.

During the parley, the Taliban reportedly offered a cease-fire and support to the Karzai administration in exchange for the Taliban being allowed to live in peace. It was a shrewd move—Karzai had no political constituency in Afghanistan. The Taliban could provide that as well as become a counterbalance to the conquering Northern Alliance.

There were other Pashtun rivals who could make Karzai's life difficult. Abdul Rasul Sayyaf, a leader of one of the Soviet war–era mujahideen parties, had accrued a significant following in and around Kabul. Gulbuddin Hekmatyar, leader of Hizb-i-Islami Gulbuddin (HiG), another mujahideen party, had been in exile in Iran and could begin angling for power. His party was powerful in the east as well as in the Pashtun enclaves of the north.

As Karzai tells the story, he agreed in principle to the Taliban's offer but knew he needed to gain American approved before proceeding.

It seemed like a win-win. Karzai would gain important southern Pashtun backing. Mullah Omar would openly support the new Afghan government. Provided sensible guarantees could have been made, the United States might have gained the Taliban's support for America's primary objective: locating Osama bin Laden and al Qaeda's high command. But the United States did not see a distinction between al Qaeda and the Taliban.

American Defense Secretary Donald Rumsfeld flatly refused Karzai's proposal. "There are a lot of fanatical people," he concluded. "And we need to finish the job."[2] The United States meant to eradicate both the Taliban and al Qaeda.

US forces continued hunting Taliban senior leaders. Those that turned themselves in or were captured were sent to military prisons in Bagram, Afghanistan, and Guantanamo Bay, Cuba.

The Taliban fled to Pakistan. Like the Taliban, the best early chance for a favorable and durable outcome to the conflict in Afghanistan vanished.

3

Light Footprints to a Long War

Overview

Operation Enduring Freedom seemed an outright success. By the end of 2001, the Taliban regime had disintegrated, with many senior Taliban and al Qaeda leaders fleeing across the border into Pakistan.[1] Northern Alliance forces moved into and beyond Kabul.[2] The Bonn Conference in November 2001 began the process of forming a new government and political order in Afghanistan. Hamid Karzai was selected to lead the transitional administration; he would be elected president in 2004 and again in 2009. All that was left, the Bush administration believed, was to hunt down the remaining al Qaeda and Taliban senior leaders while the international community rebuilt Afghanistan, so the United States could move to the next phase in the Global War on Terrorism.[3]

After the United States rejected the peace overture in December 2001, however, the Taliban began reorganizing.[4] They would continue to grow well into 2006, largely due to such factors as external financial backing, sanctuary in Pakistan, and local support in Afghanistan.[5] At the same time, the Afghan government was transforming into a predatory kleptocracy that fomented widespread resentment and fueled the return of the Taliban.[6] By 2019, according to the Department of Defense, the Taliban controlled or contested over half the country.

Afghanistan had the potential for a successful outcome. The Taliban in 2001 was among the world's most maligned and incompetent governments.[7] The September 11 attacks generated global antipathy toward their brutal regime. The state of ruin in Afghanistan after over 20 years of continuous conflict and misrule engendered international support for ousting the Taliban. The regime collapsed quickly after the US-led attack. Although Afghanistan is a polyglot of several ethnicities and a small, ethnically distinct Shi'a minority, no ethnic or sectarian conflict stirred in the wake of the Taliban's fall.[8]

How, then, did things deteriorate? *New York Times* reporter and longtime regional expert Carlotta Gall blames Pakistan for "driving the violence in Afghanistan for its own cynical, hegemonic reasons" and criticizes the ineffectual US responses to it.[9] Journalist and regional expert Ahmed Rashid adds American neglect of nation-building.[10] Afghanistan expert Barnett Rubin cites additional US mistakes, such as an overly militarized focus, bureaucratic dysfunction, failure to utilize early opportunities for reconciliation, and torture at American detention facilities.[11] Former ISAF and Pentagon senior advisor Sarah Chayes points to US acquiescence and unwitting promotion of corruption in the Afghan government.[12] Political scientist Stephen Walt and retired US Army lieutenant general Daniel F. Bolger underscore overly ambitious aims and the underestimation of the means required to achieve them.[13]

These views have merit. In many ways they are a product of a more fundamental problem. The administration failed to examine various war termination outcomes to determine which held the best prospects of a favorable and durable result: one that prevented Afghanistan from once again becoming an al Qaeda safe haven, and permitted the United States to pivot to address other threats. Instead, the administration assumed that a decisive military victory that ousted the Taliban and was followed by helping Afghan allies develop a new government would lead to a favorable and durable political outcome.

US policymakers should have examined multiple war termination outcomes. Decisive victory, as the Bush administration envisioned it, amounted to the Taliban's elimination, to include their acceptance of the new government, imprisonment, submission to interrogation and war crimes trials, and provision of intelligence that led to the capture of Osama bin Laden and his top lieutenants. Success under this construct required the support of neighboring and regional countries, especially Pakistan, to apprehend and turn over Taliban senior leaders who fled from Afghanistan. It also required the formation of an Afghan government that could win the support of people who had backed the Taliban. Once the key requirements for a decisive victory outcome were established, policymakers should have determined the necessary ways and means to bring the outcome about and then war-gamed the likelihood of success compared to other options.

A second option, for example, might envisage a negotiated outcome with the Taliban after the latter's military overthrow. The Taliban leverage would be very low, so they might accept major concessions in exchange for being allowed to participate politically in Afghanistan. This approach would likely engender greater support from neighbors and regional actors. It would also bring a defeated Taliban into a new Afghan political fabric. The resources needed for this option would probably be lower than the decisive victory and would rely more heavily on diplomacy than military force. International peacekeepers could be needed, too.

There is no evidence, however, that war termination options were ever examined.

Reluctant to Get Too Involved

The intellectually lazy presumption that the overthrow of the Taliban would result in a decisive victory was reinforced by two other currents at play in 2001: antipathy toward so-called nation-building and a belief that information technology and precision munitions would revolutionize war and reduce the need for sizeable land forces.

The September 11, 2001, terror attacks by al Qaeda on the American homeland were psychologically dislocating. The day was deadlier than the Japanese attack on Pearl Harbor on December 6, 1941, and left Americans with a chilling sense of vulnerability. President Bush felt compelled to respond but carefully sought to limit the scale and duration of the American commitment because of perceived terrorist threats across the globe.

The Bush administration's approach to the war in Afghanistan was rooted in a rejection of nation-building and a desire to avoid getting bogged down in a remote, landlocked country. As a presidential candidate, George W. Bush had campaigned against nation-building to distinguish himself from President Bill Clinton and from his election opponent, Vice President Al Gore. The peacekeeping mission in Somalia, begun during the George H. W. Bush administration, had ended disastrously early in the Clinton administration.[14] Similar missions to Haiti, Bosnia, and Kosovo were unpopular with many military officials and critics, who viewed such efforts as a distraction from the primary mission of fighting and winning the nation's wars.[15] Although the Balkan missions had kept the peace, the NATO military footprint was sizable. Bush argued during the campaign that they had drained resources, sapped readiness, and undermined the military's morale.[16] As Defense Secretary Donald Rumsfeld later put it in December 2001: "Nation building does not have a brilliant record across the globe."[17]

The fear of getting bogged down was reinforced by Afghanistan's history as the so-called "graveyard of Empires." Afghanistan had earned a reputation—justified or not—for forcing occupying powers into quagmires.[18] The Soviet defeat, having occurred less than 15 years earlier, was still in the memories of senior officials such as Vice President Richard Cheney and Rumsfeld. The administration wanted to avoid the same trap.[19]

A related current in the Bush administration's thinking was Secretary Rumsfeld's belief that the Pentagon needed to lose its fixation on large-scale ground forces. Information technology, he believed, had revolutionized military affairs and permitted wars to be fought and won with far fewer "boots on the ground." The new defense secretary had challenged what he considered to be an antiquated view of war by the Pentagon's brass, particularly the army. Enthusiastic about concepts of "network-centric warfare," which promised "dominant battlespace knowledge," Rumsfeld envisioned wars of the future fought from the air and by small special operations forces teams calling in precision fires on massed enemy formations.[20] In a war with global dimensions, harnessing such promise would be critical for success. Afghanistan was to be a proving ground for his vision of future war.

Operation Enduring Freedom thus envisioned US special operations forces teaming up with the Northern Alliance—a collection of armed militias—to oust the Taliban regime. The daring military campaign was a spectacular success. Special operations forces on horseback called in American firepower that obliterated Taliban positions and annihilated the Taliban's efforts to move reserve forces. Having never experienced such devastatingly accurate firepower, many Taliban leaders complained that their forces became demoralized and began to disintegrate. Northern Alliance forces poured through the breaches. The Taliban were collapsing. The race to Kabul was on. Meanwhile, other special operations forces sought local allies in the south and east. Hamid Karzai returned from exile to lead resistance forces in his native Kandahar—a critical province in the Taliban's heartland of southern Afghanistan.

By using special forces and local partners instead of a large-scale conventional attack, the United States could conserve valuable resources, maintain strategic agility, and, ideally, prevent a feared backlash against foreign presence.[21] Experts on Afghan history reportedly reinforced the administration's fears of getting bogged down, leading Rumsfeld to believe that a very limited military footprint that continued hunting al Qaeda and Taliban leaders would best prevent reported Afghan hostility toward foreigners from developing into armed conflict.[22] The United States wanted to win using minimal resources, then exit as quickly as possible—an approach later dubbed as "light footprint."[23]

One of Many Fronts

The Bush administration, moreover, was seeing the conflict as part of the Global War on Terrorism rather than as a war in Afghanistan alone. "This military action is a part of our campaign against terrorism," Bush told the American people on the eve of the Afghanistan invasion, "another front in a war that has already been joined through diplomacy, intelligence, the freezing of financial assets and the arrests of known terrorists by law enforcement agents in 38 countries. Given the nature and reach of our enemies, we will win this conflict by the patient accumulation of successes, by meeting a series of challenges with determination and will and purpose."[24]

Terrorist attacks on the Indian Legislative Assembly on October 1, 2001, and on the Indian Parliament on December 13, 2001, for which Pakistan was deemed culpable, seemed to add credibility to the claim that the threat was global.[25] Meanwhile, the Bush administration was convinced that Saddam Hussein's Iraq was a state sponsor of terrorism.[26] If the latter provided al Qaeda with weapons of mass destruction, future attacks on the United States could be far more damaging than those of September 11.

"We're a peaceful nation," Bush argued, "Yet, as we have learned, so suddenly and so tragically, there can be no peace in a world of sudden terror. In the face of today's new threat, the only way to pursue peace is to pursue those who threaten it."[27] In his State of the Union address on January 29, 2002, Bush famously called out Iraq, Iran, and North Korea as an "axis of evil" that threatened America and its allies with weapons of mass destruction through "a terrorist underworld—including groups like Hamas, Hezbollah, Islamic Jihad, Jaish-i-Mohammed—[that] operate[s] in remote jungles and deserts, and hide[s] in the centers of large cities."[28] The threat was shadowy, dispersed, and deadly. From this perspective, the United States could not afford to get fixated on Afghanistan.[29]

An Unrealistic Approach

The problem was not so much the overall aim of defeating al Qaeda and preventing their return to Afghanistan, nor the desire to retain strategic flexibility to address the unpredictable challenges of a larger Global War on Terrorism, but the contradiction between a decisive victory outcome and the minimalist ways and means devoted to it. Returning to the critical factors framework, the Bush administration needed to ensure that the post-Taliban government could win and retain popular legitimacy and that the Taliban would be unable to foment a viable insurgency.

There is no evidence that the Bush administration thought in detail about these challenges nor the war termination outcome that would best address them, given the strategic aims and selected means. There seems to have been an implicit belief that the Taliban were a regime without any internal constituency and thus incapable of organizing resistance after being overthrown. The United States also assumed implicitly that once the Taliban were overthrown, all Afghans would work together peacefully for the common good to bring about an inclusive, democratic government. Likewise, the Bush administration seems to have assumed that Afghanistan's neighbors, Pakistan in particular, would welcome whatever outcome emerged. None of those beliefs proved valid.

The failure to test these assumptions by examining other war termination outcomes blinded the Bush administration to a damaging contradiction: the pursuit of decisive victory increased the likelihood of a Taliban resurgence from sanctuary in Pakistan and a long-term, large-scale military commitment to deal with it. A negotiated outcome after the Taliban's overthrow, on the other hand, was more likely to avoid the quagmire in which the Bush administration soon found itself.

4

Plans Hit Reality

A Recent History of Bad Neighbors and Worse Governance

The standard rational model of economics assumes that people take risks when odds are favorable and avoid risk when they are not. As intuitive as this may sound, it might not be true. Daniel Kahneman and Amos Tversky, two of the world's most prominent scholars on decision-making in risk and uncertainty, discovered that problems such as cognitive bias, availability heuristics, and intuitive decision-making consistently lead to decisions that fail to achieve value maximization.[1] The individual tests they performed can be criticized for involving math more complicated than most individuals perform on a daily basis, basing conclusions on answers involving trivial sums of money, and observing decisions made in isolation.[2] Nonetheless, the aggregate results reveal that decisions are not always rational—at least from a standard economic model.[3]

The tendency for plans and forecasts to be unrealistically close to best case scenarios is what Kahneman and Tversky call the planning fallacy. Kahneman notes that executives routinely take on risky projects because they are overly optimistic about the odds of success.[4] Their decisions, he argues, could be improved by consulting the statistics of similar cases.[5]

War is unpredictable. The prospects for success can defy precise modeling and probabilities.[6] But knowledge of such probabilities can be useful in checking for overly optimistic assumptions and forecasts. The critical factors framework, for instance, can help policymakers test the likelihood that a particular conflict has the potential for a decisive victory outcome. If a power intends to intervene on behalf of a government of damaged legitimacy to achieve a quick, decisive victory against a sustainable insurgency, the historically low probability of success should force policymakers and strategists to explain why they believe this situation is sufficiently different from historical norms.

Were the emerging problems in Afghanistan foreseeable? A brief review of Afghanistan's recent history shows obvious red flags.

Regional politics weigh heavily on landlocked Afghanistan. Pakistan, India, Iran, and others have historically co-opted certain Afghan constituencies to secure their interests and to check or undermine the interests of their rivals. Frictions between Afghanistan and Pakistan since the latter's founding in 1947 created enduring tensions, leading Afghans to view malign activity from their eastern neighbor as the root cause of most of their problems. Support from India and Iran (and often Russia) has been a historic counterbalance.[7] The Indo-Pakistan rivalry in Afghanistan has been intense and bloody, as both countries have sought controlling influence.[8] Iran has mainly sought to secure political and economic interests in their near abroad of western Afghanistan and with their Shi'a coreligionists (the Hazaras in central Afghanistan), and to prevent both Pakistan and Saudi Arabia from gaining too much influence in Afghanistan.

The Durand Line and Tensions with Pakistan

Pakistan and Afghanistan have had a particularly difficult history. An 1893 *Memorandum of Understanding* between Mortimer Durand, the foreign secretary of British India, and Afghanistan's Emir Abdur Rahman Khan demarcated the border between British India and Afghanistan. The so-called Durand Line separated various Pashtun tribes between the two empires and placed Baluchistan in British India.[9] These were once the lands of the Afghan empire that stretched to the Indus river in the east and the Indian Ocean in the south.[10] By 1893, these areas were locally controlled but under the influence of British India. Afghanistan was quick to deny the validity of the agreement.

Taking advantage of British exhaustion after the First World War, Afghans sought to test the boundary during the Third Anglo-Afghan War of 1919. Afghanistan was defeated, forcing Emir Aminullah Khan to reaffirm the border in Article 2 of the peace treaty that ended the conflict.[11]

Afghanistan, however, has been persistent in denying the legitimacy of the border. When British India was partitioned in 1947 to create a primarily Hindu India and a primarily Muslim Pakistan, Afghanistan was the only state in the United Nations that refused to recognize Pakistan.[12] Since then, tensions have existed between the two countries. They found themselves on opposing sides during the Cold War. Afghanistan sought and received material and economic support from India and the Soviet Union to resist Pakistan.[13] The

United States and Pakistan, meanwhile, became partners in the Southeast Asia Treaty Organization (SEATO) and the Central Treaty Organization (CENTO), parts of a system of alliances designed to contain the Soviet Union and communist expansion. Pakistan tried twice to call on US military assistance in its wars with India in 1965 and 1971, only to be disappointed.[14] This history helped perpetuate a new narrative in Pakistan that would have reverberating effects in later years: American abandonment.

The Soviet-Afghan War

In 1973, Mohammad Daoud Khan, former prime minister and first cousin to Afghanistan's King Zahir Shah, overthrew his cousin, the king, and established a republic with himself as president.[15] Daoud Khan was subsequently overthrown during the USSR-backed communist-led 1978 Saur Revolution, which created the People's Democratic Republic of Afghanistan (PDRA). This new USSR- and India-aligned Afghanistan created alarm in Pakistan. In Pakistan's view, a two-front war to dismantle the Pakistani state was now a clear and present danger. Pakistan began to support various mujahideen parties that arose and rebelled against communist Afghanistan.

Seven Pakistan-backed mujahideen parties emerged. The so-called Peshawar Seven (Peshawar is a city in western Afghanistan) consisted of Jamiat-e-Islami (led by Burhanuddin Rabbani), Hizb-i-Islami Gulbuddin (Gulbuddin Hekmatyar), the Islamic Union for the Liberation of Afghanistan (Abdul Rasul Sayyaf), the National Islamic Front for Afghanistan (Pir Gailani), the Afghanistan National Liberation Front (Sibghatullah Mojaddedi), and the Revolutionary Islamic Movement (Mahammad Nabi Mohammadi).[16] Alarmed by the burgeoning insurgency, the Soviets installed Babrak Kamal in 1979 and began to provide large-scale military support. Hundreds of thousands of Afghans fled to neighboring countries, mostly Pakistan. Seeing an opportunity to bloody the Soviet nose, the United States began a large-scale covert program, funneled through Pakistan, to support the rebels.[17]

By 1985, the Soviets were withdrawing from Afghanistan. The 1988 Geneva Accords signed by the USSR, Pakistan, and the United States called for a cease-fire and an end to external support for the Afghan mujahideen.[18] Pakistan, contrary to the accords, continued providing support and sanctuary to the Afghan mujahideen. The Soviets kept military advisors in Afghanistan to support the communist regime led by President Mohammad Najibullah.

Meanwhile, Soviet-backed communist governments in Eastern Europe began to fall in late 1989 and the Warsaw Pact military alliance crumbled. With the Soviets out of the Afghan war and their grip on Eastern Europe failing, American interest in Central Asia faded.

The Afghan Civil War

The loosely aligned mujahideen parties continued fighting against the Najibullah regime from 1989 to 1992. As the Soviet Union collapsed and could no longer fund their Afghan clients, the Afghan state imploded under the combined weight of insurgency and fiscal crisis.

The two largest mujahideen parties, Jamiat and Hizb-i-Islami Gulbuddin, vied for control of Kabul. Gulbuddin Hekmatyar, Pakistan's closest mujahideen ally, was poised to invade the capital from the south. Ahmad Shah Massoud's Shura-e-Nazar (Supreme Council of the North) army, which supported Jamiat, raced from the north. Control of the capital meant leverage in forming a post-communist government. Osama bin Laden reportedly tried but failed to broker a peace agreement between the two sides.[19] Poised to seize the capital in April 1992, Hekmatyar hesitated. Reportedly, he wanted to organize his forces to enter Kabul in triumph the next morning. Massoud was quicker: he seized the capital that night.

With the communist government overthrown and the mujahideen in control, Pakistan must have believed it could finally secure a friendly—even client—government in Afghanistan. They would be disappointed. Pakistan hosted the Peshawar Accords of 1992 that created the Islamic State of Afghanistan (ISA). This was a power-sharing agreement among six of the seven mujahideen parties, effectively salami-slicing government into warlord-controlled fiefdoms. Hekmatyar refused to sign the agreement because he believed that he deserved to be in charge. His powerful Hizb-i-Islami Party began fighting the ISA, with support from Pakistan. The Afghan civil war had begun.

Pakistan attempted to broker another peace deal in 1993 (the so-called Islamabad Accords). This agreement installed Hekmatyar as prime minister, with Burhanuddin Rabbani, head of the overwhelmingly Tajik Jamiat Party, as president. The cease-fire lasted barely 24 hours. The new prime minister's forces began shelling the capital in a renewed bid to take Kabul by force.[20]

With Pakistan backing Hekmatyar, the main parties of the ISA sought funding elsewhere. Sayyaf's party was reportedly bankrolled by Saudi Arabia, Abdur Rashid Dostum's Uzbeks by Uzbekistan, and the Shi'a parties by Iran. ISA president Burhanuddin Rabbani's Jamiat-e-Islami (to include Massoud's Shura-e-Nazar) were funded mainly by India for the purpose of ensuring Afghanistan did not become a client state of Pakistan.[21]

Back in America, the Clinton administration struggled to find a solution. Ambassador Robin L. Raphel was one of America's foremost South Asia experts. She spent many years in Iran, marrying another American diplomat there in 1972 just prior to the revolution and the fall of the shah. Raphel continued serving in the region, which included postings in Islamabad and New Delhi. In 1993, as the United States began to recognize the growing importance of the increasingly unstable region, President Bill Clinton appointed her as the first US assistant secretary of state for South Asia.

Raphel raced to Afghanistan in November 1993, braving the chaos to talk with key leaders in an effort to end the civil war. She saw President Rabbani and Massoud in Kabul and met a new group—the Taliban—in Kandahar. She talked with Hekmatyar at his field headquarters just outside of Kabul. Her message was the same to all: there is no military solution; a political resolution is needed for Afghanistan to move forward. They refused. Years later, in June 2018, Raphel and I met Hekmatyar at his home in Kabul (he was given the place after making peace with the Afghan government). "Our last conversation," he recalled to Raphel with a smile, "was not so pleasant."

Afghanistan descended even further into chaos. Kabul was rubbled as factions fought in the streets and neighborhoods. Murder, mayhem, and massacres were common. Throughout the countryside, warlords and local strongmen ran amok, murdering, raping, and pillaging. Their crimes against the Afghan people were staggering.[22] Seeking advantage, Sayyaf and President Rabbani invited al Qaeda leader Osama bin Laden to Afghanistan. Bin Laden had fought with the mujahideen parties during the Soviet-Afghan War. He based al Qaeda, an international terrorist organization, in Sudan. Following pressure from the United States, bin Laden was forced to leave.[23] Sayyaf and Rabbani believed the terrorist leader's money, connections, and military support would prove valuable in the Afghan civil war.

Enter the Taliban

The Taliban arose in 1994 in opposition to the wanton lawlessness that had become daily fixtures of life in Afghanistan.[24] Warlords on all sides of the civil war set up checkpoints on main roads to control movement, extract tolls, and reportedly rape boys and women. These abuses outraged Afghans. In Oruzgan, Kandahar, and Helmand, self-styled "students" banded together under the leadership of a one-eyed mullah named Mohammed Omar—a veteran mujahideen of the Soviet war—to put a stop to the warlords. Support surged behind them. Afghans wanted an end to the bloodshed and began to believe the Taliban could deliver peace and justice.

The movement gained momentum. As Hekmatyar looked increasingly unlikely to be successful, Pakistan began to move support toward the Taliban. On September 27, 1996, the latter seized control of Kabul and established the Islamic Emirate of Afghanistan.

The Taliban came to power under the banner of being anti-government. The group's leaders had little to no experience in government—and the government they knew (ISA) was corrupt and predatory. They believed that minimalist governance according to Islamic principles should bring peace to Afghanistan.

Such hopes vanished in the face of reality. The needs of a desperate population after nearly 20 years of war were overwhelming. The Taliban had no idea what to do. The economy and infrastructure were shattered by the civil war. Homeless and displaced Afghans numbered in the millions. There was no functioning system of government, no institutions that could address such staggering problems. Afghanistan desperately needed international aid. Only three states, however, had recognized the Taliban government: Pakistan, United Arab Emirates, and Saudi Arabia, and none of them provided much aid. "We were a government once," Mullah Omar's former secretary Tayyab Agha told me in 2011, "But we were cut off from the international community. We did not have the knowledge or the means to govern properly." Al Qaeda, according to Taliban officials I spoke with, was the Taliban's primary source of funding.

Without the money or expertise to govern, the Taliban focused on what they knew—enforcing their Deobandi version of Sharia law. They organized so-called Virtue and Vice squads to ensure compliance. They undertook the bizarre decision to destroy the thousand-year-old giant Buddha statues in Bamiyan. They harbored al Qaeda. They staged public executions of women and men in the Kabul soccer stadium. The Northern Alliance factions, some of whom had representations in Washington, D.C., and other western capitals, needed little help in encouraging the view that the Taliban was uniquely evil.

"My life as a woman was no different under the Taliban than under the warlords," an Afghan civil society leader told me in 2018. "They all had the same views and the same practices." But by that point, the history of the ISA and the civil war had been forgotten in the West; the Taliban had become a pariah in the eyes of the international community.

By 2001, the Taliban controlled 90 percent of the country. The Northern Alliance, led by Massoud's Shura-e-Nazar, resisted the Taliban from their remaining strongholds in the Panjshir valley and Badakhshan province. These forces were funded by India, Iran, and others who challenged Taliban control of (and wanted to prevent Pakistani hegemony over) Afghanistan.

If Pakistan had hoped the new regime would become a client state, or at least recognize the Durand Line, they were disappointed. During the five years of misrule by the Taliban, no agreement was made to ratify the border.

To the west, Iran had opposed the Taliban regime and supported the Hazara factions fighting it. Iran strove to prevent the rise of a Saudi Arabian client state in Afghanistan and to reduce narcotics trafficking across its borders.

Misreading a Complex Situation

On September 9, 2001, two al Qaeda operatives posed as reporters to interview Shura-e-Nazar leader Massoud at his camp in the Panjshir valley. Reportedly vain with a keen hunger for public acclamation, Massoud agreed to sit down with them to tell his story. The camera was full of explosives. It detonated, killing Massoud and throwing the Northern Alliance into chaos.

Perhaps this assassination was designed as a favor to the Taliban. Western officials had been pressuring the Taliban to arrest and turn over Osama bin Laden and his chief lieutenants after the terrorist attacks on the American embassies in Kenya and Tanzania in 1998 and the attack on the warship USS *Cole* in 2000. According to former senior Taliban officials, bin Laden had reportedly assured Mullah Omar that al Qaeda was not planning any more attacks against the West.

Then came the attacks of September 11, 2001.

The Taliban joined virtually every state in the world in condemning the attacks. As it became clear that al Qaeda was responsible, America demanded the Taliban hand over bin Laden. The Taliban prevaricated, asking for evidence that bin Laden was behind the attack. Frustrated and out of patience, the United States rapidly planned the military campaign to oust the Taliban regime and kill or capture bin Laden.

That Afghanistan was riven with internal tensions exacerbated by external rivalries was information easily available to American policymakers in 2001. These challenges should have suggested that establishing a post-Taliban regime that would be accepted by Afghan elites and Afghanistan's neighbors would be extremely difficult. Failure to manage these tensions had the potential of leading to a renewal of violence that could bog down American forces.

The Bush administration recognized as early as October 2001 that some of these regional frictions and interests could be problematic but did not think through how their decisive victory approach might affect the calculations of Afghan and international actors. They presumed a spirit of cooperation in the aftermath: that Afghan elites would sacrifice self-interest for the common good, and that Afghanistan's neighbors would abandon their intense rivalries and help the new government succeed. They never considered the likelihood that such internal and external frictions could morph into a new and different conflict. A US State Department cable noted optimistically, "We do not see any irreconcilable conflict among these interests as long as Afghans and outside interests are flexible."[25]

The Bush administration fell victim to the planning fallacy. In presuming decisive victory, the Bush administration underestimated the post-Taliban regime challenges as internal and external rivalries unfolded in predictable ways.

5

The Fall of the Taliban and the Bonn Conference

Scientist and bestselling author James Gleick has been fascinated by the science of chaos ever since being introduced by Edward Lorenz to the so-called butterfly effect. Lorenz, a meteorologist, wondered about the unpredictability of weather systems. He noted that minor changes to what he called the initial conditions of the system could produce disproportionately large effects. He famously described that a butterfly flapping its wings in Indonesia could create a hurricane in the Caribbean.[1]

Linear relationships can be captured on a graph. They are very attractive, Gleick writes, "because you can take them apart, and put them back together . . . the pieces add up."[2] The weather, by contrast, was what Lorenz called a nonlinear system; elements within the system expressed relationships that were not strictly proportional. Large inputs to a system might result in little to no change. Relatively small or subtle inputs to initial conditions, by contrast, could produce significant differences.

Physicist M. Michael Waldrop turned these insights into a larger study of complexity. A system is complex when "a great many independent agents are interacting with each other in a great many ways."[3] The Latin roots of complex (com = together; plex = woven) mean interwoven. This is in contrast to complicated, which means layered together. A complicated system is linear; that is, the pieces add up. Real life, as Waldrop, Gleick, and others in the field suggest, is more complex than complicated. Independent actors interact in a great many ways, and the outcomes are uncertain.

The danger for policymakers and strategists is mistaking a nonlinear system for a linear one—trying to turn a complex situation into a complicated one. The urge to do so is tempting, as it is much easier to explain a way forward in a linear manner, where the pieces add up, than to make sense of a complex situation and draw out the strategic implications. In Washington, D.C., where cabinet-level officials are bombarded with myriad global problems and difficult domestic policy challenges, and have major departments to run, there is limited bandwidth for nuance. Linear explanations and solutions can be far easier to grasp and accept.

The problem is that the linear approach is normally at odds with reality. This problem frustrated General Stanley A. McChrystal, the former commander of America's Joint Special Operations Command and, later, of the International Security Assistance Forces—Afghanistan (ISAF). In his bestselling book *Team of Teams,* McChrystal describes the challenges his elite special operations forces had keeping pace with the enemy in Iraq, despite the fact that their adversaries tended to be poorly trained and resourced. The core problem: "We were using complicated solutions to attack a complex problem."[4]

Things that are complicated, he explains, have many parts but they are joined in relatively simple ways. The workings of an aircraft engine might be confusing, but they can be broken down into a clear set of relationships. Complexity occurs when the interactions create multiple interdependencies.[5] The elements cannot be broken down into a clear set of linear relationships.

Once McChrystal's team began to appreciate complexity, they were able to design strategies that ultimately defeated al Qaeda in Iraq. "The solution we devised," he recalls, "was a 'team of teams'—an organization in which the relationships between the constituent teams resembled those between individuals on a single team: teams that traditionally resided in separate silos would now have to become fused to one another via trust and purpose."[6] Open communications, shared consciousness, common purpose, and empowered execution made McChrystal's team of teams successful.

The US and international efforts in Afghanistan, unfortunately, were being plagued by the same problems McChrystal was encountering—the tendency to use complicated, linear solutions to address complex problems. We see this materializing in the arrangement of political, diplomatic, military, and economic milestones for Afghanistan, and in US strategies that employed bureaucracy-centric efforts (military campaign plans, capacity-building programs, economic development initiatives, etc.) that were unprioritized and poorly integrated.

The whole became less than the sum of its parts. Multiple, independent entities interacted with one another to exploit the chaotic situation and turn major milestones to their advantage. Local actors sought to secure power and influence while reducing that of their rivals. Afghanistan's neighbors vied for influence to secure their interests and keep others off balance. The efforts of US bureaucracies and international actors often worked at cross-purposes. As

US and international officials fixated on achieving certain milestones, they grew blind to the deterioration of the broader political and security situation that was happening in plain sight. They were implementing a complicated solution to a complex problem.

A Technocratic Approach

As the Northern Alliance and coalition forces advanced against the Taliban and al Qaeda, US and international diplomats began discussions about post-Taliban Afghanistan. The United Nations convened an international conference outside of Bonn, under the leadership of veteran diplomat Lakhdar Brahimi. They aimed to gather together Afghan leaders who could find a political solution.

The conference participants established key milestones for the development of a new government: creation of an Afghan Interim Transitional Administration; convening an *Emergency Loya Jirga*[7] in 2002 to select an Afghan Transitional Administration, which would develop a new constitution and govern the country until the 2004 elections; a 2003 *Constitutional Loya Jirga*; a presidential election in 2004, and parliamentary elections a year later. Follow the road map, they assumed, and Afghanistan would have a government supported by the Afghan people.

The diplomats crafted a complicated solution to a complex problem. While western diplomats tended to see the way forward in technocratic milestones and processes, Afghan and regional actors viewed the situation as an opportunity to gain power and influence—and to deny the same to their rivals. Whoever won the looming scramble would gain control of the government as well as western largesse; they could also use western backing to take down their rivals. Authoring the so-called facts on the ground—namely, seizing Kabul and assuming control of security ministries—would give the possessor control of the game.

Facts on the Ground

By November 10, 2001, Massoud's Shura-e-Nazar forces, now commanded by Mohammad Qassim Fahim, were racing toward Kabul. Fahim appreciated the importance of seizing the capital, as his leader had done in 1992. Control of the capital in 2001 would put Fahim in control of events.

Western officials, on the other hand, wanted Kabul to remain an open city. They recognized that control of the capital would give the holder controlling influence at the Bonn Conference. Diplomatic and military officials urged Fahim to stop north of Kabul.

Fahim used Massoud's playbook instead. Ignoring US requests that he hold north of Kabul until the Bonn Conference was completed, Fahim captured Kabul on November 13 and began taking control of the city.[8] US officials watched it unfold, unprepared for such defiance. Many Afghans perceived that the United States backed Fahim's seizure of the capital. By mid-December 2001, the Taliban forces had collapsed and their leaders either returned home or fled to Pakistan for safety.[9] The vacuum created by the removal of the Taliban was filled by Northern Alliance forces and local militias, primarily those associated with Shura-e-Nazar.[10]

As Fahim's forces seized control of Kabul, they gained control of the security ministries by fait accompli. They immediately organized police and military forces and directed allied warlords and strongmen to do the same in their local areas.[11] US envoy James Dobbins recalled that diplomatic efforts were unable to keep pace with the rapidly unfolding events.

With Kabul firmly under their control, Fahim and other Northern Alliance leaders dominated the political situation in the lead-up to the Bonn Conference. Fully entrenched in the de facto security ministries, Fahim and his allies gained significant leverage. Any effort to make the security architecture of the new government more inclusive would mean ousting a Fahim-placed incumbent. He could extract a high price for any such concessions. Although the design of the conference was to invite a diverse array of non-Taliban leaders to chart a new political future for Afghanistan, the Northern Alliance and warlords from the former Islamic State of Afghanistan wielded significant influence over the composition of the Bonn Conference and Interim Transitional Authority.[12] Control the guns, control the situation.

The selection of who would lead the Transitional Authority was a delicate one. The person (and faction) chosen would have significant advantage for the 2004 elections—and would have the backing of the United States and the West. The United States grew increasingly concerned that the divisive former President Burhanuddin Rabbani would try to get himself installed as president.[13] With Northern Alliance leader Ahmad Shah Massoud assassinated by al Qaeda operatives on September 9, 2001, no natural consensus leader among the Northern Alliance was available. Even had he been alive, Massoud was a Tajik in a country traditionally ruled by the plurality Pashtuns.

Zalmay Khalilzad, an Afghan émigré to the United States and a rising star within the Republican Party ranks, was the US special envoy for the Bonn Conference. He was a natural choice to lead the American effort at Bonn. As expected, he dominated the international actors present and played a decisive role shaping the new Afghan government.

Khalilzad and others recognized the advantages of having a Pashtun head of the interim administration. Pashtuns had ruled Afghanistan since the creation of the Afghan empire under Ahmad Shah Durrani in 1747. Some favored the return of the king, Zahir Shah, who lived in exile in Rome after being deposed in the 1973 coup. Zahir Shah appeared to be willing to be selected as head of state. Sensing that America wanted Afghanistan to become a democracy rather than a monarchy, Khalilzad went to Rome and reportedly arm-twisted the former ruler. Zahir Shah dropped out and endorsed Khalilzad's choice: a tribal leader from Kandahar named Hamid Karzai.[14]

Consensus on who would lead the interim administration thus began to form around Hamid Karzai. Fiercely anti-Taliban and without blood on his hands from the civil war, Karzai proved an attractive option.[15] Because he had no party, militia, or following of his own, he was unobjectionable to the warlords, who traditionally favored weak leaders they could control and manipulate. Karzai spoke perfect, almost poetic English, which made him attractive to the United States and the western powers. He had spent much of his life in Afghanistan and in the region, rather than in the United States or Europe. Karzai, in fact, was reportedly arrested and tortured in 1994 by Fahim Khan, now the de facto leader of Afghanistan's military.[16]

The factors that made Karzai an attractive choice to lead the new government would leave him beholden to warlords. These powerful and well-armed figures could make Karzai's life miserable or could even kill him if he threatened their interests. Preventing Karzai from consolidating Pashtun support would enable the Northern Alliance leaders to maintain the upper hand.

The most powerful and organized Pashtun party was Hekmatyar's Hizb-i-Islami Gulbuddin (HiG). Gulbuddin Hekmatyar, a power-obsessed warlord who was once considered America's favorite mujahideen leader during the Soviet war, had run afoul of the United States for his role in the Afghan civil war. He welcomed the removal of the Taliban but opposed the presence of international forces in Afghanistan.

Seeing the HiG as a threat to their political hegemony, the Northern Alliance factions sought to marginalize Hekmatyar and his party and keep them out of the Bonn negotiations. They suggested that HiG was allied to the Taliban and al Qaeda, thus encouraging the American military to target Hekmatyar and his key leaders. Leaflets reportedly appeared in refugee camps in Pakistan claiming Hekmatyar had joined al Qaeda. Hekmatyar was later accused of an April 2002 assassination attempt on Karzai, which he vehemently denied.[17] Such manipulative efforts were successful in keeping the HiG out of Bonn, out of power, and in the United States' crosshairs.

In February 2003, the United States designated Hekmatyar as a global terrorist. They reportedly fired on his vehicle in May of that year.[18] Hekmatyar declared jihad on the Americans afterward. His reported efforts to join the Karzai government in 2004 were rebuffed.[19] Like the Taliban, the HiG turned to insurgency.

The Warlords Are Back

During the Bonn process the former warlord government known as the Islamic State of Afghanistan seemed to reemerge.[20] Shura-e-Nazar, Massoud's party of Panjshiris now dominated by Fahim, controlled the key security ministries: Defense (Fahim), Interior (Younis Qanooni), National Directorate of Security (Muhammad Arif Sarwari), and Foreign (Dr. Abdullah Abdullah). Other warlords such as Uzbek leader Dostum, Tajiks Ismail Khan and Atta Noor, and Hazaran leaders Muhammad Mohaqqeq and Mohammad Karim Khalili were given positions of power and influence.

Fahim promoted himself from general to "Marshal" and made it clear to Karzai who controlled the guns (and hence the real power) in Afghanistan. When Karzai arrived in Kabul in December 2001 as the newly appointed head of the Afghan Interim Authority, he was greeted on the tarmac by Fahim with his forces in military formation. Karzai's former jailer pointedly asked him, "Where are your men?" Karzai pointed to Fahim's militia and deftly replied, "You are my men." Karzai would have been unlikely to miss Fahim's not-so-subtle message.[21] Politics was blood sport in Afghanistan. Karzai would either need international military support to curb the influence of the warlords or he would have to co-opt them. This political calculation would decisively shape the rule of Hamid Karzai.

Consolidating Power in Kabul

The Bonn process had called for several steps in the political formation of a new government. The warlords and elites craftily exploited each one for personal advantage. The stakes were high for warlords and strongmen to consolidate their positions and influence. Their significant wealth and ability to use militia forces for intimidation, due in part to the scarcity of international forces, enabled these elites to ensure their representatives were always present, to buy votes as necessary, and to intimidate opposition.[22]

They were enabled in this unwittingly by the United States. Seeking to remain focused on counterterrorism operations against al Qaeda, the United States wanted to assign responsibility for security and reconstruction in Afghanistan to the international community as soon as possible.[23] This left a vacuum of responsibility for the overall international effort and no checks against the warlords.

The Bonn process began to unfold as planned. An Emergency Loya Jirga was convened on June 7, 2002, with the purpose of establishing the Afghan Transitional Administration (ATA), to include filling the presidency and cabinet positions. A proposal to invite the Taliban was floated and quickly rejected by the Northern Alliance factions.[24] Southern Pashtuns had little representation in the Emergency Loya Jirga. Feeling marginalized during the proceedings, many of them walked out.[25] Vote buying and intimidation by the warlords ensured the votes went the way they wanted.

As expected, Karzai was chosen as the leader of the ATA. The Northern Alliance retained control of all but two ministries. The ATA was responsible for drafting the new constitution, which would be ratified in the December 2003 Constitutional Loya Jirga.

The new constitution, partially modeled on the 1964 Afghan constitution, created a highly centralized government with little to no provisions for accountability. The Afghan parliament was given the authority to approve and to impeach ministers, a process that would later be used to extract enormous bribes.[26] The failure to implement a census, called for in the Bonn Agreement, and a poorly organized voter registration process left elections open to widespread fraud.[27] A method known as the Single Non-Transferable Vote system was selected for parliamentary and provincial council elections. This favored the better organized parties, which happened to be those of the warlords.[28]

The international community, Barnett Rubin concludes, reempowered leaders that the Afghan people had rejected when they sided with the Taliban in 1996.[29] Efforts by international organizations to address the history of human rights abuses by these warlords during the Afghan civil war were repeatedly repressed by both the United States and the Afghan government.[30] "The Bonn conference did not reflect the interests of the Afghan people," a former Afghan government senior advisor reflected. "It reflected the interests of those deposed by the Taliban."[31]

Gathering Storm

The gathering storm within Afghanistan and the region largely escaped the Bush administration's notice. The Americans set up bases primarily in the southern and eastern parts of the country to support the counterterrorism efforts. The military command restricted international forces to Kabul to avoid complicating the hunt for al Qaeda and Taliban leaders.[32] "General Franks is very much in charge of everything, and he doesn't want to worry about a multinational force," explained a US military spokesman. "The US has one goal: Attack AQ and get the job done. And they're not too worried about the rest of it right now."[33]

The absence of international forces in the countryside, combined with the US military's exclusive focus on hunting al Qaeda and Taliban senior leaders, left America and the international community blind to the predatory behavior of Afghan militias on the Afghan people. The Northern Alliance and local strongmen used their militias to secure power, land, and de facto authority. Civilian casualties mounted.

American forces and Afghan militias, meanwhile, closed in on Osama bin Laden and the remainders of al Qaeda in the Tora Bora region of eastern Afghanistan, but, in the end, were unable to prevent bin Laden's escape to Pakistan.[34] Despite all efforts, the world's most wanted terrorist was still at large.

Across the border in Pakistan, the situation was deteriorating just as quickly. Pakistan had supported the Taliban's rise to power because the Islamic State of Afghanistan had been closely tied to India. The rapid overthrow of the Taliban, coupled with the marginalization of both the Taliban and Hizb-i-Islami during the Bonn process, must have been disconcerting to Pakistan. India quickly became a close partner of the new Afghan government, in part because so many former partners—ex-communists and ISA leaders—were back in power, but also because India sought all opportunities to thwart Pakistan's designs.[35]

Threats from the Bush administration to Pakistan's President Musharraf were successful in gaining a measure of Pakistani support against al Qaeda, but Afghanistan's eastern neighbor was unwilling to turn on its Taliban allies.[36] The myriad Afghan refugee camps still in Pakistan gave the Taliban plenty of places to hide, plan, recruit, and reconstitute.[37] Pakistan came to believe that India and Afghanistan were supporting the insurgency in Baluchistan and aimed to partition much of Pakistan between them.[38] The Taliban and Hizb-i-Islami Gulbuddin became Pakistan's insurance against what the Pakistani military perceived to be a looming existential threat. This anxiety underpinned Pakistan's belief that instability in a hostile Afghanistan was preferable to a stable, hostile Afghanistan.[39]

US officials were tone deaf to these concerns, so Pakistan began to play a double game. To keep the United States happy, Pakistan supplied information on and even arrested al Qaeda leaders. Meanwhile, Pakistan allowed the Taliban to establish sanctuary and create an insurgency. To the United States officials, al Qaeda and Taliban were equivalent—terrorists. But Pakistan viewed these groups differently and insisted on being seen as a good partner on terrorism (al Qaeda) while silently enabling the Taliban's insurgency.

US relations with Iran, another of Afghanistan's neighbors, had become strained, too. Iran first proved helpful in the Bonn process.[40] That support ended abruptly, however, in 2002, following President Bush's declaration that Iran was part of the axis of evil. Nonetheless, Iran continued backing the Hazara factions and promoting economic development in Herat and western Afghanistan.[41]

The United States could not keep pace with the scrimmage for political power among Afghan and international actors, and most likely failed to understand the intricacies of the politics and issues. Without a governing strategy that looked beyond the ouster of the Taliban, the United States found itself unable to adapt to a dynamic situation. In Kabul, a Northern Alliance-dominated government formed with a politically weak and isolated Hamid Karzai at its head. Pakistan grew alarmed by the dominant influence competitors like India and Iran wielded over the new Afghan government. America was backing these arrangements, unwittingly feeding the fears of the southern and eastern Pashtuns in Afghanistan, as well as those of the Pakistani military.

By adopting a technocratic approach to a complex issue, the international community squandered tremendous leverage without realizing the strategic implications. Western officials were not ignorant of the broad challenges; they were, however, naive in believing they could finesse the behavior of other actors. Failing to address how the likely actions of domestic, regional, and international actors might undermine US plans was a salient mistake.

Western officials celebrated progress as the Bonn process milestones were reached, but they did not recognize that the manner in which they had been achieved damaged the legitimacy of the new government. The security services were dominated by Shura-e-Nazar and other Northern Alliance factions, military and development contracts were captured and pilfered by powerful warlords and their backers, government positions began to be brokered by warlords and sold to the highest bidder. In the eyes of many Afghans, the despised Islamic State of Afghanistan had returned in a different guise. Across Afghanistan's eastern border, the Pakistani military grew alarmed at the emerging India-allied Afghan government. These changes, many in Afghanistan and the region assumed, had international backing.

To understand why highly talented western officials were unable to see the gathering storm, let's examine America's bureaucratic way of waging war.

6

America's Bureaucratic Way of War

Historian Russell Weigley famously characterized the American way of war as the use of industrial mobilization, grinding attrition, and firepower-centric methods to crush an adversary under the weight of American might.[1] In some ways, the invasions of Afghanistan and Iraq seemed to usher in a new, bolder, swifter way of war characterized by precision munitions, special operations forces, and the seamless integration of land, sea, and air power.[2] Known as Joint Operations, the American military's ability to overcome interservice rivalries among the army, navy, and air force and operate as a team was on powerful display. Terms such as "shock and awe" described the battlefield effects of this new military synergy.

These important advances in military operations have not been matched by the US interagency. The United States has no analogous way to integrate elements of national power—diplomatic, military, political, economic, intelligence—into a coherent whole. While the US military has learned to fight wars as an integrated team, America's national security agencies have not established how to wage war as an integrated team.

Instead, America wages war by bureaucracy.

Before the war begins, diplomats try to prevent hostilities and may attempt to gather allies in the event that war breaks out. Once war is declared, uniformed militaries fight one another until one side wins, or a stalemate ensues. Following the military contest, diplomats negotiate a peace treaty, with their negotiating leverage largely determined by the outcomes of the military engagements. Once peace is concluded, aid and development agencies repair the damage.

Conventional war is typically fought in these four broad stages: (1) diplomacy, (2) clash of military forces, (3) diplomacy, and (4) aid and development. Think of a relay race: the diplomats get the baton and do their job, they hand the baton to the military to fight, the military hands that baton back to the diplomats to negotiate peace, and then the diplomats give the baton to those repairing the damages.

The relay race sequencing has the added convenience of aligning with the major national security agencies: State—Defense—State—USAID. Certainly, there is military activity occurring in support of the State-led phases (deploying forces, defending territory, securing populations, etc.). Diplomats will continue seeking support from allies or encouraging the enemy's allies to defect during the military-led phase. Their actions are normally supporting efforts to the military contest. The bureaucratic lines are very clear: each agency optimizes its own efforts in its lead phase while the other agencies support. The priorities are also well-established.

Does this sequencing and focus on military objectives work for irregular war? As noted earlier, outcomes are more greatly impacted by the insurgency's ability to sustain tangible internal and external support and by the government's measure of political legitimacy in insurgent-contested areas than by battlefield engagements.[3] This means that the political and diplomatic efforts to undermine the insurgency and to bolster the credibility of the host nation government are as (or even more) critical for success than are the military efforts—even during the war.

Because the clash between fielded forces is so central to success in conventional war, political and military leaders work carefully to "define a military objective that, if achieved, can deliver the political objective."[4] Military objectives such as the defeat of the enemy's forces, the capture of the capital, or the seizure of territory are common ways to seek a decisive military victory. Interventions against insurgencies, however, don't tend to hinge upon the clash of opposing forces; there may not even be a realistic military objective that could deliver the political aims. Moreover, the nature of guerrilla warfare renders the annihilation of insurgent forces very difficult—and nearly impossible if they have sanctuary across a national border. Furthermore, given that there is no insurgent capital, seizing territory might not make any difference. This is especially true if the population provides clandestine support to the insurgency.

The nature of the struggle between a host nation and an insurgency tends to require an intervening power to employ elements of national power concurrently rather than sequentially. Rather than a relay race that features one agency at a time, irregular war tends to be more of a team sport, like basketball. Every agency is on the court at once and must work together as a team so that the whole is greater than the sum of its parts.

The host government's political legitimacy—its ability to govern justly and effectively—is the foundation for success (State and USAID). International support and recognition can boost perceptions of legitimacy (State). To defend the people and government, recruits need to be organized and trained (Defense). The insurgency needs to be fought (Defense). Economic development and services can boost popular satisfaction with the government (USAID). Greater host nation political legitimacy leads to better information from the people about the insurgency (State and Intelligence). Better information leads to more effective military operations (Intelligence and Defense). More precise military operations damage the capability of the insurgency and prevent civilian casualty incidents that could undermine the new government (Defense).

The team aspect is critical. Like in basketball, the players cannot simply afford to stay in their individual comfort zones. Overemphasis on a single player will enable the opponents to adjust and to find and exploit weaknesses in the team overall. The 2004 US Olympic basketball team was bursting with the best professional athletes in the game. They should have won another gold for Team USA. Instead, they barely salvaged a bronze medal after losses to Puerto Rico, Lithuania, and Argentina. The obvious lesson: a group of less talented individuals who play together as a team can be more effective than top talents who play as individuals. The point guard can have the most assists, the shooting guard can have the most points, one forward can have the most rebounds, the other can have the most steals, and the center can have the most blocks—and the team can still lose the game. Talent silos will doom a group of individually exceptional players to defeat at the hands of a team of less individually gifted players who work well together.

Bureaucratic silos can create the same effect. Key national security agencies deploy their forces and capabilities to the host country. While they are encouraged to work together, the lines of authority extend from each agency lead in country directly back to Washington, D.C. There is no American official in the combat zone who has the authority and responsibility to manage all of the deployed elements of national power and who is held accountable for achieving US objectives. Each agency can optimize its performance even as the team performs poorly overall.[5]

America's bureaucratic way of war reinforces agency-centric silos and undermines the teamwork necessary for success. In Afghanistan, the absence of a coordinating authority and governing strategy led to disjointed efforts between the military campaign and diplomatic actions to organize a post-Taliban government. As military and diplomatic leaders focused on their comfort zones, major gaps developed and were exploited by others, to the detriment of US interests. The Bush administration did not put a single US official in charge of managing American efforts on the ground, so no one was empowered to integrate the efforts of various agencies and make decisions that addressed the emerging problems in the dynamic conflict. International silos added to the coordination challenges. The result was a slow failure because the whole was less than the sum of its parts.

Believing that victory was won after the Taliban's ouster, the Bush administration aimed to minimize its commitment in Afghanistan by outsourcing reconstruction to other international actors.[6] President Bush initially focused American assistance on building Afghanistan's army and national police.[7] He urged the international community to support other major needs: "We know that true peace will only be achieved when we give the Afghan people the means to achieve their own aspirations." Although the United States was not prepared to do nation-building, Bush later reflected, "We had liberated the country from a primitive dictatorship, and we had a moral obligation to leave behind something better."[8]

The administration remained divided on the type and extent of support America should provide.[9] On April 8, 2002, Defense Secretary Donald Rumsfeld discouraged Secretary of State Colin Powell from making a further commitment of American resources to Afghan reconstruction, arguing that "The U.S. spent billions of dollars freeing Afghanistan and providing security. . . . There is no reason on earth for the U.S. to commit to pay 20 percent for the Afghan army. I urge you to get DoS [Department of State] turned around on this—the U.S. position should be zero."[10] Powell replied eight days later and invoked the president's decision to support Afghan reconstruction. The United States, he argued, needed to do its "fair share," particularly regarding the Afghan military and police.[11] Rumsfeld eventually conceded, having accepted the view that by accelerating the development of Afghan security forces the United States could withdraw its own forces more quickly.[12]

Institutionalizing Silos: The Lead Nation Concept

The result of America's efforts to get the international community to support nonmilitary efforts was a commitment by the Group of Eight (G8) countries (the world's top eight economic powers) to provide a broad range of development support for Afghanistan. In April 2002, the G8 announced a "lead nation concept" for the Afghan security sector, which entailed five interdependent efforts. Each one was assigned to a lead donor nation: Afghan National Army (US), Afghan National Police (Germany), counternarcotics (UK), judiciary (Italy), and disarmament, demobilization, and reintegration [DDR] (Japan).[13]

The lead nation concept was quickly and aptly exploited by Afghan elites to promote personal and political power. With no overarching strategy or decision-making body that could set priorities, integrate efforts, and mitigate unintended consequences, the efforts of each lead nation were open to manipulation.

By November 2002, UN Security Council Resolution 1444 called for the establishment of "fully representative, professional and multi-ethnic army and police forces."[14] For the US-supported Afghan National Army, Minister of Defense Fahim wanted a large, 200,000-man conscript force raised by provincial levy (where each province provides a designated number of recruits) to secure the country and defend its borders.[15] Fahim planned to handpick the cadre, thus keeping the military under his control.

Provincial levy was the traditional method of raising an Afghan army, and it was also an effective way of ensuring every part of the country shared the burden of the military commitment. The levy would, over time, create a large body of people who had experienced military service. Such a method of raising and maintaining forces often leads to an army becoming a so-called "school of the nation." But the advantages would come with disadvantages: this type of manning system would draw recruits from an overwhelmingly illiterate population and would need to rely on simple weapons and tactics.

The United States rejected Fahim's concept and limited the size of the Afghan Army to a volunteer force of roughly 50,000.[16] US military officials believed that Afghanistan needed only a small military to dissuade warlords from using their militias to destabilize the government. According to a RAND study on the development of the Afghan security sector, "When U.S. [Security Force Assistance] efforts in Afghanistan were first being planned in 2002, a driving assumption was that the Taliban had been decisively defeated and would not reappear. Instead, the major threat envisioned was a return to the warlordism that had plagued the country in the 1990s and had given rise to the Taliban."[17] A smaller, professional force, US officials argued, would be less likely to fracture along ethnic lines.

The levy system, however, would have the same benefit as the smaller, professional force while also providing a stronger deterrent to Afghanistan's neighbors. Moreover, the levy system would enable the Afghan military to expand more easily in times of threat and contract in times of peace. Overall, such a force would be less expensive and more flexible than the smaller, professional force.

US officials remained wedded to the idea of the volunteer force, likely because of their own experiences: conscription during the Vietnam War had profoundly negative effects on both American society and the military effort. It disproportionately relied on low-income and poorly educated groups to fill the ranks; the wealthy had many ways to opt out of the system. Morale plummeted. The United States went to an all-volunteer force in the late 1970s, ultimately creating the highly educated and sophisticated military that crushed Saddam Hussein's army in the 1991 Gulf War and the Taliban in 2001. Such a military relied upon intelligent recruits who were highly motivated to join and capable of handling the sophisticated technology. As noncommissioned officers and junior officers, recruits were expected to exercise judgment and initiative.

The general population of Afghanistan, after two decades of war, was very different from the general population of the United States. Over 90 percent of the population was illiterate. The country, driven apart by decades of war, was badly in need of cohesion. Armed militant groups had already begun preying on the population. The military leaders had largely been trained in the very hierarchical Soviet system. The ingredients that made a high-tech volunteer force possible in the United States were simply not present in Afghanistan.

Mirror imaging likely provided the rationale for American insistence on a small, volunteer force. The US military officials advising the effort had no recent experience with conscript or levy forces. They had no doctrine to draw from in developing such a force. The manning, equipping, training, sustainment, and training systems required substantially different approaches than that to which the American officials were accustomed. The small, volunteer military was ill-suited for Afghanistan.

Whatever its reservations, the Afghan government had to accept what it could get. Defense Minister Fahim moved quickly to manipulate recruiting and training efforts to lock in Shura-e-Nazar's control of the army. Ninety of the 100 general officers were from Panjshir—a tiny province that was Fahim's home.[18] As insurance, Fahim kept his own large militia while using the DDR process to disarm his rivals (more on that later).[19] Many Afghans perceived US support for Fahim's actions.

The Ministry of Interior, meanwhile, was responsible for the police, subnational governance, and counternarcotics. Police development formed a silo under the Germans, and counternarcotics was supported by the British. No one bothered to address subnational governance.

German efforts to build the police forces proceeded very slowly. For a country of nearly 30 million people, the Germans planned for an Afghan national police force of 1,500 officers to be trained over the course of a five-year program.[20] The small force was to be ready by 2007. Notably, the plan for local policing was unaddressed. Rumsfeld began to get frustrated at the slow pace of forming the army and police, demanding ways to accelerate the efforts. "There is not a sufficient sense of urgency on the part of anybody," he complained to Under Secretary of Defense for Policy Douglas Feith and Chairman of the Joint Chiefs General Richard Myers.[21] The Germans and Americans began to pull the police in different directions—the former toward basic western-style law enforcement, the latter toward a paramilitary role.[22]

US counterinsurgency doctrine suggests a country should have roughly one security official for every 50 people. Hence, a country of 30 million would theoretically need 600,000. While the ratio could be smaller during peacetime, the original plan for the size of Afghan forces was barely one-tenth of the standard size.

Perverse incentives grew. Police chief positions mostly went to favored local strongmen or the highest bidder. The police chiefs used their official position to consolidate power, marginalize their rivals, and engage in illicit economic activities.

Governorships were often allocated the same way. This would create perverse incentives for officials to turn a profit on their positions through misappropriation of customs revenues, aid dollars, or even extortion (such as land theft and kidnapping for ransom).[23] Afghans believed these corrupt officials were backed by the US military. The absence of any ombudsman or government watchdog, and perceptions of international acquiescence or complicity, meant that the Afghan people had no avenue to register complaints or air grievances. No lead nation was established to promote subnational governance, so the international community had little information on how well or poorly the new government was connecting to the people outside Kabul.

The British counternarcotics efforts, meanwhile, were exploited by local strongmen, who directed international eradication efforts at their rivals while keeping unwanted attention away from their own crops.[24] The Rule of Law sector did not develop beyond an Italian-written plan that was never implemented. The justice sector positions were often allocated in ways like those used for police chiefs and governors.[25] Open to bribery, justice, too, would often go to the highest bidder.

Powerful warlords like Defense Minister Fahim and others manipulated the disarmament, demobilization, and reintegration (DDR) efforts by the Japanese to bring their militias into the army and police while disarming rivals.[26]

Each ministry developed a strong, informal chain of command tied to the warlords, who controlled key appointments. Successive ministers were never able to assert proper authority over the army or police.[27] Warlords and powerful elites used threats of violence and the ability to block policy or legislative action to prevent reforms that could damage their interests.

To make matters worse, the military efforts were divided into silos. The coalition military (International Security Assistance Forces—Afghanistan [ISAF]) was small, Kabul-based, and commanded by a series of generals who rotated in rapid succession. US combat forces existed under a separate command that supported the counterterrorism mission. These forces focused on killing or capturing senior al Qaeda and Taliban leaders. Comfortable in a military silo, the US combat forces remained largely uninformed about local political dynamics or the damaging effects of being manipulated by local power brokers.

Setting the Stage for Taliban Resurgence

Many Afghan officials and strongmen used this ignorance for personal advantage—with devastating impact. Afghan elites played on the Americans' dangerous combination of aggressiveness and naivete and began manipulating intelligence to use unwitting American forces to settle scores with rivals or to consolidate power.[28] Such military operations, combined with the predatory actions of warlords, strongmen, Afghan police, and military and government officials, as well as reports of torture in American prisons such as Bagram and Guantanamo, caused significant civilian harm.

These problems began to foster a sense of alienation and unwanted military occupation—themes the Taliban would deftly exploit in their recruiting and propaganda. When directed against community leaders, civilian harm had disproportionately large effects in driving the people away from the government and often into the embrace of armed opposition groups.[29]

Hamid Karzai, the head of the Afghan Transitional Administration, recognized the damage the warlords and strongmen were having on the population and on the legitimacy of his own government. Karzai's best hope to keep them in check was political support—with military backing—from the international community and, more specifically, from the United States.

But no such support was given. The United States refused to support him in a 2002 effort to take on a threat from eastern warlord Pacha Khan Zadran.[30] A year later, Rumsfeld was adamant to CENTCOM commander John Abizaid, Zalmay Khalilzad, and others, "We do not want him making moves [against warlords] under the mistaken belief that we are going to back him up militarily."[31] International officials would continue to press Karzai to make reforms and keep the warlords in check, but they provided no clear backing to help him manage the fallout.

Some American officials sent very different messages. In an April 2003 meeting with Karzai, US Congressman Dana Rohrabacher, who had longstanding relationships with the Northern Alliance warlords, encouraged the Afghan president to "integrate" these "ethnic leaders" into the government before devolving power to them. Karzai reportedly rejected the idea, suggesting that actions of that nature would only recreate the conditions of 1994 that originally catalyzed the development of the Taliban. Rohrabacher allegedly replied that the "wild West" in America was secured by local strongmen and their militias. He contended that radical Islam and Pashtun nationalism were in league with one another, which seemed to imply that the United States was at war with both. In yet another stunning statement, the congressman reportedly told Karzai that the United States, Saudis, and Pakistanis "created" the Taliban to bring law and order to Afghanistan.[32]

Karzai must have been deeply troubled by the conversation. He may have perceived that Rohrabacher was expressing true US policy, one of empowering the warlords or carving up Afghanistan, waging a secret war on Pashtuns, and having a role in the creation of the Taliban. Karzai would, in the future, levy charges against the United States on all three fronts. American officials would dismiss them as paranoid conspiracy theories.

These kinds of problems, which arose from the combined effect of silos and the absence of an overall authority to direct and manage the international efforts, undermined the foundations of the new government. Predatory

Afghan officials and warlords manipulated the disjointed efforts to their personal advantage. With only a counterterrorism-focused US military presence outside Kabul, international officials had almost no view on what was occurring in the provinces and how predatory Afghan officials and strongmen were alienating significant parts of the population. The perception that post-Taliban Afghanistan was peaceful was an illusion. The violence was being perpetrated by the very people who were supposed to be providing security and reporting incidents.[33]

As the abuses by predatory actors and militias grew in scale and scope, the Taliban began to reorganize.[34] The "harassment and persecution [of former Taliban officials and Afghan civilians] by pro-government groups and by US forces," political scientist Antonio Giustozzi writes, "was a major factor in the original mobilisation of the Taliban insurgency . . . and continued to drive recruits to the Taliban even in 2014–15."[35]

In the face of US resistance, UN envoy Lakhdar Brahimi continued to push in 2002 and 2003 for greater international military presence to protect Afghan civilians from predatory militias. "Skirmishes between local commanders," he observed in July 2003, "continue to cause civilian casualties in many parts of the country where terrorism is no longer an issue. . . . There are daily reports of abuses committed by gunmen against the population . . . all too often—while wielding the formal title of military commander, police or security chief."[36] Predatory activities by the government and civilian harm by Afghan and coalition forces drew disaffected groups to the Taliban cause.

Across the border, Pakistan military officials, alarmed by their diminishing sway and India's growing influence on Afghanistan, supported the Taliban's use of Pakistani territory to foment and sustain an insurgency.[37] Prolific military businesses and logistics companies in Pakistan, as well as a secretive intelligence service that operated outside civilian control, gave the Taliban ready access to supplies, logistics, and expertise. The vast refugee camps there offered substantial numbers of recruits.

The absence of any US official managing the American efforts on the ground enabled the former Northern Alliance leaders to more effectively influence the United States in blocking Karzai's efforts toward peace. Jim Dobbins and Carter Malkasian describe how the Taliban sent several peace overtures in the early years, even offering to surrender in December 2001.[38] Negotiations were rejected by the United States and the Northern Alliance factions in government.[39] Secretary of State Colin Powell's suggestion that the United States sought talks with moderate Taliban was ridiculed as naive by Foreign Minister Abdullah Abdullah.[40]

Rumsfeld flatly rejected Karzai's December 2001 idea that the Taliban be allowed to "live in dignity" in retirement.[41] He told Karzai that al Qaeda and Taliban leaders still needed to be hunted down. "There are a lot of fanatical people," he concluded, "And we need to finish the job."[42] Brahimi later described this decision to be a fundamental error in the Bonn process.[43] Taliban leaders who turned themselves in were sent to prisons in Bagram and Guantanamo, and some reportedly were tortured. Others were killed or captured in raids. The rest fled to Pakistan.

The preamble of the new Afghan constitution placed the resistance against the Taliban on equal footing with the jihad against the Soviets.[44] The message to the Taliban was quite clear—they had no place in the new order.

America's bureaucratic way of war played an essential role in these disconnects—problems that would continue as the war in Afghanistan grew more Americanized. They would play a damaging role in Iraq as well.

Conclusion to Part II

The lack of strategic thinking on war termination—how to achieve a favorable and durable outcome—heightened the risk that the successful overthrow of the Taliban would turn into a quagmire. The Bush administration had given detailed attention to the military campaign but wished away the challenges presented in the aftermath. Such neglect damaged the United States' ability to address the exploitative actions taken by Afghan and international actors during the chaotic post-Taliban period. These activities, many of which were hidden in plain sight, undermined the legitimacy of the nascent government in the eyes of Afghans and aroused suspicions among Afghanistan's neighbors—especially Pakistan.

American officials attempted to simplify the complex challenges of post-Taliban Afghanistan by creating a step-by-step blueprint for political, military, and economic development. This became known as the Bonn process. Officials crafted a detailed plan that arrayed key milestones along a timeline. Each milestone had its particular plan and deliverables. Afghanistan achieved each one. But the situation in the country slowly deteriorated.

American officials in Kabul and Washington found themselves completely unprepared for the intense and sometimes violent struggle for control among Afghan elites. "Local elites know how to consolidate power," observed Lieutenant General Terry A. Wolff, "but not how to build a country."[1] Each milestone was manipulated for advantage, mainly by Northern Alliance figures who enjoyed almost uncritical US backing. The result was a set of rules and systems that seemed to marginalize southern Pashtuns.

America's (and NATO's) bureaucratic way of war added to the challenges. Each agency worked hard within its silo, but the whole became less than the sum of its parts. The absence of a senior official to manage US efforts in Afghanistan hurt America's ability to set the new Afghan government and security forces on solid ground, prevent abuses that could give rise to an insurgency, and address Pakistan's concerns about Indian influence. The so-called lead nation concept and the tendency of US agencies to operate in bureaucratic silos undermined the woefully under-resourced stabilization effort and left international civilian and military efforts open to manipulation.

The limited US interests in Afghanistan, and the Bush administration's desire to minimize security and reconstruction commitments, should have led to a more thoughtful discussion with Karzai about the Taliban's offer to capitulate. The Taliban had little to no leverage—negotiations by competent Afghan and international authorities were highly likely to result in a favorable outcome and probably could have prevented or limited the scale of the insurgency. The Bush administration, however, viewed the Taliban and al Qaeda as one and the same—international terrorist organizations bent on attacking the United States. Given the September 11 attacks and the way American political leaders framed the conflict, an effort to negotiate with the Taliban in 2001 or 2002 probably would have entailed huge audience costs.

Strategic empathy is the ability to place yourself into the mind of another so that you can see yourself and the situation from their point of view.[2] Sun Tzu counseled generals to know the enemy and yourself.[3] Knowledge of the enemy is more than assessing their military forces and capabilities. You need to understand another's perspective—and their pressures and motivations—to gauge their intentions and anticipate their moves. Knowing yourself, particularly your strategic empathy limitations, will motivate you to close gaps in your understanding.

Strategic narcissism, by contrast, is the tendency to view others only in relation to your interests and to believe that their success relies upon their adopting your point of view.[4] Problems such as confirmation bias and mirror imaging indicate strategic narcissism, which can result in leaders believing what they want to believe about partners and enemies while presuming that others aspire to their point of view and strategies. Empathetic competitors manipulate narcissists to gain advantage.

Lack of strategic empathy undermined America's ability to understand the objectives and motivations of the key actors in the early stages of the conflict—a problem that persisted throughout this war and the Iraq war, too. The upshot was a US approach that unwittingly created incentives for the conflict to widen. The Afghan government, as we will see in Part III, turned into a predatory kleptocracy, the Taliban mobilized into an insurgency, and Pakistan played the double game of supporting some US counterterrorism objectives while allowing the Taliban to use Pakistani soil as an insurgent sanctuary.

In 2004, according to a former Taliban official who was part of the delegation, the Taliban made another peace overture.[5] But the Bush administration still refused. Standing alongside President Karzai in February 2004, Rumsfeld said, "I've not seen any indication that the Taliban pose any military threat to the security of Afghanistan." Karzai, noting that he was being contacted daily by Taliban leaders seeking to be allowed to return home, surmised, "The Taliban doesn't exist anymore. They're defeated. They're gone."[6] The 2004 overture would be the Taliban's last for many years.

Had the Bush administration considered multiple options for war termination, they still might have come to the view that negotiations with the Taliban were unacceptable. Transition-and-withdraw was another option, and one that the Bush administration backed into as the insurgency fomented. A strategy that considered in advance the requirements for a successful transition would likely be based on factors such as establishing and maintaining political legitimacy and inclusiveness, deploying sufficiently capable peacekeeping forces until credible local forces were built and trained, ensuring civilian protection, and emphasizing regional cooperation. The United States, however, was myopic in its approach; to the detriment of its long-term goals, the United States continued to invest most of its time, energy, and resources into hunting down Taliban remnants—efforts that were manipulated by local actors.

By 2006, an insurgency that had durable internal and external support was fighting against a predatory, kleptocratic host nation government that was losing legitimacy. The stage was set for quagmire.

Part III

Persisting in a Failing Approach

Overview

The US government had not developed a clear and coherent strategy for the war in Afghanistan and was having difficulty understanding and adapting to a deteriorating situation. The Taliban's insurgency, with a sanctuary in Pakistan and growing Afghan support, expanded.[1] As we saw in Part I, the civilian harm caused by pro-government predatory militias and coalition forces contributed significantly to the growth and sustainability of the Taliban. High-profile civilian casualty incidents were driving deep wedges between the US and Afghan governments, undermining the legitimacy of both, and alienating the Afghan population.[2] The Afghan government, meanwhile, had become a predatory kleptocracy, which was driving more people into the arms of the Taliban.[3] America's bureaucratic way of war unwittingly reinforced rather than diminished these problems. American military and civilian officials, however, continued citing myriad examples of progress even as the security situation deteriorated.[4]

The ineffective strategy remained in place until 2008 when Barack Obama was elected president of the United States. He campaigned that America needed to withdraw from Iraq and focus instead on Afghanistan and made good on this pledge. He directed a thorough review of the Afghan conflict and boosted military, civilian, and diplomatic efforts to reverse a deteriorating situation. To encourage Afghan government reform, Obama put a timeline on American presence. The design was to build Afghan government and security force capacity, gain Pakistani support in closing insurgent sanctuaries, and degrade the Taliban into a residual insurgency. Despite impressive examples of progress, however, the Afghan government and security forces remained corrupt and unable to win the battle of legitimacy in contested and insurgent-controlled areas, Pakistan proved unwilling to close insurgent sanctuaries, and the Taliban sustained high levels of violence. Obama clung stubbornly to the withdrawal timeline as the situation deteriorated. By 2015, he felt compelled to slow and eventually stop the drawdown.

Why did both administrations persist in strategies that were not succeeding? This question has received little attention. Confirmation bias, political and bureaucratic frictions, and patron-client difficulties offer insights on the persistence of ineffective strategies in both wars for both administrations. These problems impeded their ability to recognize that critical strategic factors, such as insurgent sustainability and the host nation government's inability to win the battle of legitimacy, undermined the prospects of success. Although both administrations would make changes on the margins, they never determined if transition was a realistic outcome.

7

Accelerating Success, 2003–2007

As the situation in Afghanistan deteriorated, why did the United States cling stubbornly to bad strategies? Confirmation bias is a part of the answer. Political scientists Elizabeth Stanley and John P. Sawyer argue that losing or ineffective strategies can become "sticky" when decision-makers have insufficient or faulty information, or when bureaucratic or organizational filters and biases prevent leaders from accessing or using available data.[1] Information overload can magnify the problem.[2] Policymakers can get bombarded with information of varying reliability and contradictory assessments and forecasts. Making sense of it can be overwhelming. Confirmation bias can result.

Confirmation bias is the tendency to seek data and to interpret information in ways that conform to currently held beliefs.[3] Decision-makers can place higher credibility on confirmatory information or assessments while discounting the reliability or value of different ones. More facts or better information might not correct the error. Decision-makers often dig in their heels when challenged. "Arguing the facts doesn't help [when confronting confirmation bias]," cautions behavioral scientist Christopher Graves, "in fact, it makes the situation worse."[4] Confirmation bias may thus entrench the status quo.[5] Confirmation bias, if present, results in officials giving confirmatory information excessive weight while discounting the value of contradictory information or interpreting such data to support pre-existing beliefs. The effect is a tendency to resist changes to the status quo.

America's bureaucratic way of war reinforced the status quo. Seams and fault lines between silos can exacerbate confirmation bias, and, with it, strategic damage. Seams denote gaps between silos that can be exploited by host nation actors or adversaries; fault lines describe problems in which efforts in one silo undermine efforts in others. Actions by one agency may damage the efforts of others and undermine overall national objectives. These problems are often missed in the assessments because each agency measures progress within their respective silos. The tendency to aggregate in-silo metrics and milestones to assess overall progress can create a misleading strategic picture. The result of these challenges is that the whole can be less than the sum of its parts.[6] Senior officials, many conditioned by decades of working in bureaucracies, could not see this problem.

After more than a year of bureaucratic infighting over the scale and scope of US assistance to Afghanistan, Rumsfeld finally relented in mid-2003 to a plan drafted by Zalmay Khalilzad called "Accelerating Success." Khalilzad was soon chosen to be the US ambassador to Afghanistan and would have the opportunity to implement his ideas. Overall, the so-called Marshall Plan for Afghanistan amounted to $1.2 billion in aid (Congress would eventually approve $1.6 billion).

Although agreeing to higher development expenditures, Rumsfeld continued to resist further troop increases. He reportedly accepted "Accelerating Success" and a larger NATO-run ISAF to extricate the United States from Afghanistan more quickly.[7] To enact Khalilzad's plan, the United States and NATO began fielding Provincial Reconstruction Teams (PRTs) across Afghanistan to implement local assistance projects. USAID was to manage infrastructure development.[8]

"Accelerating Success" gave a much-needed boost in resources but did not sufficiently address the factors that posed the highest risks to success: an insurgency with durable internal and external support and a host nation government unable to win the battle of legitimacy in contested areas. The plan focused on key milestones and developmental efforts. A National Security Council memorandum to principals (cabinet secretaries) on January 18, 2005, assessed the progress of Accelerating Success. "The President's 2004 'Accelerating Success in Afghanistan' initiative," the note explained, "led to *transformative changes* in governance, security, and reconstruction in Afghanistan." The memo lauded the president's leadership, highlighted the activity metrics, and explained that a new working group on Afghanistan had improved interagency coordination. It cited as examples of progress the 2004 election, the creation of 19 PRTs, the fielding of a 16,000-strong Afghan Army and a police force of 25,000, efforts to check warlords, over 25,000 militia troops demobilized, and completion of the Kabul to Kandahar highway. It noted challenges with counternarcotics, police training, and donors making good on pledges. All metrics were "Yellow," which indicated reasons for caution, but there were no red flags.[9]

The NSC was grading its homework based on inputs from agency silos. They measured government legitimacy positively because of the voter turnout in the 2004 election and the self-assessment that warlord influence was "progressively undercut." The latter, the memo remarked, were now turning to peaceful politics. Although noting the violence, the NSC memo did not measure the state of the insurgency. Curiously, it claimed that the UN reported improved countrywide access in September 2004 compared to the previous year. A US Government Accountability Office report, however, cites UN maps that show a significant deterioration in security for the period covered by the memo.[10] The NSC document did not assess regional malign activity, despite the existence of insurgent sanctuaries in Pakistan. In essence, the White House measured specific agency-centric inputs (e.g., 19 PRTs created) and outputs (e.g., 16,000 Afghan Army fielded) but not outcomes, such as insurgent strength and government legitimacy.

Sufficiently confident were they that the Pentagon suggested at the Berlin Conference in September 2005 to withdraw up to 4,000 US forces and replace them with NATO troops.[11] "It makes sense that as NATO forces go in," General John P. Abizaid, the head the US Central Command, reportedly told the *New York Times,* "that we could drop some of the U.S. requirements."[12] The Pentagon eventually scrubbed the withdrawal plan due to NATO objections.

The tendency to measure progress within each silo reinforced confirmation bias that the war was on track and masked the emerging strategic risks. The achievements to date were impressive, given the catastrophe in Afghanistan after over 20 years of war. Other indicators, however, were present that should have led the Bush administration to question its optimistic assessments.

Chapter 5 noted the widespread violence on the part of various warlords who aimed to consolidate and expand their control, including their use of predatory militias and manipulation of international forces and officials.[13] Karzai was in an extraordinarily difficult position. Nearly every Afghan leader over the past century had been overthrown or murdered, and Karzai's political rivals controlled the security forces. International military forces, in a bit of mirror imaging, mostly found themselves incapable of conceiving a civil-military relationship other than objective control, in which militaries are professional and apolitical. Some Afghan military leaders undermined Karzai's political authority.[14] Warlords could also mobilize well-armed militias rapidly and organize street protests. Both the Pentagon and the State Department continued to underestimate the importance of these patron-client issues.[15] Karzai often felt, even in the last years of his presidency, that the Afghan National Security Forces (ANSF) were not under his control.

International officials continued to press Karzai to make reforms and keep the warlords in check but provided no clear backing or muscle to help him manage the fallout. Nonetheless, Karzai took the bold step of selecting Ahmad Zia Massoud rather than Fahim as his first vice presidential running mate for the 2004 elections, which caused significant tensions.[16] After winning the election in 2004, Karzai removed several warlords and their lieutenants from ministerial positions, including replacing Fahim at the Ministry of Defense with his deputy, Rahim Wardak.[17]

The 2005 parliamentary elections, however, would offer the warlords new opportunities for political influence. In May 2005, protests erupted across Afghanistan, instigated by a *Newsweek* article that alleged interrogators at the US military prison at Guantanamo had desecrated the Koran.[18] Due in part to this instability, the September 2005 parliamentary elections were a significant victory for warlords and local strongmen.[19] They could block legislative efforts that might undermine their influence.

Rather than turning their attention to peaceful politics, the warlords used the elections to expand their means of control.[20] The United States had made it clear that Karzai could not count on the international military to help him impose his will on the warlords. Karzai had to undertake efforts that kept the international community happy while balancing the interests of dangerous elites. His efforts to co-opt them, to gain subjective control (i.e., to co-opt the military into armed political supporters), were partially successful but also heightened internal frictions and corruption.

These factors, Sarah Chayes argues, began to provide the incentives that would turn the Afghan government into a predatory kleptocracy.[21] A kleptocracy is a government by people who seek primarily to use their position for personal gain at the expense of the governed. In a kleptocracy, the elites exploit the country and people to gain power and influence.

Government officials, in this case, often purchase their positions for significant sums. These payments are not patriotic contributions to the national treasure. Officials purchase license to use their position for private gain. They use time-tested techniques such as extortion of customs and business revenues, kidnapping for ransom, land theft, misappropriation of international

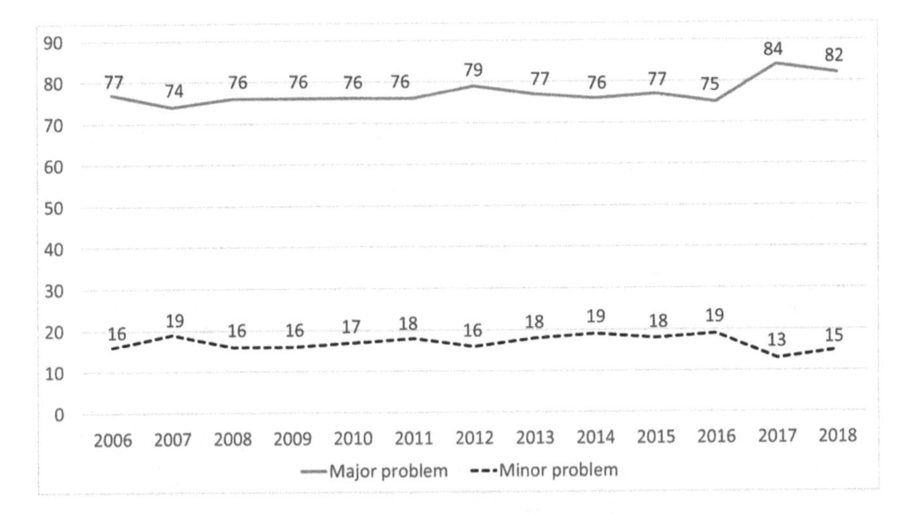

Figure 2. Perception of Corruption as a Problem in Afghanistan

Question: "Please tell me whether you think corruption is a major problem, a minor problem, or no problem at all in the following areas. (a) In your daily life. (b) In Afghanistan as a whole." Graphs show percentages of people who say "major problem" or "minor problem" regarding (b).

Source: Afghanistan in 2018: A Survey of the Afghan People, The Asia Foundation, 2018, 117, reprinted with permission.

aid and investment, robbery, and participation in illicit economic activity to recoup the sale price and turn a profit.

To be sure, some kleptocrats are more predatory than others. Some might even aim to support the people while helping themselves. Over time, the competition to participate in the system raises prices and thus requires more predatory actions to turn a profit. This cycle drives out the well-intended and attracts unsavory characters.

Western officials assume a government is of the people, by the people, and for the people. Mirror imaging helps to explain why western officials had difficultly recognizing the Afghan government's legitimacy problems. Large-scale military support, massive amounts of aid, and seemingly uncritical support to Afghan officials unwittingly enabled the kleptocracy.

Corruption was one of the main issues voiced by the Afghan people, and the United States never managed to hear it properly. The Asia Foundation surveys of the Afghan people show a constant dissatisfaction with the level of corruption in Afghanistan in general, and the number of those indicating

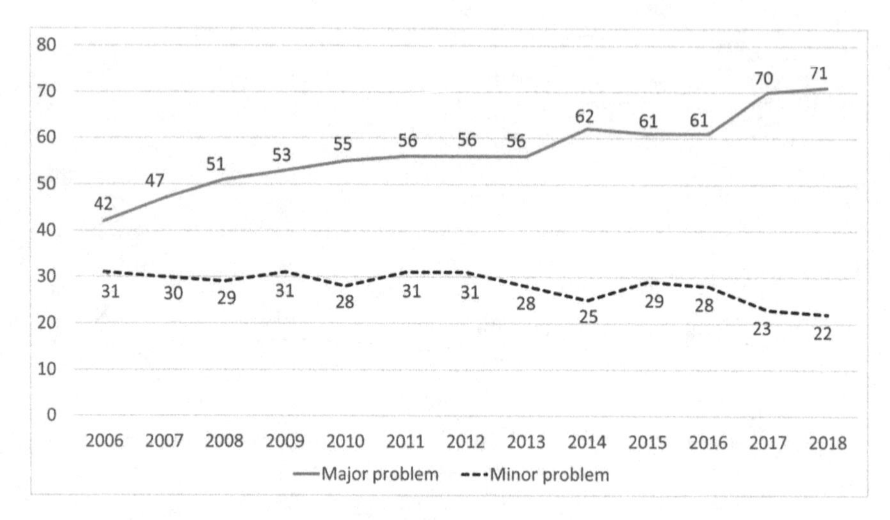

Figure 3. Perception of Corruption as a Problem in Daily Life

Question: "Please tell me whether you think corruption is a major problem, a minor problem, or no problem at all in the following areas. (a) In your daily life. (b) In Afghanistan as a whole." Graphs show percentages of people who say "major problem" or "minor problem" regarding (a).

Source: Afghanistan in 2018: A Survey of the Afghan People, The Asia Foundation, 2018, 118; reprinted with permission.

that corruption was a problem in their daily life increased between 2006 and 2018.[22]

The milestone-centric approach to political and economic development was easily captured by Afghan elites, who quickly froze out their rivals while the international community lauded the progress. Elites super-empowered by their proximity to the United States gained significant political and economic power, backed up by the military muscle of local militias, police, or unwitting coalition forces. Government, police, and military positions were increasingly for sale at exorbitant prices. Many officials purchased their jobs from elites in Kabul in return for license to turn a profit. International aid and development dollars, customs revenues, and black market racketeering were all eligible for extortion. American and NATO officials lacked the strategic empathy to understand these problems and their implications.

A 2008 report by Oxfam found that barely 15 percent of international aid dollars made it to the local levels.[23] More problematic, some officials would engage in land theft, kidnapping for ransom, and other forms of extortion, seemingly backed by coalition military might. "You Americans are either too stupid to realize you are being used in this way," an Afghan elder explained to me in 2009, "or you are complicit."[24] Well-meaning efforts by agencies or nations self-synchronized in damaging ways. The whole was perpetually less than the sum of its parts.

It is no wonder, then, that the political instability and Taliban offensives in 2006 came as such a shock to the Bush administration. Karzai's tone, as seen in reporting cables from the American diplomats, grew deeply suspicious and pessimistic that year. He suspected some of the warlords were fomenting violence to destabilize his government.

Angry anti-US demonstrations in Kabul erupted on May 29, 2006, after a traffic accident in which an American military vehicle plowed into a dozen civilian ones. Fourteen civilians were reportedly killed and 90 injured during the riots.[25] The demonstrations came on the heels of an incident in Kandahar in which 35 civilians perished.[26] "If it [the perpetrators] had been Taliban or al-Qaeda," Karzai explained darkly, "the bombs [in Kabul] would have been more effective."[27]

Other anti-American protests erupted in February and September 2006 over cartoons in a Danish newspaper and an American film, respectively, that mocked the Prophet, Mohammad.[28] Corruption and predatory behavior by Afghan officials continued.[29] Karzai worried that his government and the coalition were losing the support of the people.[30] To reduce the internal tensions, he began to slow down disarming some militias and to bring more warlords into the cabinet.[31]

Meanwhile, the Taliban initiated large-scale military offensives, particularly in southern Afghanistan.[32] Retired US general Barry R. McCaffrey, in a June 2006 assessment for Rumsfeld, was alarmed by the deteriorating security situation.[33] An August 2006 security assessment on southern Afghanistan noted, "The Taliban are becoming increasingly willing to defend key terrain with large, sophisticated, well-armed groups of fighters."[34] Afghan and Indian officials had been warning American officials since 2005 that Pakistan was actively supporting the insurgency.[35]

The reporting cables did not detail the scale and scope of Pakistani complicity. Still, the existence of a sanctuary in Pakistan from which the Taliban were able to plan, coordinate logistics, train, and recruit was evident.[36] By 2006, Karzai's tone on Pakistan had moved from conciliation to hostility.[37] The Afghan government grew closer to India, reinforcing Pakistan's fear of a hostile Afghanistan.[38] In November 2006, the Defense Intelligence Agency's director, Lieutenant General Michael D. Maples, forecasted that insurgent attacks in 2006 were likely to be twice as high as in 2005. He also revised an earlier optimistic assessment of Afghan governance. "Nearly five years after the Taliban's fall," Maples testified, "many Afghans expected the situation to be better by now and are beginning to blame President Karzai. These unrealized expectations are likely contributing to an erosion of support for his administration."[39]

Such problems should have led Bush administration officials to question the assumptions underpinning their strategy. In response to the growing challenges, however, the US and Afghan governments and international community sought to improve coordination and levels of support. The London Compact of February 2006 called for an expansion of the Afghan Army from 34,000 to 70,000 by the end of 2010.[40] By that time, the army and police were to be able to secure the country and allow international forces to withdraw. Governance, anticorruption, the rule of law, and other reforms were to be achieved by then as well, but no conditions were attached to the benchmarks.[41]

The commitments for reform were aspirational. To improve coordination on the ground, Karzai agreed in August 2006 to the "Afghan Development Zone" (ADZ) concept, developed by the ISAF commander, British general David Richards. The ADZ, the American ambassador noted, "attempts to unify security and development efforts from the GOA and IC . . . and should expand to encompass ever-widening territories."[42] The underlying assumption was that projects and "service-delivery" rather than good governance would restore public confidence in the government.

In the absence of clear guidance from Washington, the US embassy and military command in Afghanistan published a "Strategic Directive for Afghanistan" on September 11, 2006, signed by Ambassador Ronald Neumann and Lieutenant General Karl Eikenberry. The intention was to integrate American civilian and military efforts in Afghanistan. The country team outlined its strategic goal as "[c]reate a self-reliant Afghanistan that provides effective governance, is self-securing, committed to representative government, economically viable, and rejects narco-production." The military identified its primary task as "defeat insurgent and terrorist threats and establish effective, adequate security." They defined success as: "insurgents are defeated and no longer threaten internal development nor the safety and security of Afghanistan," and "the ANSF are capable of effectively securing their territory against insurgent and criminal threats with only limited coalition support."[43] The document specified agency-centric lines of effort related to security, governance, development, and strategic communications and portrayed how they all point to a self-reliant Afghanistan.

The directive noted that the 34,000-strong Afghan National Army (ANA) would double in size to 70,000 and be capable of operating independently by fiscal year 2011. With more resources, they argued, the ANA could achieve that milestone by 2009. A police force of roughly 80,000 would be merely one year behind.[44] The military command made these aggressive forecasts even though the international training mission for the Afghan Army and police were, as McCaffrey described in a June 2006 assessment read by Rumsfeld, chronically and "miserably" under-resourced.[45] The commonly used rule of thumb for the size of security forces needed to combat an insurgency was 20 military or police for every 1,000 people.[46] With a population of roughly 30 million, Afghanistan would require a security force 600,000 strong. Barely one-third of that number of Afghan and international forces was forecasted to be available by 2009.

The small army and police were justified by US military officials based on comparisons with the security forces of neighboring states, none of which were in the grip of an insurgency.[47] Despite General McCaffrey's alarming report, Rumsfeld only asked Chairman of the Joint Chiefs of Staff General Peter Pace to look into the reported lack of small arms for the ANA and the size of the ANSF.[48] The Pentagon reportedly forecasted in 2006 that the Afghan government would not be able to fund even an army of 50,000 until 2063.[49]

8

Failing to Keep Pace with the Insurgency, 2007–2009

An insurgency battles a host nation's government for the right to rule all or part of the country.[1] An intervening power generally aims to help the host nation government defeat the insurgent threat. US counterinsurgency doctrine presumes alignment between the host government and the external counterinsurgent.[2] Political scientist Thomas Grant argues that good allies are rare among host nation governments that are fighting insurgencies because effectively governed countries tend not to provide "inspiration or excuse for guerilla war."[3] Adverse selection is the political science term for this problem. "The same governmental shortcomings that facilitate the emergence of an insurgency," notes Walter Ladwig, "also undercut the effectiveness of the counterinsurgent response."[4] These kinds of conflicts, Stephen Biddle argues, are "strongly associated with weak states and corrupt, unrepresentative, clientelist regimes."[5] A counterinsurgency is thus unlikely to succeed without reforms by the host government, Daniel Byman contends, "but these regimes are likely to subvert the reforms that threaten the existing power structure."[6]

Adverse selection is part of a broader phenomenon known as principal-agent theory. This theory, developed by economists to explain interactions by parties to a contract, describes challenges in which one actor (the principal) delegates authority to another actor (the agent) to carry our actions on its behalf.[7] I will use Ladwig's term, patron-client relationship, because of the equal status in international law between the host country and external power.[8] The patron, in this case the United States, supports a host nation government (the client) with the primary aim of advancing American strategic interests. The faster a host nation can govern and secure itself, the quicker the United States can withdraw its troops and reduce capacity-building, aid, and assistance expenditures.

This aspiration has proven difficult in practice. Three challenges in patron-client relationships are interest misalignment, information asymmetry, and moral hazard. These problems can impede the intervening power's ability to change a losing or ineffective strategy, damage capacity-building efforts, and undermine the prospects of a successful outcome. Interest misalignment occurs because the primary objectives of the intervening power (defeat the insurgency, enable the host nation to secure and govern itself, and return home as quickly and inexpensively as possible) compete with powerful incentives for the client (maintaining power and international largesse). This difference results in the host government promoting political, economic, sectarian, ethnic, or other arrangements that benefit its core supporters—even if these same measures are inspiring the insurgency.[9] Many of the standard prescriptions for counterinsurgencies, such as political and economic reform, a professional military, greater political inclusion, and reconciliation, can be more threatening to the regime than the insurgency itself.[10] "The regime's interests are thus typically focused less on external enemies than on internal threats from rival elites," observes Stephen Biddle, "and especially the state military itself, which is often seen as a threat at least equal to that of foreign enemies."[11] This situation complicates efforts to develop host nation security forces. Because armed elites can pose a much more significant threat than the insurgency does, regimes tend to undertake various forms of appeasement, cooptation, or enfeeblement of the security forces.[12] In Samuel Huntington's framework, they opt for subjective control of the military to prevent a coup.[13] US military advisors raised on the principle of objective control, in which the military agrees to be apolitical in return for substantial professional autonomy, can have difficulty recognizing the difference in their host nation partner. "The kind of powerful, politically independent, technically proficient, non-corrupt military the United States seeks," notes Biddle, "is often seen by the partner state as a far greater threat to their self-interest than foreign invasion or terrorist infiltration."[14]

Information asymmetry occurs because the host nation is unlikely to divulge its corrupt practices. The client is likely to manipulate the patron to maintain military and economic support while maximizing autonomy. Hiding information, disguising intentions and interests, and paying lip service to the patron's demands are typical parts of the playbook. "In reality," Ladwig suggests, the patron "has, at best, only indirect control over its client's economic, political, and military policies."[15]

A moral hazard occurs when the patron is committed to the client's survival, and the client does not bear the full consequences of its actions. This situation may create incentives for the client or rival elites to engage in high-risk behaviors, knowing that the patron will not allow things to go too far.[16] To encourage reform, patrons may reassure clients that the aid and support will be forthcoming if the regime undertakes actions the patron deems necessary. To maintain domestic support for the ongoing assistance and troop presence, the patron may paint the client's survival as a vital interest. This combination could raise doubts in the client's mind that the patron will halt the intervention or support if the client does not comply with the reforms. Why risk internal instability by enacting painful and potentially destabilizing reforms if the insurgency threat will be met by the intervening power anyway?[17]

Conversely, publicity of the client's problems may reduce the patron's public support for the intervention. The client may avoid enacting what it deems to be risky reforms if the intervening power is going to leave anyway or not make good on its promises.[18] In short, a client has the incentive to resist changes that may heighten the risk of domestic instability even if such reforms would improve its chances to defeat the insurgency or bring it into a peace process. These problems undermine the prospects of transition strategies.

How can patrons sway clients? Walter Ladwig examines two influence strategies patrons frequently use: inducements and conditionality.[19] Inducement seeks to persuade a client to change behavior with promises of aid and support. This approach, Ladwig notes, tends to be preferred by US policymakers. Conditionality, on the other hand, uses rewards and punishments to affect a client's behavior and reduce moral hazards. To know if the client is enacting the reforms or shirking, the patron must use intrusive monitoring.[20] Such oversight can be very resource-intensive and unpopular with the host nation. Conditionality increases the likelihood of frictions in the patron-client relationship. The client may highlight or amplify these challenges to reduce the willingness of the patron to enact such measures. Inducements are the path of least resistance for the patron, but Ladwig and Biddle show that conditionality is more likely to be effective.

In September 2006, as Karzai was heading to Washington to meet with President Bush, Ambassador Ron Neumann noted that the Afghan president "is at the lowest point of public confidence in his government. A deteriorating security situation, coupled with rampant corruption at all levels, has sapped confidence and feeds public perceptions of a weak government and governance system." The ambassador also cited the problems with Pakistan and the nexus of corruption, narcotics, and insecurity. He advised Bush to improve funding for Afghanistan, increase the size of the security forces, and urge Karzai to "do more" on corruption and narcotics. "We, in turn, must recognize he is not strong enough and Afghanistan not stable enough for him to do these things without our encouragement, our occasional pressure, and a lot of our money and force to back him up."[21]

Bush took stock of the wars in Iraq and Afghanistan in the fall of 2006. He fundamentally changed the strategy in Iraq, as chapter 27 will show. For Afghanistan, he simply provided more resources. "Today, five short years later, the Taliban have been driven from power, al Qaeda has been driven from its camps, and Afghanistan is free," President Bush announced in a February 2007 speech on Afghanistan. "That's why I say we have made remarkable progress."

Nonetheless, he noted the significant Taliban offensives in 2006 and decided to increase support to Afghanistan even as he was surging military forces in Iraq. He outlined five capacity-building efforts: increase the size of the Afghan Army from 32,000 to 70,000; strengthen the NATO forces in Afghanistan (to include an increase of US forces); build provincial government capacity and develop local economies; reduce poppy cultivation; and help Karzai fight corruption, especially in the justice sector. He added a sixth effort, which was to work with Pakistan's President Musharraf to defeat terrorists and extremists in Pakistan.[22]

Confirmation bias reinforced the Bush administration's tendency to overestimate the legitimacy of the Afghan government and underestimate the growing strength of the Taliban. They misdiagnosed the problem in Afghanistan as one of inadequate capacity. Warning signs of an increasingly capable insurgency with external sanctuary and a predatory, corrupt Afghan government were met with more resources but not a new approach. Exploitable bureaucratic silos and interest misalignment between the US and Afghan governments meant that more resources were at high risk of elite capture, as was discussed in the preceding chapter. Without addressing these problems, more resources might even make corruption worse.

Relations with Pakistan, which was designated by the United States as a major non-NATO ally in 2004, were misaligned, too. A Pakistani military offensive to eliminate Afghan Taliban sanctuaries could theoretically bring about an end to the insurgency. Still, Musharraf was not about to go to war with a group that was advancing Pakistan's interests in Afghanistan, even if they undermined American interests and conducted operations that killed

American soldiers.[23] Musharraf, instead, used US funding to support operations against the Pakistani Taliban.[24]

Despite Bush's desire to increase support, by August 2007 the US training teams fielded only 1,000 of the 2,400 trainers required. NATO provided only 20 of their agreed 70 teams. The ANA had only 53 percent of the equipment deemed critical by the Combined Security Training Command—Afghanistan (CSTC-A).[25] By June 2008, only 2 of 105 Afghan Army units had a fully capable rating.[26] The Afghan police were in even worse condition.[27] CSTC-A revised its readiness forecast by only one year in 2008, noting the army and police would be "fully capable" by the end of 2011.[28] When Lieutenant General James Dubik conducted a CENTCOM-directed assessment of the ANSF in 2008, he found that their slow rate of growth was unable to keep pace with the growing insurgency.[29]

As the Taliban threat and US resources grew, so did the scale of corruption. In 2005 Afghanistan ranked 117 out of 158 in the Transparency International Corruption Perceptions Index. By 2009 it vaulted to 179 out of 180, behind only Somalia.[30] Afghanistan expert Astri Suhkre blames the growth in corruption on international community largesse: "The money flow simply overwhelmed the country's social and institutional capacity to deal with it in a legal and socially acceptable manner."[31] This argument presumes that international aid that was aligned to Afghanistan's "absorptive capacity" would have been used for its intended purposes, and excess would not have been available for misappropriation.

Patron-client challenges cast doubts upon this premise. Sarah Chayes, for instance, illustrates that the government had become a vertically integrated kleptocracy. As noted earlier, government positions were sold by power brokers in Kabul at increasingly high prices. In exchange, officials had license to recoup the money and turn a profit—many using predatory practices to do so. An official who did not pay the power brokers would lose the job.[32] Even if international aid and contracts had aligned to Afghanistan's "absorptive capacity," the problems would persist. The main incentive for many government officials was sustaining the kleptocracy, not serving the people.[33]

The Department of Defense's semi-annual report to Congress in January 2009 cites data showing governance getting worse from 2006 to 2008.[34] By 2010, according to a United Nations report, corruption had eclipsed insecurity and unemployment as the most significant concern among Afghans.[35] Annual surveys showed that an overwhelming majority of Afghans viewed corruption as a significant problem in their everyday lives (see figure 3).[36] A highly sophisticated kleptocracy and international largesse became mutually reinforcing. Officials' and power brokers' insatiable demand for money and power was sapping the legitimacy of the government and the international community. The predatory nature of the government and associated warlords led more Afghans to withhold their support or even transfer it to the Taliban.[37] Civilian casualties by international forces, meanwhile, continued to undermine the mission. The United Nations Assistance Mission in Afghanistan (UNAMA) attributed 39 percent of the civilian fatalities in 2008 to foreign forces.[38]

Bush changed the strategy in Iraq, but not in Afghanistan. In 2007 Iraq was unraveling rapidly. Afghanistan was a slow, less perceptible failure. Confirmation bias had led to a dangerous complacency as American officials fixated on examples of progress.[39] The tendency to operate in bureaucratic silos created seams that were being ably exploited by Afghan elites and fault lines that were interacting in destructive ways. Military efforts were damaging governance and legitimacy; elites captured political milestones; diplomatic efforts resulted in Pakistan being awarded major non-NATO ally status despite their support for the Taliban, which reinforced Karzai's cynicism and India's alarm. Patron-client challenges reinforced these problems. The United States had not developed a coordinated strategy with the Afghan or Pakistani governments, so interests and incentives remained dangerously misaligned. Political instability and kleptocracy in Afghanistan and double-dealing by Pakistan were undermining US interests. The harder the international community tried to fight and to spend its way out of the problems, the more deeply it was sucked into the quagmire.[40]

9

The Good War Going Badly

If the first step to making war a rational instrument is to determine the political goals, the second is to understand the character of the conflict.[1] This assessment includes estimating the aims of the different belligerents; evaluating the importance of those goals to them (the "value of the object" for Clausewitz[2]); assessing the strengths and weaknesses, capabilities, and limitations of each actor; calculating comparative advantages; and weighing potential risks and opportunities.[3] This process, political scientist Richard K. Betts argues, has a high risk of error as the policymakers struggle to use intelligence reports and assessments that can be vague, incomplete, misleading, or contradictory to make strategic decisions.[4] At the same time, he notes, policymakers can ignore, discount, or misinterpret accurate intelligence.[5] Intelligence producers and consumers are imperfect.

External powers face the additional problem of competing risks. Interventions do not occur in isolation of world events and other national interests. When interests clash, countries generally prioritize the more important ones, which may be to the detriment of an ongoing war. This risk calculus helps to explain why the Soviet Union did not invade Western Europe or start a nuclear war with the United States over its tangible support to Afghan insurgents in the 1980s. Likewise, the United States did not invade Iran over its support to Shi'a militants in Iraq and has not conducted a ground invasion of Pakistan to eliminate Taliban sanctuaries. Such actions could have increased the costs of supporting an insurgency. But their consequences to broader national security interests were far too high. Careful management of risk and uncertainty by the intervening power across an array of national security concerns may result in choices that reduce the prospects of decisive victory even further. The new Obama administration grappled with these tough choices as it faced a wide array of domestic and international challenges.

As is often the case, a change in administration was necessary for a change in strategy. Obama's changes to the Afghanistan strategy, however, were more significant in scale than approach. Cognitive bias, bureaucratic frictions, and patron-client problems continued to impede the US government's ability to recognize the limited prospects for a successful transition. Obama campaigned that Afghanistan was the "good" war—a war of necessity—while Iraq was the war of choice.[6] One of his first acts as president was to order an interagency review of the war in Afghanistan, led by former CIA official and South Asia expert Bruce Reidel. The process served to concentrate the minds of senior administration officials and bring a shared appreciation of the challenges.

Reidel's report outlined that the situation in Afghanistan was worse than expected. Afghanistan and Pakistan, it argued, should be a single integrated theater. The United States needed a stronger relationship with Pakistan to change its strategic calculus so they would stop using militant groups to advance their interests. The report called for a more effectively resourced counterinsurgency campaign in Afghanistan, to include holding President Karzai accountable for dealing with corruption. It also supported an Afghan-led reconciliation effort. President Obama outlined these findings in a March 27, 2009, speech and approved ISAF commander General David McKiernan's request for 17,000 more American troops.[7]

Obama also made personnel decisions. He directed Secretary of State Hillary Clinton to create a special representative for Afghanistan and Pakistan (SRAP) and selected veteran diplomat Richard Holbrooke for the position. Holbrooke's mission was regional diplomacy and prospective talks with the Taliban. Soon, other international partners created similar posts. The director of the Joint Staff, Lieutenant General Stanley A. McChrystal, created the Pakistan-Afghanistan Coordination Cell (PACC) to improve the effectiveness of the Joint Staff's efforts. Robert Gates, who Obama had kept on as secretary of defense, picked a former deputy chief of mission to Afghanistan, David Sedney, to lead an upgraded Afghanistan-Pakistan-Central Asia office. He also appointed a former CIA operative during the Soviet-Afghan War, Michael Vickers, as the assistant secretary for special operations. Not satisfied with how McKiernan was responding to the new direction, Gates relieved him in favor of McChrystal. As the new commander was leaving for Kabul, Gates directed him to provide, within 60 days, an assessment of the war considering the president's new direction and to inform him of any additional resources required.[8]

McChrystal's assessment described the situation as "serious and deteriorating."[9] The Taliban and other insurgent groups had sanctuary in Pakistan and were threatening Kandahar and several critical locations across the country.[10] They were tightening their grip in the provinces and districts around Kabul. The Afghan government was weak and corrupt, while the international effort was disjointed, ineffective, and creating animosity among the population. ISAF, the assessment argued, needed to address two critical threats: a growing insurgency and a crisis in confidence in the coalition and the Afghan government. Success was achievable, McChrystal noted, but not by "doubling down" on the same ways of doing business. A fundamentally different approach was needed, which included protecting the civilian population from harm by the Taliban *and* the Afghan government. He emphasized that responsible and accountable governance should be on par in priority with security. Without the former, success was not possible. He called for much greater unity of effort, both within the military coalition and with the international civilian efforts, and he recommended conflict resolution initiatives, such as reintegration and reconciliation.[11]

The assessment outlined five strategic risks: (1) the loss of coalition will and support; (2) lack of Afghan government political will to enact needed reforms; (3) failure by ISAF partners to provide adequate civilian capabilities to support good governance and economic development; (4) significant adaptations by insurgent groups; and (5) external malign activity from Pakistan and Iran.[12] All five were clear and present problems; the magnitude of risks 2, 3, and 5 was becoming apparent.[13] These are issues that tend to fall in the seams between bureaucratic silos. At no time was any American official held responsible or accountable to address them. The assessment did not address the question of war termination directly. Still, it argued that helping the government win the battle of legitimacy in the eyes of Afghans would reduce the Taliban's ability to recruit local fighters and control population, thus forcing them to seek an end to the war.[14]

McChrystal provided periodic updates on the assessment to Gates, Chairman of the Joint Chiefs of Staff Admiral Michael Mullen, and Central Command's General David Petraeus. Recognizing that any discussions over troop increases would be contentious, Gates directed McChrystal not to provide a resource recommendation until the NSC reviewed the assessment.[15]

The assessment, however, was leaked to the *Washington Post* soon after being sent to Gates.[16] The issue of "more resources" was already being discussed by pundits in Washington, including some civilian members of the assessment team after they returned home.[17] The White House, already suspicious of the military, believed the Pentagon was using leaks to force the president to approve a troop surge. The leak, which was not by McChrystal or his staff, exacerbated the suspicion into mistrust.[18] These internal bureaucratic and principal-agent problems set the stage for a civil-military crisis in

June 2010 when a *Rolling Stone* article reported disparaging remarks by McChrystal's staff about the president.[19]

As the NSC reviewed the assessment, Gates directed McChrystal to develop resourcing options. McChrystal offered low-, medium-, and high-risk options: 80,000, 40,000, and 20,000 troops, respectively.[20] Equally important, he urged, was a significant increase in intelligence capabilities, civilian expertise, development assistance, and a doubling of the size and capability of the Afghan National Security Forces. The administration deliberated from August until late November.[21]

These discussions took place in the wake of the August 20, 2009, presidential elections in Afghanistan. Convinced that good governance was impossible under another Karzai term, SRAP Holbrooke actively promoted Karzai's rivals, particularly former Foreign Minister Abdullah Abdullah.[22] Karzai suspected this and began accusing Holbrooke of interference.[23] Karzai won 55 percent of the initial tally to Abdullah's 28 percent.[24] Abdullah refused to accept the results and accused Karzai of widespread fraud.[25] Holbrooke and UNAMA deputy Peter Galbraith pushed hard for the internationally staffed Electoral Complaints Commission to investigate. Galbraith was fired after accusing UNAMA chief Kai Eide of a cover-up.[26] The Electoral Complaints Commission declared roughly 1 million ballots to be fraudulent, just enough to put Karzai's percentage below 50 percent and trigger a run-off between the top two candidates. Karzai refused Abdullah's entreaties and American suggestions for a power-sharing deal.[27] By November, a frustrated Abdullah declined to participate in the run-off, ceding the election to Karzai.[28] Obama reportedly called Karzai afterward and lectured him about corruption.[29] In his inaugural speech, Karzai called for a transition to Afghan-led security to begin within two years, for Afghan forces to take over security responsibility by the end of 2014, and for international forces by then to be reduced and limited to training and support roles.[30] Criticized for getting nothing out of the elections dispute in terms of lucrative jobs, Abdullah and his supporters would increase their brinksmanship in the disputed election of 2014.[31]

The mess surrounding the elections intensified the debate in Washington over McChrystal's requests for more resources.[32] The US ambassador in Kabul, Karl Eikenberry, cabled that a surge would be a mistake because Karzai was an "unreliable partner" who lacked the political will to reform.[33] Vice President Joe Biden continued to advocate for a smaller-footprint counterterrorism mission (dubbed "CT-plus").[34] Obama was reportedly upset that the military only offered options in terms of troop numbers.[35] He believed the Pentagon was trying to box him in to approve a significant troop surge. Only the military, though, was asked to provide options.[36] No one sought—and no agency provided—strategies that placed diplomatic efforts (such as reconciliation) or political efforts (such as addressing governance and corruption) as the top priority with the military in support. America's bureaucratic way of war offered only military-centric options to the president.

Afghanistan, moreover, was not being considered in isolation. Obama wanted to wind down wars in Iraq and Afghanistan so that he could fund his domestic agenda, which included a major national health care initiative, addressing massive budget deficits, and bringing the economy out of recession after the 2008 banking crisis. Expanding the war in Afghanistan even further than he already had could inhibit those priorities.

Seeking to limit the US commitment to Afghanistan and to send a signal to Karzai to get serious about reform, the Obama administration began debating a timeline to withdraw troops. The administration did not address conditionality or other options to tackle corruption. The military deferred to its recent experience in Iraq as a gauge for when the campaign would show results.[37] Based on advice from the uniformed military, Gates suggested that areas cleared of the Taliban could transition to the Afghan government within two years.[38] The White House thus decided to withdraw the surge forces beginning in July 2011.

The NSC did not discuss methods besides transition-and-withdraw for achieving a favorable and durable outcome, nor did they examine the feasibility of transition.[39] The questions for debate were limited to the scale of the surge and its timeline. The military was to attrite the Taliban and expand the ANSF while the civilians were to build government capacity. Diplomats were to convince Pakistan to pressure and eventually shut down insurgent sanctuaries.[40] The Obama administration believed these efforts would reduce the Taliban to a residual insurgency by the end of 2014. At that point, the ANSF could defeat them. The NSC discussions did not address the probability of tackling government legitimacy and insurgent sustainability. Beyond the Iraq example, the NSC did not examine readily available studies to assess the validity of the transition theory and the likely amount of time required.[41] The White House reportedly rebuffed suggestions to review comparative examples.[42] "History," former Under Secretary of Defense for Policy Michèle A. Flournoy recalled, "never had a seat at the table."[43]

10

Surging into the Good War

Audience costs is a political science term that refers to the public support penalties a leader suffers for backing down after escalating a foreign policy crisis.[1] Leaders could be reluctant to change a strategy over fears about constituent reactions. Michael Tomz conducted a wide range of experiments to determine whether and to what extent audience costs are real and affect national security choices. He found that constituents tend to disapprove of leaders who make threats and then back down and that leaders regard disapproval as a liability.[2] To be sure, situations exist in which people might approve of backing down after making threats, as Stephen Walt suggests the French and British public did during the 1936 Rhineland crisis and the Munich agreement in 1938.[3] With the end of the Great War scarcely 20 years before, public sentiment remained rooted in keeping the peace. Tomz's experiments also suggest that the actual use of force increases the intensity of audience costs.[4] These rose even further when US casualties were involved.[5] Political leaders are sensitive to audience costs because high approval ratings are considered an essential source of presidential power.[6] Thus, leaders may avoid changes that have high potential audience costs—such as seeking negotiations rather than decisive victory if the leader has painted the adversary in good versus evil terms.[7]

President Obama's updated approach showed the administration's limited understanding of the strategic challenges and his waning patience with the war. The new plan increased the scale of the effort but failed to examine some problematic assumptions about the insurgency and the Afghan government. Bureaucratic frictions and problems with the Karzai government continued to impede progress. Periodic assessments by each agency within its respective silo reinforced views that the war was on track, even as the situation kept deteriorating.

Obama announced his decision in a speech at the US Military Academy on December 1, 2009. Outlining the importance of success in the war, Obama reiterated the March 2009 aims of disrupting, dismantling, and defeating al Qaeda and preventing its reemergence in either Afghanistan or

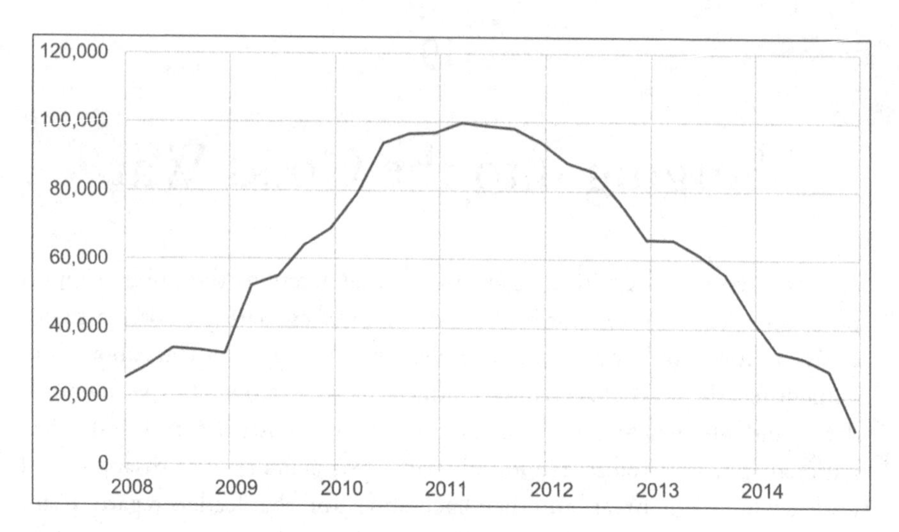

Figure 4. US Troop Numbers in Afghanistan
Source: Based on data in Heidi M. Peters, Moshe Schwartz, and Lawrence Kapp, "Department of Defense Contractor and Troop Levels in Iraq and Afghanistan: 2007–2014," Congressional Research Service, July 22, 2015, 2.

Pakistan. The US government aimed to achieve a secure, stable, sovereign Afghanistan that could defend itself and prevent the reemergence of terrorist safe havens.[8] He approved a three-fold military, political, and diplomatic surge. He would commit an additional 30,000 US forces, bringing the American total to nearly 100,000 troops. He requested an additional 7,000 from coalition partners. The surge forces would begin to withdraw by July 2011, and the transition to handing over security to the Afghans would begin. Not stated in the speech, the administration had set December 2014 as the end of the combat mission and transition to full Afghan security responsibility.[9] The larger ANSF was to be able to handle a residual insurgency that had been substantially degraded by the military.[10] The president also approved a significant civilian increase to build Afghan government capacity.[11] He directed diplomatic efforts to improve support from Pakistan and to promote Taliban reconciliation.

The Obama administration did not deliver a clear and coherent strategy. It was, instead, a highly detailed policy from which the departments and agencies wrote individual plans and then executed them within bureaucratic silos. There was no effort to develop with Afghan partners a coordinated strategy for the war or to address the misaligned aims and incentives.[12]

Despite Obama's concerns about reform, there was no conditionality tied to US support and resources. The military campaign plan was supposed to be a joint endeavor with the Afghan Army and police but was written by ISAF and translated into *Dari*.[13] The capacity-building programs often followed the same pattern. The State Department did not develop a governance strategy and had no requirement to explain how they would tackle corruption problems.[14] Capacity-building—training civil servants and providing resources to government officials—was assumed to lead to good governance. Despite approving a much-needed uplift in civilian support, the Obama administration had enormous difficulty finding qualified people with the right expertise. Of those, most went to the US embassy in Kabul rather than into the provinces and districts.[15]

State did not develop a strategy for changing Pakistan's strategic calculus or for reconciliation.[16] Holbrooke arranged US-Pakistan strategic dialogue, which was a series of meetings and an ambitious list of working groups, but it was all carrot and no stick. There was little connective tissue between these carrots and changes to Pakistan's national security assumptions and approaches.

Holbrooke's staff also wrote concept papers on reconciliation, but they did not show how to integrate these ideas into the Obama strategy.[17] This omission was unfortunate because a negotiated outcome had the potential to achieve US war aims. Holbrooke's SRAP office saw reconciliation as an alternative to the military campaign, not a complementary effort. Like their Defense Department counterparts, they could not envision how coordinated politico-military actions could boost the prospects of negotiations toward a successful outcome. The soldiers and diplomats stayed in their silos, and no one in the NSC thought to examine the merits of alternative war termination approaches.

Finally, the Obama administration did not fix civil-military integration. The starkly different views presented by McChrystal and Eikenberry, for instance, should have raised serious questions as to whether unity of effort was possible unless one of them was in charge of coordinating the American efforts in Afghanistan. No US official in Kabul was responsible for achieving US aims, so Obama had no one to hold accountable for success. Political, diplomatic, military, and intelligence efforts continued operating in bureaucratic silos directed from Washington. ISAF commanders and ambassadors worked to improve on-the-ground coordination, but these could neither override demands from the top nor manage in-theater priorities and trade-offs.[18] The

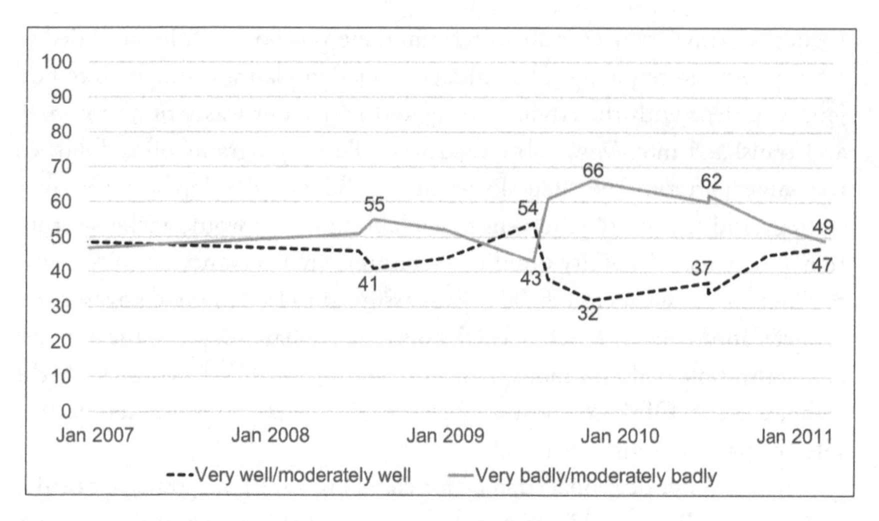

Figure 5. US Public Optimism on the Afghanistan War
The chart shows the results of a Gallup poll, asking: "In general, how would you say things are going for the U.S. in Afghanistan?"

Source: Based on data from Jeff Jones and Lydia Saad, "Gallup News Service, March Wave 1, March 25–27, 2011," https://www.gallup.com/file/poll/147002/Afghanistan_110408.pdf (accessed July 9, 2020).

Obama administration said that civilian and military efforts would be mutually reinforcing, but, as we will also see in the Iraq case study, declaring something to be so is not enough. Siloed actions continued to damage the prospects for success, and Afghans exploited the seams between the silos. The whole was less than the sum of its parts.

Obama's speech sent mixed messages on war termination, which reflected conflicting views within the administration. The announcement of both a surge and a withdrawal timeline heightened the risk that the Taliban could simply play for time—a problem that could undercut US leverage if talks with the Taliban came about. Vali Nasr, one of Holbrooke's chief advisors, believes that the White House was engaged in a concerted effort to block the issue and to prevent Holbrooke from having any role in policymaking.[19] Holbrooke, however, had yet to coalesce his views into a coherent form, so the NSC could not consider the merits of an approach that emphasized a negotiated outcome.[20] Had they done so, the implications of announcing a withdrawal timeline would have come into question. Participants in the discussions recalled neither Holbrooke nor Clinton raising the concern.[21]

Why did the Obama administration select an approach riven with such obvious problems? Limiting audience costs is part of the explanation.

Obama had to satisfy multiple audiences. He wanted to make good on his pledge to improve the situation in Afghanistan and to refocus on his domestic agenda. Gates assessed that Obama was growing ambivalent and wanted to wind down the war as quickly as possible without creating a rift in the administration.[22] Besides, the president needed to show he was listening to the advice of his military commanders while not giving them everything they wanted. He wanted to send a signal to the Afghan government and to his base of support in the Democratic Party that US commitment was not open-ended.[23] Finally, according to a Gallup poll, US public optimism on the war effort had just taken a sudden dip (see figure 5)—a trend that might have further encouraged Obama to bring the war to a quicker end.[24]

Obama wanted Karzai to take ownership of the war and reform his corrupt government. McChrystal and Eikenberry received more resources. Obama's ambivalence, however, did not result in demands for more details about reconciliation or Pakistan. In short, a new president, who campaigned on Afghanistan as the real central front in the war on terror, felt compelled to give the military, the diplomats, his vice president, and his political base enough of what they wanted to keep them on board. What suffered was strategic coherence.[25]

11

More Shovels in the Quicksand

Cognitive bias, bureaucratic frictions, and patron-client problems reinforced Obama's stubborn insistence on the withdrawal timelines and his unwillingness to examine the assumptions underpinning transition. Obama had lost patience with the war.

The Obama administration told McChrystal that he was not going to get the resources and authorities for an outright win. They changed the military's mission from "defeat" the Taliban to "degrade."[1] Even if the military campaign went well, though, without good governance and significant interdiction of Taliban sanctuaries in Pakistan, the war would drag on. These problems were evident during Operation Moshtarak, which aimed to secure the Taliban stronghold of Marjah and Taliban-held areas in the northern parts of Nad-e Ali district in Helmand province. This operation was to be followed months later by Operation Hamkari, an offensive into Taliban-held areas of Kandahar.[2] The two actions, if successful, would take parts of the southern Pashtun heartland away from the Taliban.

The initial push into Marjah by a combined force of US Marines and Afghan troops was successful in wresting control of the area. Still, the hoped for "government in a box" (district-level officials with economic resources at their disposal) that was to arrive and begin earning legitimacy was disappointing.[3] Abdul Rahman Jan, the former Helmand chief of police, was a predatory actor who reportedly controlled much of the Noorzai tribe-dominated police in the province. He allegedly had let the Taliban take control of Marjah in 2007 in retaliation for his removal as the provincial chief of police. Retribution, residents feared, would be more likely than reconciliation and population security if he returned to the position.[4] Marjah, supposed to be a signature offensive in a new counterinsurgency campaign, soon became, in McChrystal's words, "a bleeding ulcer."[5] By the spring of 2010, McChrystal had come to believe that a negotiated outcome might hold the best prospects for durable success.[6]

The intellectual work toward that conclusion began in the fall of 2009. McChrystal asked retired British lieutenant general Sir Graeme Lamb, who engineered reconciliation efforts with Sunni tribes in Iraq, to come to Kabul and develop a reintegration program.[7] Lamb and a small team of two US and two UK officers talked extensively with Masoom Stanekzai, a former communist official with a clean reputation, whom Karzai designated to lead the effort for the Afghan government, and a host of Afghan officials and elders from across the country.[8] The team reviewed previous efforts, including some localized initiatives that had been effective and the failed Afghan government's Peace Through Strength (PTS) program. Successful efforts tended to focus on dispute resolution and inclusive governance.[9] PTS, however, was a scheme to bribe insurgents to defect and was marred by corruption.

Lamb and his team determined that good governance and dispute resolution were critical for a reintegration program. ISAF would need to support these efforts and develop ways to avoid targeting people engaged in substantive discussions. Local development projects that created jobs within affected communities could reinforce, but not substitute for, dispute resolution and improved local governance.[10] Reintegration would probably hit a glass ceiling at local levels, they determined, due to the kleptocratic and predatory nature of the government and its resistance to reform. A corrupt governor or chief of police could undermine any initial success.

Furthermore, isolated efforts were vulnerable to Taliban disruption. Without significant support from the top down, the bottom-up approach would have only limited impact. Reconciliation, they determined, was critical for successful reintegration and vice versa.[11]

The team outlined the problem to McChrystal, who asked Lamb to study reconciliation further and advise him on the results.[12] Lamb and the small group held meetings with diplomats, current and former Afghan officials, and former Taliban senior leaders. The group discerned that the Taliban had sustainable support within Afghanistan and sanctuary in Pakistan. The latter was not about to turn against the Taliban or throw them out of the country. The predatory corruption within the Afghan government and security forces was so profound that it undermined any realistic prospect of an Afghan government victory in the foreseeable future.[13]

Opportunities existed, too. The Taliban's "Code of Conduct" and 2009 *Eid* messages had indicated that many Taliban public political positions were not dissimilar from public statements made by the Afghan government. There was a potential basis for dialogue, but the competition for power would be the most challenging issue. An effort at a brokered deal would be problematic and potentially destabilizing—like the peace deals that fell apart in 1992 and 1993 and led to the Afghan civil war. A peace process, Lamb's team suggested, would need to be a coordinated effort, akin to the Northern Ireland process. Finally, the group raised the problem with the timeline. There was a significant risk that the Taliban could simply wait out the United States. The best time to begin talks, therefore, was before all the surge forces arrived: take advantage of the uncertainty in the Taliban's mind and get the effort moving. If the Taliban withstood the surge and US forces began to drawdown, American leverage would decline substantially. The Taliban could play for time. I briefed this issue to a Deputies Committee small group meeting (a meeting of cabinet-level deputies) in late January 2010, after discussing it with Holbrooke's staff.

To be successful, reintegration (as outlined by Lamb) and reconciliation efforts had to work hand-in-glove.[14] Lower-level fighters leaving the ranks would put increasing pressure on the senior leadership. At the same time, high-level talks could induce local commanders to bide their time and perhaps participate in district-level conflict resolution. By the late spring of 2010, facilitated by their respective staffs, Holbrooke and McChrystal had agreed on this view.[15] McChrystal believed the time was right to begin facilitating discussions between Karzai and Pakistan's chief of army staff, General Ashfaq Kayani.[16] He was keen to avoid stepping into Holbrooke's lane but needed to help the effort gain traction.[17] Such collaboration between the military command and SRAP ended when Obama accepted McChrystal's resignation in June 2010, the day after the *Rolling Stone* article was published.[18]

12

Misapplying the Iraq Formula

Heuristics are rules of thumb that people use to find adequate, albeit imperfect, answers to difficult questions.[1] These efforts to simplify complex and difficult-to-understand issues sometimes work very well, but at other times they lead to significant decision-making errors. The availability heuristic, for instance, is the reliance upon recent or very vivid memories to make judgments.[2] This rule of thumb can be useful in many cases. A person who burned their hand on a stove one day will think twice before touching the stove again. It can also lead to errors in judgment. Although flying remains by far the safest mode of travel, ticket sales tend to drop significantly after a plane crash or terrorist incident involving an aircraft. The vivid image of the event creates an impression of risk that is far greater than the statistical reality. Heuristics can simplify complex national security decisions but may lead to significant errors.

The perceived success of the surge in Iraq affected the Obama administration's decision-making about the troop uplift and withdrawal timeline in Afghanistan. The availability heuristic and confirmation bias kept the White House fixated on the withdrawal timeline, undermined assessments of the war's progress, impeded efforts to explore a negotiated outcome, and heightened civil-military tensions. General David Petraeus, the CENTCOM commander, was selected to replace McChrystal. Still recovering from prostate surgery, Petraeus brought with him vast experience and a brilliant reputation from Iraq. Petraeus's view was that enough military pressure on the Taliban, plus a reintegration effort that could turn former insurgents into local police, could work in Afghanistan as it did in Iraq. Petraeus tweaked his "Anaconda" concept from Iraq for application to Afghanistan and brought in many members from the Multi-National Force—Iraq (MNF-I) and CENTCOM teams to execute his plan.[3] His reliance on the Iraq experience to guide his approach to Afghanistan has come under some criticism.[4] A joke around the ISAF staff that summer was that every sentence began with the same two words: "In Iraq. . . ." Petraeus recognized, though, that time was short and that he needed to produce unambiguous results if there was any hope of convincing Obama to delay the withdrawal timelines and allow more time for his theory of success to work. He needed everyone on board, not questioning the logic.[5]

There were good reasons, however, to doubt that the formula used in Iraq could be replicated successfully in Afghanistan.[6] The prevailing interpretation of the Iraq experience was that the surge in forces plus application of the new counterinsurgency doctrine and a reconciliation program convinced the Sunni tribes to turn against al Qaeda in Iraq (AQI).[7] Reductions in violence were evident within a year.[8] The command assumed that a similar effort in Afghanistan should produce like results.

No other cases, however, were used for comparison, nor was the Iraq thesis examined critically.[9] Escalations of the war in Vietnam or the Soviet war in Afghanistan, for instance, failed to achieve success. The sharp downturn in violence in Iraq was more complicated than portrayed by the military at the time and relied on political and social factors that were not present in Afghanistan. It would also turn out to be temporary. Stephen Biddle examined the cause and effect relationships between the surge and the Awakening. He cautioned against a formulaic belief that more troops plus the new doctrine would lead to success.[10]

The conflict in Afghanistan differed substantially from the one in Iraq. The former was atomized and localized. Tribal cohesion had broken down during 30 years of war and efforts by the communists and the Taliban to replace tribal identity with political or religious conformity.[11] Convincing a Pashtun tribal leader to switch sides and bring thousands of people with him would not work in Afghanistan because hardly any tribal leaders commanded such a following. The Taliban, since 2009, were placing increased emphasis on governance and were using the full range of coercion and persuasion to gain control and support.[12] Their political program was far more sophisticated than AQI's. The Taliban had a strong constituency in many rural areas of the south and east, and among Pashtun enclaves in the north and west. Unlike in Iraq, the conflict did not break down into distinct ethnic and sectarian lines. Turning former insurgents into local police was likely to produce government-sanctioned predatory actors.[13]

To add to the challenges, the Afghan government ran the reintegration program. The Afghan government had no intention of using the program as a forcing function for good governance and local dispute resolution. Reintegration councils were often cronies of the governor, police chief, or local power broker, the same actors who drove people to fight the government in the first place. The program soon resembled the bribery schemes of the past.[14] It was absorbed by the kleptocracy, as government officials and reintegration councils extorted money and made false reintegration reports. Curiously, the majority of "reintegrees" were ethnic Tajiks from the north and west, where the Taliban had no presence whatsoever.

In some cases, local elites co-opted predatory militias to declare themselves to be Taliban so they could "reintegrate" and become a coalition-funded Afghan local police that continued their abusive practices.[15] ISAF could not demonstrate a correlation between reintegration numbers and lower levels of violence.[16] Nonetheless, military officials continued to claim it was taking Taliban fighters and leaders off the battlefield.[17] Such confirmation bias meant that ISAF made no efforts to reform a program that the command viewed as central to success. There would be no Sons of Iraq equivalent in Afghanistan. And yet, US officials continued to argue that the war was on track, citing examples of progress within their bureaucratic silos.

13

Assessments and Risks

Financial columnist Gillian Tett reported extensively on the 2008 financial crisis. The more she learned about organizational behavior, the more she wondered why smart and capable people working in modern institutions collectively act in ways that sometimes seem stupid. The answers inspired her groundbreaking book *The Silo Effect*. Nearly everywhere she looked, tunnel vision and tribalism appeared. "People were trapped inside their little specialist departments, social groups, teams, or pockets of knowledge."[1] Silos in big banks, for instance, often created perverse incentives that encouraged employees to do things that made sense at a micro level but look very foolish from a macro perspective.[2]

The silo effect undermined management's ability to assess these problems and understand the risks. When viewed within silos, performance may be satisfactory, and dangers may be manageable. The macro perspective must deal with issues such as compounding risks. These occur when actions in one silo affect other silos. Simply presenting individual assessments of risk and performance from each silo to create an overall evaluation is flawed because it ignores the effects of interaction and masks compounding risks. The Union Bank of Switzerland made this error during the financial crisis.[3] Robert W. Komer's classic study *Bureaucracy Does Its Thing* outlines how on-the-ground turf battles, bureaucratic infighting, institutional inertia, and bureaucratic silos undermined US efforts in Vietnam.[4] Such challenges persistently undermined the quality of the US government assessments in Afghanistan and Iraq, too.

The administration's new strategy required an annual assessment, which offered the opportunity to assess the strategic direction of the war and the likelihood of success. This process began in October and lasted until December 2010. Violence hit record highs that year. The Afghan security forces were growing in size and capability but had serious leadership and corruption problems and were progressing more slowly than forecasted toward being able to operate independently. The Kabul Bank collapsed late that same year, after over $900 million had been looted by corrupt officials, warlords, and other power brokers. Mahmoud Karzai, the Afghan president's brother, and Qaseem Fahim, the vice president's son, were implicated.[5] Afghanistan remained at the very top of the world's most corrupt governments.[6] These problems led Under Secretary of Defense for Policy Michèle A. Flournoy to conclude that the Afghan government was not winning the battle of legitimacy.[7] Diplomatic efforts to induce Pakistan to "change its strategic calculus" and to turn on the Afghan Taliban, meanwhile, had yet to achieve any tangible results.[8]

Interagency battle lines formed quickly. The intelligence community assessed that the Taliban had strengthened; the military command countered that the increases in violence were due to ISAF taking the fight into more Taliban-controlled areas.[9] The Defense Department insisted the counterinsurgency campaign was on track and simply needed more time to work.[10] After all, the surge forces had only been entirely on the ground for a few months. Gates famously noted, "The sense of progress among those closest to the fight is palpable."[11] Holbrooke, who died suddenly of a heart attack during the process, was convinced that COIN (shorthand for counterinsurgency) had already failed and wanted a significant push on reconciliation. Members of the White House staff, skeptical of the surge in the first place, were convinced the military campaign was unlikely to produce results. Some reportedly pushed for a quicker drawdown.[12] The Departments of Defense and State both accused the White House staff of being policy advocates rather than honest brokers and of placing the highest weight on the most cynical interpretation of events.[13]

Confirmation bias was damaging the objectivity of assessments. Individual agencies and the White House were making selective use of information to bolster their cases. The intelligence community assessments about the strength of the Taliban, endemic corruption in the Afghan government, and sanctuaries in Pakistan competed with optimistic claims of progress from the field.[14] The conflicting assessments had plenty of facts on their side. What the NSC did not adequately consider was that the positive results that were happening on the ground were not nearly sufficient to alter the factors that were undermining strategic success. A dispassionate view of the evidence suggested that a successful transition in 2014 was highly unlikely. Interagency disagreement and an inability to develop and assess strategically relevant metrics, however, reinforced a bias toward maintaining the status quo.

Overall, the contentious 2010 Afghanistan-Pakistan Annual Review refined the war aims and specified five lines of effort: a civil-military campaign to degrade the Taliban and build Afghan capacity; strategic partnership; transition to full Afghan sovereignty (security, economic, political); regional diplomacy; and reconciliation.[15] The lines of effort were coequal and unprioritized but assumed to be mutually reinforcing toward a successful outcome. The updated approach did not address the unity of effort problems, so each agency continued to optimize its silo. The NSC handled disagreements by watering down issues toward consensus.[16] Reconciliation became a friction point, as we will discuss in the next part.

The United States had a final opportunity at the end of 2011 to assess whether the transition strategy remained viable, the last chance to alter course before the 2012 US presidential elections and the troop drawdowns in 2013. Administration officials and the military command in Kabul continued highlighting progress while noting that government corruption and insurgent safe havens in Pakistan were critical risks to success. At no point, however, did any senior official testify that these risks were insurmountable or that success required a significant course correction.[17] The October 2011 semiannual report from the Department of Defense to Congress, for instance, summarized the situation in optimistic terms:

> [T]he International Security Assistance Force (ISAF) and its Afghan partners have made important security gains, reversing violence trends in much of the country (except along the border with Pakistan), and beginning transition to Afghan security lead in seven areas. . . . Although security continues to improve, the insurgency's safe havens in Pakistan, as well as the limited capacity of the Afghan government, remain the biggest risks to the process of turning security gains into a durable, stable Afghanistan. The insurgency remains resilient, benefitting from safe havens inside Pakistan. . . . Nevertheless, sustained progress has provided increased security and stability for the Afghan population and enabled the beginning of transition in July of security responsibilities to Afghan forces in seven areas, comprising 25 percent of the Afghan population.[18]

The military believed its campaign plan would be successful in its specified tasks. However, the military component, although dwarfing other agencies in resources, comprised only a small part of the administration's five lines of effort. The command was responsible for the military portion of the civil-military campaign to degrade the Taliban and build Afghan capacity (in this case, the ANSF). The governance and corruption problems, which belonged to State, were outside their authorities.[19] For strategic partnerships, the military played a supporting role in negotiations with the Afghan government. On transition, the military oversaw the security transition, but not the political and economic ones. The military likewise played only supporting roles in regional diplomacy and reconciliation. Chairman of the Joint Chiefs of Staff Admiral Michael Mullen's congressional testimony in September 2011 captures the matter: "The military component of our strategy, to the extent it can be separated from the strategy as a whole, is meeting our objectives. Afghan and ISAF forces have wrested the initiative and the momentum from the Taliban in several key areas. The number of insurgent-initiated attacks has for several months been the same or lower than it was at the same time last year. And we are on a pace and even slightly ahead of our end strength goals for the Afghan national security forces."[20]

Senator Carl Levin, the chairman of the Senate Armed Services Committee, asked Mullen if the effort was on course to meet the Obama timetable. "As far as I can see, yes, sir . . . while the risk is up, I think it's manageable and that there's no question that we can get there and sustain the military success and the military component of the campaign."[21]

America's bureaucratic way of war encouraged military officials to stay in their bureaucratic lane and focus on military progress and risks.[22] Officials from other agencies reported metrics from their silos. What no one questioned was whether a successful military campaign could still result in strategic failure. After all, the military could not make a dent in the Taliban's sustainability and the government legitimacy problems.[23] No one addressed those damaging cross-cutting issues, because no person below the president was accountable for strategic success and thus forced to confront them.

Despite the security gains, the insurgency remained sustainable. Violence levels in the second half of 2011 were indeed lower than in the second half of 2010, but characterizing that as a trend was misleading, as 2011 showed a higher level of violence in the first half of the year (see figure 6). The 2010 parliamentary elections counted for much of the spike in attacks that summer, as the Taliban aimed to discredit the election and candidates used violence to suppress voting in areas where people supported their opponents, so the year-on-year comparison was misleading. Compared to 2009, the number of violent incidents in 2011 was three times higher. The military was

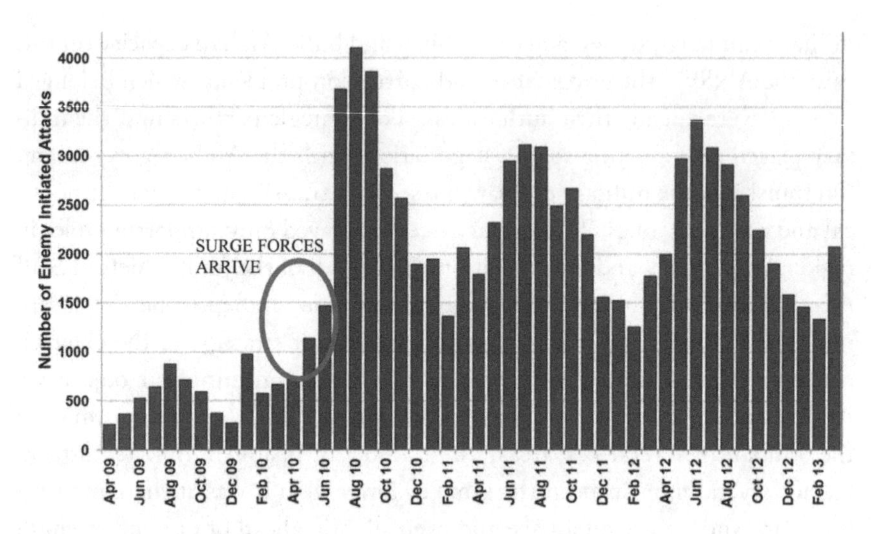

Figure 6. Enemy-Initiated Attacks, Afghanistan, 2009–2013
Source: US Department of Defense, *Report on Progress Toward Security and Stability in Afghanistan*, July 2013, A-3 (circle and annotation added).

supposed to hand over a residual insurgency to the ANSF by 2014, but there were no standards that defined what residual meant or even if insurgent strength in 2009 was the benchmark.[24]

The military offered no reason in its 2011 semiannual reports to Congress, or in future assessments or testimonies, to be overly concerned whether the ANSF could handle the insurgency by the end of 2014. "The ANSF are on track to assume full security responsibility by the end of 2014," the Department of Defense assessed, "after successfully securing the presidential and provincial council elections and performing well during the fighting season."[25] Although the Afghan security forces looked strong on paper, corruption and poor leadership were sapping their strength. Patron-client problems such as predation, sale of fuel, food, and equipment on the black market, and ghost soldiers damaged readiness faster than the coalition could build it. "How long would you stand and fight if your commander is stealing your food and equipment?" asked a former Afghan military officer.[26] The Afghan local police, despite some positive examples, was particularly corrupt and predatory.[27] The fielding of western systems and equipment, meanwhile, was increasing ANSF dependency on western advisors. A June 2011 Senate Foreign Relations Committee report warned of Afghanistan's growing dependence and corruption.[28] ANSF readiness to assume security responsibility was eroding even as their numbers and resources were increasing.

Adding to the problems, the insurgent sanctuary in Pakistan remained intact. The efforts to build a strategic partnership with Pakistan that would change the latter's strategic calculus were ineffective. In May 2011, Obama approved the raid into Abbottabad, Pakistan, that killed al Qaeda leader Osama bin Laden. Pakistan was outraged by the breach of its sovereignty.[29] A month earlier, a US contractor had killed two Pakistani citizens in Lahore and was detained by Pakistani authorities despite American objections about diplomatic immunity.[30] A final blow came in a border incident in which US aircraft killed 25 Pakistan Frontier Corps soldiers who had fired on a nearby ISAF-ANSF patrol.[31] At that point, Pakistan cut the ISAF logistics line leading from Karachi to Afghanistan, forcing the coalition to move supplies through Russia and Central Asia instead. In his final testimony to the Senate Armed Service Committee, Admiral Michael Mullen, who worked diligently to build a productive relationship with Pakistan's chief of army staff, General Kayani, decried Pakistan's use of insurgents as a "veritable arm" of foreign policy.[32] A frustrated Senator Carl Levin, the chairman of the Senate Armed Services Committee, reportedly asked the Pakistani prime minister, Yousaf Raza Gilani, why Pakistan had not condemned attacks on US forces by groups operating from Pakistani soil. Gilani did not answer him.[33]

Finally, the Afghan government was not showing any evidence of reform. According to Transparency International's Corruption Perceptions Index, Afghanistan was tied with Myanmar as the third most corrupt country in the world, behind North Korea and Somalia.[34] Karzai resisted efforts to address corruption, faulting the United States for the problem.[35] Mullen carefully explained, "If we continue to draw down forces apace while such public and systemic corruption is left unchecked, I believe we risk leaving behind a government in which we cannot reasonably expect Afghans to have faith."[36] Some Defense officials pressed hard for the US government to develop a credible anticorruption strategy, mainly Mullen's senior advisor Sarah Chayes. Still, State was unable to put one together and feared the risks of breaking the relationship with Karzai. The Department of Defense eventually dropped the matter.[37]

The White House was unwilling to address these problems or examine their implications. Part of the issue was bandwidth. Senior officials complained that getting any time on the national security advisor's or president's calendar was nearly impossible.[38] By 2011, although deputies meetings were frequent, NSC meetings on Afghanistan were rare.[39] Events such as the Arab Spring, Iraq withdrawal, bin Laden raid, Libya intervention, and domestic challenges competed for NSC attention. When the war in Afghanistan did come up, it generally involved crisis management.[40] Little time or energy was available for considering highly complex strategic issues.

The reconciliation effort, as we will discuss in the next part, was in disarray by the fall of 2011 and was not available to serve as a credible alternative to transition. Believing that the "occupation narrative" was the driving force behind the Taliban's ability to recruit, some senior White House and State officials clung to the argument that the drawdown would reduce the strength of the insurgency on its own.[41] In this view, winding down the war would lower the threat of the Taliban and ease the ANSF's job of securing the country. That Afghans had been fighting other Afghans from the Soviet withdrawal in 1988 to the Taliban overthrow in 2001 did not factor into this fanciful theory.

In some ways, Defense's assurances about the military campaign plan were self-fulfilling for drawdown advocates in the White House. As long as the administration believed that ANSF development was on track to secure the country by the end of 2014, there was no reason to take the political risks of shifting priority to reconciliation, tackling corruption, or reexamining the withdrawal timeline. Confirmation bias, reinforced by evidence of progress within bureaucratic silos, kept the withdrawal timeline off-limits to serious examination.

The administration decided not to conduct a thorough review in 2011 as it did at the end of 2010. Although some Defense officials believed they could convince the president to "put more time on the clock" (delay the drawdown), Obama held firm. He announced in June 2011 that the withdrawal had begun and that the combat mission would end in 2014, at which time the Afghans would be responsible for their security.[42] The advocates of drawdown had won the argument, and there was no reason, from their standpoint, to open an interagency process that would invariably try to rehash the issue.

In the event, the security situation declined as international forces handed over security responsibility to the ANSF in mid-2013.[43] Although the surge achieved temporary effects where foreign forces concentrated, the Afghan government proved incapable of winning the battle for legitimacy in contested areas.[44] The Afghan National Security Forces, ISAF asserted, was to be the "defeat mechanism" of the Taliban.[45] Problems, however, continued to materialize. Although rated by coalition advisors at the end of 2014 as capable of independent operations to secure the province, nearly the entire Helmand-based 215th Afghan Army Corps collapsed a year later.[46] The 215th was the newest corps in the Afghan Army, but the post-2014 advisory mission did not cover them. The British refused to take on the task, and the Obama administration declined to raise its troop presence to make up the difference. An Afghan military officer reported finding a Taliban warehouse of abandoned corps equipment during a 2015 operation in Marjah.[47] By September of that year, the Taliban temporarily captured Kunduz, a provincial capital, and the first major city they had controlled since 2001.[48] Al Qaeda's presence reportedly began to grow in 2015, and an Islamic State affiliate established a presence in the strategic border province of Nangarhar.[49] In early 2017, more than 30,000 suspected ghost soldiers were dropped from the rolls—nearly 10 percent of the ANSF's reported strength.[50] Instead of fighting a residual Taliban insurgency, as the US policy had forecasted in 2009, the ANSF were facing the most muscular Taliban force since 2001 and a growing terrorist presence.[51] ISAF handed off a strengthening insurgency to an ANSF that was severely compromised by corruption, poor leadership, and political pressures.

Conclusion to Part III

Why did the Bush and Obama administrations fixate on ineffective strategies? Confirmation bias reinforced the Bush administration's belief that the war was over, that a small military effort could defeat the remnants of the Taliban, and that the international community could pick up the burden of reconstruction. Officials emphasized reports of progress within political, military, and economic silos and discounted evidence of mounting problems. The Obama administration's belief that the Taliban would be unwilling to fight other Afghans and its frustration over the Karzai government's endemic corruption, coupled with the Pentagon's narrative about the ANSF's readiness, reinforced the withdrawal timeline. Evidence mounted that the Afghan government was losing legitimacy, and the ANSF was corrupt and poorly led. The Taliban remained resilient with their sanctuaries in Pakistan and local support in Afghanistan intact. Still, the Obama administration would not reexamine the drawdown timeline until the disaster became apparent at the end of 2014. Poor strategic empathy compounded the effects of confirmation bias in both administrations. Civilian and military officials, focused in their silos, never addressed the cross-cutting issues that jeopardized US aims. Defense officials had to use subtle language about risks to avoid running afoul of the White House, but no one in Congress picked up the nuances.

The United States' national security architecture discourages placing a civilian or military official in charge of the full range of American political, diplomatic, military, intelligence, and economic capabilities deployed to war zones like Afghanistan and Iraq. The result is a bureaucratic way of war in which no one below the president is accountable for overall success. This absence of authority gives leaders on the ground opportunities—wittingly or otherwise—to operate in ways that might impede the president's goals.

Chapter 6 outlined early bureaucratic frictions that delayed counterinsurgency and nation-building efforts and allowed predatory actors at local and national levels to seize control of key centers of power and run roughshod over parts of the population. Elites continued to manipulate American forces and officials into advancing their personal and political agendas—sometimes by targeting their rivals. ISAF grew wise to this by 2010, but the strategic damage was not reversible. As both administrations operated in military, political, diplomatic, and economic silos, strategic risks were emerging along the seams and fault lines. Each effort was making progress, but the whole was less than the sum of its parts. Military operations conducted with inadequate information or manipulated by elites undermined governance and legitimacy. Predatory corruption damaged government legitimacy and motivated people to fight back, thus deteriorating the security situation. The inability of US government agencies to agree about reconciliation, as will be explored in detail in Part IV, left the Obama administration without a viable alternative to a failing transition plan. The tendency by both administrations to measure success within silos reinforced the narrative of progress and undermined their ability to understand and address the emerging strategic risks.

Severe patron-client problems emerged for both administrations that impeded strategic decision-making and coordination, masked the magnitude of challenges, and undermined capacity-building efforts. The US and Afghan governments never developed a common strategy for the war. While the United States wanted to win quickly and leave, the new Afghan government focused on consolidating power and extending international presence and financial support. Karzai grew increasingly suspect of US intentions during the Obama administration. As corruption grew under the Karzai administration, so did Obama's desire to withdraw.

As Obama stood firm on the timeline, some senior officials fixated on changing the president's mind instead of addressing the issues that were undermining transition—such as the corruption and poor leadership in the government and ANSF. Even if officials addressed the main challenges to transition, the timeline was probably still unrealistic. They could have had a sounder basis for asking for more time or making a more serious effort at negotiations.

Both administrations, meanwhile, provided funding to the Pakistani military in the hopes of inducing them to shut down Taliban sanctuaries. The Pakistanis fought the Pakistani Taliban instead, leaving the Afghan Taliban and its Haqqani network untouched. This support to Pakistan increased Karzai's cynicism about American intentions, alarmed India, and reinforced Pakistan's incentives to support instability in Afghanistan so as to maintain the flow of American cash. The US approach "became a *de facto* military attrition campaign," Douglas E. Lute, the Obama administration's policy lead for South Asia from 2009 to 2013, recalls, "the political and diplomatic efforts never materialized."[1]

These challenges also undermined the use of the transition method to achieve a favorable and durable outcome. By 2010, when the Afghanistan mission finally had significant resources and Washington's attention, the deck was stacked against a successful transition. The US government did not understand the predatory kleptocracy that had emerged or appreciate the strategic risks it entailed. No one was responsible or accountable for addressing the problem. The State Department was unwilling to take on the task of addressing the kleptocracy. Efforts by ISAF to do so were too limited to have much effect.

When the matter finally reached the NSC, the Obama administration punted. Even the staggering Kabul bank crisis was not enough to motivate the interagency to address the problem. Meanwhile, the Afghan government and political elites had long since co-opted the ANSF leadership. Army and police positions were bought and sold at exorbitant prices. The reported price for a Tier 1 province police chief could go as high as $3 million to $5 million; a customs official allegedly paid $5 million for his position. The primary motivator for too many leaders was feeding the kleptocracy.[2] For them, fighting and winning the war was a lesser consideration. The selling and buying of offices, a cabinet-level Afghan official said, "is our biggest problem."[3] US officials, fixated on their silos, failed to grasp the overall picture.

The predatory behavior was creating more support for the Taliban. Sufficient support within Afghanistan, combined with the sanctuary in Pakistan, kept the Taliban sustainable. Corruption and poor leadership in the government and security forces undermined the Afghan government's ability to win the battle of legitimacy in contested and Taliban-controlled areas. Perverse incentives drained readiness and performance so fast that the sizeable capacity-building programs of the US embassy and the NATO training mission never caught up.

To be fair, the risk factors in Afghanistan were subtler than they were in Iraq. The Iraqi government's predatory sectarianism was easier to deduce and address than Afghanistan's predatory kleptocracy. Pakistan was less evident in their support for the Taliban (as was India for anti-Pakistan factions) than was Iran for the Sadrist militias. Al Qaeda in Iraq, wanting to provoke a sectarian civil war, never had a governance strategy and could not care less about winning the battle of legitimacy. The Taliban did. While Iraq hurtled into a disaster, Afghanistan crept incrementally but relentlessly toward failure.

Part IV

Ending the War in Afghanistan

Overview

President Obama's December 1, 2009, announcement of both a surge of forces and a timeline for withdrawal limited American leverage during the exploratory talks with the Taliban. They undermined the potential to secure a favorable and durable outcome by the end of his administration. The Taliban, thus, was interested in an arrangement with the United States but not a peace process to resolve the broader conflict. They aimed to use exploratory talks to gain concessions that improved their legitimacy while coaxing the United States to complete the withdrawal of its forces. American bargaining power declined as the drawdown continued. By March 2012, the Taliban believed they had little to gain from the talks and little to lose by walking out of them.[1] Reconciliation (the term used for peace efforts) ended in disaster with the abortive attempt to open a Taliban political office in Doha on June 18, 2013.[2] The fallout from that event damaged US legitimacy and contributed to an increasingly toxic environment in Afghanistan and the region.[3]

The United States and the Taliban did manage to conclude a detainee exchange in June 2014, but this effort was delinked from a peace process. Various exploratory conversations between Afghan officials and Taliban figures (outside the political commission) occurred in 2015 but stalled after the Taliban announced the death of their reclusive leader, Mullah Mohammad Omar.[4] By mid-2016, a new four-party effort called the Quadrilateral Coordination Group had yet to produce meaningful results.[5] As the security situation declined, President Obama ordered an extension of the US military presence beyond 2016.[6] "When we first sent our forces into Afghanistan 14 years ago," Obama announced, "few Americans imagined we'd be there—in any capacity—this long."[7]

14

Reconciliation versus Transition

Transition is not the only alternative to a decisive victory. An intervening power might consider negotiations instead. Donald Wittman, writing in 1979 in the wake of the end of the Vietnam conflict, developed a rational bargaining model to explain how adversaries reach a negotiated settlement.[1] States, he argues, tend to begin a war with high expected utility—that the war will yield favorable results.[2] As the war progresses, combatants gain information about relative strengths and weaknesses and the likelihood of success.[3] Political scientist Harrison Wagner argues that war is not a contest to disarm one another, but a bargaining process in which states use force or the threat of force to influence others.[4] Wars thus tend to stop short of one side destroying the other.[5]

Demands from external actors for cease-fires and negotiations, however, may go unheeded until the war settles into a stalemate. I. William Zartman argues that even if the substance of proposed solutions to the conflict might be mutually acceptable, the right timing is equally necessary. A "mutually hurting stalemate" occurs when the parties cannot escalate to victory, the deadlock is painful to both parties (though not necessarily equally hurtful), and they both decide to seek a way out.[6] Provided a way out is perceived to be available and acceptable, the conflict is ripe for a negotiated outcome.[7]

How does this apply to interventions against insurgencies? The scholarship tends to focus on conventional wars with uniformed military forces that fight to win, lose, or draw. An irregular war that includes an intervening power has a different dynamic. At first, bargaining opportunities are likely limited.[8] If the insurgency seeks to bargain, but the host nation rejects, the rebels must surrender or fight for survival. The military contest is asymmetrical. The insurgent, being the militarily weaker party, generally avoids decisive battles. Guerrilla warfare aims to exhaust the stronger party.[9] The insurgency may lose every pitched battle, but that is not necessarily relevant to the outcome. An insurgent's strength grows with increased tangible support from the population and external actors. The insurgency's political and social efforts seek to improve local control. Military actions keep the group relevant and undermine the government's ability to maintain security. A sustainable insurgency has the potential to get stronger over time, especially if the government is predatory and exclusionary.

The intervening power and host nation may escalate the conflict, but the insurgency has a chance of succeeding if it survives with tangible support intact. Once the intervening power tires of the war and begins to withdraw, the rebels can play for time. The intervening power might be more open to negotiations at this point, but these could be less appealing to the insurgency. The latter will probably seek to maximize their leverage before exploring talks. They are likely better off waiting until after the withdrawal of foreign forces to test the host nation's government fighting on its own. The insurgency could be willing to negotiate concessions over the external power's departure, but not to end the conflict.[10] The United States, for example, negotiated with North Vietnam to withdraw from South Vietnam, and the Soviets negotiated with the United States and Pakistan to withdraw from Afghanistan.[11]

What if the external power elects to stay indefinitely, but just at a much lower level? The models offered by Wittman, Wagner, and Zartman can become helpful again.[12] As the intervening force reaches its sustainable level, the insurgency will likely seek to maximize battlefield gains. Once each side perceives that the costs of further advances outweigh the benefits, a mutually hurting stalemate may set in. In such cases, negotiations could begin when the situation is ripe. The challenge for an intervening power is to make its commitment to indefinite presence seem credible to the insurgency, which is easier to do at the beginning of the conflict.

Negotiations with an insurgency, however, are likely to include credible commitment challenges. During the Northern Ireland conflict, the Provisional Irish Republican Army (PIRA) was not going to be strong enough to win an outright victory or force the British military and police to leave the island. Similarly, the sustainability of the PIRA's local and external support made decisive victory over them highly unlikely.[13] Credible commitment challenges, significant differences of opinion within each party, and audience costs impeded negotiations. Nonetheless, secret talks persisted and eventually produced the Good Friday Agreement.[14]

The intended path toward a negotiated outcome may also play an important role. The very deliberate process in Northern Ireland stands in contrast to more hasty efforts to broker peace deals in other cases. Insurgencies and civil wars damage the fabric and cohesion of a society. Peace deals in a low-trust environment, even if struck, are likely to be short-lived and destabilizing.[15] The Peshawar (1992) and Islamabad (1993) Accords were efforts to create power-sharing deals among major Afghan mujahideen factions. They failed and set the stage for the Afghan civil war that would bring al Qaeda to Afghanistan and the Taliban to power.[16]

These kinds of bargaining asymmetries and challenges suggest that the ability to fall back from decisive victory to a negotiated settlement could be less available for intervening powers in irregular conflicts than in classic conventional wars.

Why did reconciliation efforts in Afghanistan fail? Former Holbrooke senior advisor Vali Nasr argues that the White House was "too skittish to try it." Still, he does not discuss why Obama was reluctant to put his eggs in the reconciliation basket.[17] Thomas Waldman, in a paper examining the intellectual history of reconciliation, suggests that the military was too closed-minded.[18] Afghanistan expert and former Pentagon senior advisor Sarah Chayes counters that this is too facile.[19] As we will discuss later, the Pentagon supported reconciliation in theory, but leaders had different notions of it. The same was true of the State Department, the Special Representative for Afghanistan and Pakistan (SRAP) office, and the White House.

Political scientists and Afghanistan experts Dobbins and Malkasian blame all US government parties for "a failure to initiate a peace process at the peak of US leverage, as NATO troops were retaking large swaths of the Taliban's heartland in Kandahar, Helmand, and nearby provinces."[20] These arguments do not explore the reasons why the US and Afghan governments, the Taliban, Pakistan, and others preferred to continue fighting an inconclusive war rather than negotiate an end to it.

Bringing an insurgency to a negotiated conclusion has been historically difficult. Political Scientist James D. Fearon argued in 2007 that only 16 percent of the 55 civil wars fought since 1955 came to a negotiated end.[21] Chris Paul and his coauthors at RAND found that only 19 of 71 conflicts against insurgencies since World War II resulted in "mixed outcomes," where neither side wins outright, and both make significant concessions to end the conflict; in other words, a mere 27 percent.[22] Fearon notes that, for the most part, the low rate of negotiated outcomes is not due to a lack of effort. "Negotiations on power-sharing are common in the midst of civil wars, as are failed attempts, often with the help of outside intervention by states or international institutions, to implement such agreements." Such efforts usually fail,

Fearon observes, due to mutually reinforcing fears: "combatants are afraid that the other side will use force to grab power and at the same time are tempted to use force to grab power themselves."[23] This problem creates a type of prisoner's dilemma in which both parties might recognize the benefits of peace, but do not trust the other party enough to take the risk to stop fighting. A third-party enforcer might make the sides abide by an agreement temporarily, but unless the warring parties build sufficient trust, the power-sharing arrangement is likely to fall apart.[24]

Scholars and Afghanistan experts have diverse opinions about reconciliation.[25] Some academics, civil society advocates, and most Afghan elites associated with the former Northern Alliance believed the Taliban to be irreconcilable.[26] They argued that any talks with the group would be akin to negotiating with terrorists, and inevitably would result in trampling of human rights and legitimizing the use of terrorism as a political weapon. Non-Pashtun elites feared that a power-sharing arrangement with the Taliban would shortchange them. Some scholars cautioned about the Taliban's tendency to use negotiations as a stalling tactic.[27] A presumption of Taliban irreconcilability reflected the views of the Bush administration, particularly in the early years.[28]

From 2008 to 2011, an increasing chorus of experts and diplomats encouraged a political solution to the conflict. Most came to this conclusion due to the intractable nature of the conflict and the unlikely prospects of a clear military victory.[29] British SRAP Sir Sherard Cowper-Coles was a consistent advocate since at least 2009. His advocacy up to that point kept reconciliation in the conversation with the United States, but he had been unsuccessful in getting the Obama administration to make it a priority.[30] The Afghanistan Study Group argued in 2010 that "the US should fast-track a peace process designed to decentralize power within Afghanistan and encourage a power-sharing balance among the principal parties."[31] By 2009, senior US commanders began to note that there was no military solution to the conflict.[32] Other experts viewed reconciliation as necessary for the endgame but believed that the surge needed time to place more pressure on the Taliban.[33] In this view, Taliban leaders would begin to opt out of the insurgency and into the political and constitutional fabric of Afghanistan.

Those supporting a negotiated outcome had to contend with the fact that neither the Afghan government nor the Taliban was willing to enter talks. Powerful constituencies in both actors believed that they could—or must—win outright. A reconciliation process needed to bring these internal groups along. With the conflict in Afghanistan raging for over 30 years by 2010, anxieties and hatreds were intense. As a veteran of the Northern Ireland peace process said to me, "Ninety percent of the negotiations is with your base, not with your adversaries."[34] The Taliban, moreover, knew they could play for time. Obama's announcement of a drawdown timeline meant that the Taliban simply needed to wait until July 2011 for the pressure to begin to ease.

Third-party actors complicated matters, creating an even broader prisoner's dilemma problem.[35] The reconciliation models that envisioned a Taliban capitulation required Pakistan to turn against the insurgent group. A loss of sanctuary in Pakistan, plus intense military pressure in Afghanistan, would have made the Taliban highly vulnerable to outright defeat.[36] The United States had proven unable thus far to motivate Pakistan to act against the Afghan Taliban. Pakistan had no interest in forcing the Taliban to negotiate.[37] China would not compel Pakistan to turn against the group.[38] India was unlikely to support a power-sharing arrangement that might put their traditional Afghan allies out of power and draw Afghanistan into Pakistan's orbit.

A reconciliation effort in Afghanistan had to face these challenges. The conditions in early 2010 were not yet ripe for negotiations to end the conflict. To bring this about required the Taliban and Afghan government and their backers to believe that neither side was likely to win outright, that the benefit of future military gains was not worth the cost, and that a path toward a peace process was compelling enough to overcome the status quo.[39] To bring about a peace process, the United States would have needed to modify its strategy from transition-and-withdraw to a negotiated outcome. The Obama administration declined to do so, and thus had little bargaining leverage with the Afghan government, the Taliban, and Pakistan.

Reconciliation never gained sufficient traction because it could not pass the credibility bar with Obama's National Security Council (NSC). Advocates for a negotiated settlement needed to outline a practical path toward a peace process that protected American interests. The game plan had to be sufficiently compelling to overcome the administration's status quo bias, fears of Taliban deceit, and the backlash the president was likely to encounter by negotiating with an al Qaeda-allied Taliban.

The absence of a conceptual apparatus for war termination within the US government made these problems even harder, undermining communication and interagency cooperation. Obama's focus on transition-and-withdraw, moreover, eroded American leverage and limited the incentives for serious negotiations. A drawdown timeline, General Petraeus reflected, "tells the enemy that he just has to hang on for a certain period and then the pressure will be less. In a contest of wills, that matters."[40] Lack of vision and strategy for reconciliation, poor coordination, and sloppy execution reinforced these obstacles and friction points, dooming the effort to failure and Afghanistan to greater violence. Success might have been a long shot, as two former SRAPs have argued.[41] Self-imposed problems, however, made the odds far steeper than they needed to be. "The United States and Afghan governments," Lute observes, "squandered their best opportunity to advance reconciliation."[42]

15

Reconciling Reconciliation

Emotional investment may complicate decisions about negotiations. Some opponents may suggest compromise means that sacrifices in lives and treasure were in vain, that talking with such evil actors is unthinkable. Fred Iklé, a former senior official in the Nixon and Reagan administrations, wrote about the intense and potentially divisive nature of strategy discussions between patriots and "traitors."[1] Stakeholders are likely to frame information in ways that support their conclusions and the interests of their constituents or departments.[2] "Those who want their country to pursue ambitious war aims will seek out the favorable military estimates and find reasons why negotiations ought to be avoided," Iklé argues. "Those who want negotiations to move ahead will select the unfavorable military estimates to argue that war aims should be scaled down."[3] In addition to misinterpreting or overlooking information, actors might simply frame information in ways that support their views. Political leaders may also want to avoid creating winners and losers in their national security teams. Lack of consensus for change thus tends to keep the status quo intact.

Internal politics contributes to strategic paralysis. States are not unitary actors that operate in lock-step or see issues from a single, shared, and objective perspective.[4] Differing assessments among decision-makers may entrench the status quo. Actors perceive proposed changes to the status quo as gains or losses of influence, prestige, or power.[5] Prospect theory, discussed in more detail in chapter 26, suggests that those deeply invested in a current strategy will fight hard to prevent change—likely harder than those advocating for a new direction. More disconcertingly, such stakeholders could manipulate information asymmetries in cynical efforts to block changes or undermine alternatives.[6]

Political scientists Stanley and Sawyer suggest that leaders may not want to seek a negotiated outcome due to personal stakes such as power, prestige, and physical or financial security.[7] They may have less prosaic motives, too. Policymakers may prefer to keep fighting if they doubt the sincerity of the adversary. Political scientist Dan Reiter analyzes examples from the US Civil War to the Korean War in which uncertainties about the adversary's credibility led decision-makers to continue fighting until the doubts subsided.[8] An adversary that offers to capitulate, for instance, removes uncertainty about their willingness to stop fighting. Alternatively, an adversary can use cease-fires and negotiations as a ploy to consolidate control, buy time to prepare for future military operations, or undermine their opponent's legitimacy. These kinds of credible commitment concerns increase when dealing with armed nonstate actors who may not feel bound by international law and the Geneva Conventions.

The Obama administration faced these challenges. Bureaucratic frictions and conflicting assessments undermined the ability to put together a coherent approach toward a negotiated settlement. As mentioned in chapter 13, reconciliation became one of five unprioritized strategic lines of effort in 2010, alongside the civil-military campaign, transition, strategic partnership, and regional diplomacy. The SRAP was responsible for regional diplomacy and reconciliation. Two factors boosted the importance of the latter. First, advisors within State and Defense were successful in convincing the NSC during the 2010 Afghanistan-Pakistan Annual Review that reconciliation was a logical adaptation of the 2009 strategy. With the Afghan government actively resisting reform and with efforts to change Pakistan's political calculus not bearing fruit, they argued, the Obama administration should investigate the prospects of reconciliation. If done carefully, exploratory measures need not conflict with the other lines of effort.[9]

Requests from the Taliban for direct talks reinforced the arguments for exploring reconciliation. Tayyab Agha, the chief of the Taliban political commission and former secretary to Taliban leader Mullah Mohammad Omar, had asked Germany in 2009 to broker a meeting with the United States. After confirming his identity, the Germans arranged the first US-Taliban meeting in Munich in November 2010.[10] The discussion covered general points but suggested that the Taliban wanted to continue the talks. The result for the United States, as Secretary of State Clinton later explained, was the "fight and talk" approach. This dual-track effort could lead toward a reconciliation process or, at worst, confirm Taliban duplicity.[11]

"Our civilian and military efforts," Clinton announced, "must support a durable and favorable political resolution of the conflict. In 2011, we will intensify our regional diplomacy to enable a political process to promote peace and stability in Afghanistan."[12] Exactly what reconciliation was supposed to achieve, however, remained vague. Officially, it was an Afghan-led dialogue with Taliban senior leaders toward a political resolution of the conflict.

Some US officials viewed reconciliation as a peace process with strong US involvement. Others believed it should focus on convincing Taliban senior leaders to defect from the insurgency and join the Afghan government. Finally, others believed that the US role was to persuade the Taliban to talk to the Afghan government. The distinctions are subtle, but the ambiguity contributed to the failure to achieve consensus in Washington and Kabul.

To build momentum for reconciliation, Secretary Clinton discussed the political and diplomatic surges at a landmark speech to the Asia Society in February 2011. "Today," she noted, "the escalating pressure of our military campaign is sharpening . . . a decision for the Taliban." The choices were to "break ties with al-Qaida, give up your arms, and abide by the Afghan constitution, and you can rejoin Afghan society; refuse and you will continue to face the consequences of being tied to al-Qaida as an enemy of the international community."[13] Clinton had just outlined what became known as the three red lines for reconciliation. Those who met the conditions could participate in the process. The United States would continue targeting those who refused. "This is the price for reaching a political resolution," she announced. "If former militants are willing to meet these red lines, they would then be able to participate in the political life of the country under their constitution." Several coalition partners who had come to the same conclusion about the need for a political settlement welcomed the speech.

This new "fight and talk" approach was complicated because the Obama administration had not decided how to achieve a favorable and durable outcome. The NSC never agreed on goals for reconciliation and how those would integrate with other efforts. Petraeus, notes Vali Nasr, was skeptical of Holbrooke's reconciliation vision.[14] When asked by the latter to discuss reconciliation, the general reportedly replied, "that's a 15-second conversation. Eventually, yes, but no. Not now."[15] Despite having only one year with surge forces in place (unless he could convince Obama to delay withdrawals), Petraeus believed talks were premature. "It was clear from the outset—due to sanctuaries they enjoyed in Pakistan," Petraeus recollected, "that we could not sufficiently pressure the leaders of the Afghan Taliban and Haqqani network to get them to negotiate seriously."[16]

The lack of war termination doctrine and clarity on reconciliation led to divergent views within the US government. Within the Obama administration, at least four different ideas competed for traction. The status quo position was that the civil-military campaign to build Afghan capacity and degrade the Taliban, plus efforts to change Pakistan's political calculus and shut down insurgent sanctuaries, would allow the Afghan government and security forces to take over responsibility for security by 2014. The Afghan troops could finish off a residual Taliban insurgency. Petraeus's view is consistent with the capitulation model and the three red lines articulated by Clinton. The Taliban could surrender and trust the Afghan and US governments to treat them fairly, or they could keep fighting. Unless the Taliban believed themselves to be on the verge of defeat, they would be unlikely to capitulate. The Afghan National Security Forces, ISAF claimed, would be the "defeat mechanism" of the Taliban.[17] In this view, reconciliation would bring surrendering Taliban leaders into the political fabric of Afghanistan. The Afghan government supported this concept, as did the military command and the ambassador in Kabul, and elements within the Pentagon and the White House. This approach was attractive because it required no political compromise, it limited audience costs, and, if achieved, it could be viewed as a clear victory. This approach could have been successful in 2001. The downsides ten years later were its conflicts with reality. The Taliban were under significant pressure, but they were nowhere near the point of contemplating surrender, and the Afghan government was hemorrhaging legitimacy.

Some members of the SRAP team argued that reconciliation should aim for a peace deal or "grand bargain" with the Taliban.[18] Vali Nasr argues that this was the view of Holbrooke and some of his advisors.[19] "COIN has failed," a senior SRAP official told me in 2011, now it was the diplomats' turn to take over.[20] The grand bargain offered the potential for a deal to end the conflict. There were downsides, however. First, peace deals in low trust environments can be destabilizing.[21] Recent Afghan experiences with peace deals were not positive ones. As noted in chapter 4, the Peshawar (1992) and Islamabad (1993) Accords led to and perpetuated the Afghan civil war. Second, there seemed little appetite within the Afghan government or polity for such a deal. To many Afghans, particularly the former Northern Alliance factions in government and civil society groups in Kabul, a peace deal with the Taliban was anathema. Senior SRAP advisor Rina Amiri voiced concerns about the potential for backlash and the need to gain Afghan buy-in.[22] Senior State and Defense officials, myself included, voiced these same concerns.

Marc Grossman, a highly accomplished and respected career ambassador, was in an awkward position as Holbrooke's replacement. He likely recognized that large-scale Taliban defections were unlikely and appreciated the downsides of the grand bargain. He consequently developed, over time, a less ambitious and potentially more achievable approach of using confidence-building measures and regional diplomacy to get the Taliban to agree to meet with the Afghan government: "To try to open the door for Afghans to talk to other Afghans about the future of Afghanistan."[23]

The upside to this approach was its limited and potentially attainable goal. It focused on playing a declining hand well rather than seeking to improve it or change the game. The United States was losing negotiating leverage but might have enough carrots to get the Taliban to agree to official talks with the Afghan government. Such a step would imply Taliban recognition of the Karzai government as a legitimate negotiating partner—a substantial achievement. This approach limited the need for interagency coordination and conformed closely to the traditional envoy-to-envoy discussions diplomats spend their careers doing. It also removed the United States from having to negotiate peace.

The downsides to this approach included the potential backlash in the United States and Afghanistan of talks with and concessions toward the Taliban—which vocal constituencies in both countries regarded as a terrorist group. Discussions with the United States would confer more political legitimacy on the Taliban than they had had to date. There was also a danger that the Taliban would get concessions from the United States, agree to the meeting with the Afghan government, and then walk out of it. This approach was unlikely to bring an end to the conflict. Unless sufficiently coordinated, it might also be viewed suspiciously by the Afghan government.

At best, this approach would set a precedent for discussions. Because this approach did not address the drawdown timeline, the Taliban could simply play for time and not engage in serious talks until after international forces completed the drawdown. At that point, the Afghan government would be in a much weaker bargaining position than it was from 2011 to 2014. In short, the approach might accomplish the narrow objective of arranging a meeting between the Afghan government and Taliban, but arguably little to support US strategic goals.

My personal view, as the Defense Department lead for the effort, together with some senior advisors at State, was that reconciliation should be a deliberate and incremental peace process (rather than a peace deal) that was coordinated carefully with the Afghan government. We based this view on our war termination research.[24] The prospects for a favorable and durable outcome that met US interests, we felt, were more likely in this approach than the other three. Reconciliation should be the priority among the five lines of effort.

We argued that a deliberate peace process held the best opportunity for the United States to achieve the most favorable and durable outcome possible at the lowest costs. Rather than continue with a strict drawdown timeline, the administration should stabilize the troop presence indefinitely to convince the Taliban it would not win through force of arms. Ideally, exploratory talks should begin before the surge forces arrived to take advantage of the uncertainty within the Taliban about how much damage the military effort would do to them. If the Taliban withstood the surge and the United States began drawing down forces, we argued, the Taliban could simply play for time. They might negotiate our withdrawal, as the North Vietnamese did during the Vietnam War, but not an end to the conflict. A stalemate in which the Afghan government controlled 90 percent of the country to the Taliban's 10 percent was a position of strength for negotiations.

A compelling vision and strategy were needed to bring about a peace process. A step-by-step approach that built trust and confidence, we reasoned, could lead within a few years to growing consensus on political principles and measures to reduce violence, and eventually to cease-fires and discussions on more sensitive political issues. This process could also test Taliban intentions at relatively low risk while assessing the integrity of their statements and the credibility of their commitments. The national-level process needed to be complemented by local reintegration efforts that focused on good governance and conflict resolution, and a regional dimension to address interstate issues. This approach was more likely than the alternatives to result in a successful resolution to the conflict but would probably take years to come to fruition. Afghanistan had been at war by then for over 30 years. A durable peace would not break out after a few meetings. The other lines of effort would need to align accordingly.

This approach had its downsides, too. It would require Obama to modify the withdrawal timelines and suffer potential audience costs.[25] Talks with the Taliban were politically risky in the United States and Afghanistan. The administration would also incur the fiscal risk to higher domestic and international priorities of having large numbers of troops in Afghanistan beyond 2014 if the negotiations had not progressed by then to entail troop reductions. This approach required strategic patience and a long-term view, both often in short supply in Kabul and Washington. Lifting the withdrawal timelines could be seen as a betrayal to Obama's political base. It would probably require lengthy interagency deliberations, and White House officials were reportedly sensitive to criticisms about drawn-out decision-making discussions.

In short, it was tough to make reconciliation the priority effort, but preferable, we thought, to losing or to post-2014 negotiations in which the Taliban had much higher leverage. State and Defense coordinated a paper along these lines just before Grossman took over at SRAP. NSC deputies approved the proposal, but it was never moved forward to principals or disseminated to the field.[26] Officials close to the effort suggested that the concept was much more ambitious than Grossman, newly arrived as SRAP, was comfortable attempting.[27]

SRAP moved forward with a more limited approach. Defense agreed that reconciliation needed to coordinate several elements of national power.[28] With 100,000 troops on the ground, the Pentagon believed their military capabilities could be helpful. They were also concerned about the security and force protection implications if negotiators made uncoordinated agreements. Conversely, military operations conducted in isolation could have very damaging effects on reconciliation. Talks could easily be derailed, for instance, if an interlocutor was killed or captured. Spoiler activity needed to be understood and managed.

Moreover, the military had several "tribes" of its own that needed to support the effort. In addition to the military and civilian officials at the Pentagon, the four-star commands at ISAF, US Central Command, US Special Operations Command, and NATO all had equities. Two ISAF commanders told me that they wanted to support the effort but needed to know what to do.[29] The military wanted to help but needed an approach that they could understand.[30]

SRAP's resistance to a reconciliation strategy might seem odd, while the Pentagon's insistence on having one might appear as a terrible encroachment. The military tends to favor deliberate strategy, while diplomats tend to embrace emergent strategy.[31] The Pentagon is very good at the former. Important issues undergo a rigorous planning effort that coordinates the activity of thousands of people, massive amounts of logistics, and lethal force and capabilities. Deliberate strategy, however, can over-engineer problems and limit flexibility. State's institutional culture views planning as a waste of time. As explained to me by former State Department officials, when a diplomat wants to slow-roll something into nonexistence, they call for a plan.[32] American diplomats tend to be more comfortable with emergent strategy—skillfully playing your hand while adapting to a dynamic situation and protecting US interests. Holbrooke appropriately likened diplomacy to jazz.[33] This approach, however, limits interagency coordination and integration and can come across as making it up as you go along—if the players do not understand music, noise results. The risks to this approach may be low if the diplomatic effort is operating on its own, but they increase significantly when other stakeholders and American lives are involved. Those who think ahead usually outmaneuver those who fixate on the next move.

Reconciliation needed an appropriate blend of both deliberate and emergent strategies. The United States never articulated its reconciliation strategy, which frustrated Defense and some in SRAP and probably made Grossman feel second-guessed and micromanaged. "There were occasions," Grossman reflected, "when some colleagues tried to micromanage the conversation with the Taliban in ways designed to make it impossible to continue, but the need to keep inter-agency representatives engaged and as supportive as possible overrode my periodic frustrations."[34] To be sure, ideas on reconciliation from Defense could be as unwelcome to some diplomats as State recommendations on military efforts would be to some generals.

The result was the lack of an agreed concept within State, within the US government, and between the American and Afghan governments. Various stakeholders were pulling in different directions and potentially sending mixed messages to other parties. "Bureaucratic silos and turf battles undermined coordination on reconciliation," reflected former Deputy SRAP Vikram Singh.[35] These paled in comparison to disconnects with Kabul, which made the issue explosive.

16

Competing Visions

Karzai, Taliban, and Pakistan

Patron-client challenges and inadequate strategic empathy among US officials further undermined the prospects of bringing the war into a peace process. Other key stakeholders, of course, had their views and interests that would need to be addressed for reconciliation to gain sufficient traction. At the risk of oversimplification, I will focus on the Afghan government, the Taliban, and Pakistan, recognizing that they are not unitary actors (or the only actors) but that their positions had to address views from an array of internal and external stakeholders. Their policies, in many ways, were designed to account for these varied inputs, pressures, and constraints.

Karzai had an altogether different view on reconciliation than did SRAP. In some ways, Karzai could not accept the fact that Afghans were fighting against his government. He believed that local fighting was the result of America bringing the war to Afghan (mainly Pashtun) villages and the associated backlash over civilian casualties and questionable detentions.[1] The Taliban leadership, Karzai believed, was wholly a creature of Pakistan. With enough money and American pressure on Pakistan, he aimed to convince Taliban senior leaders to defect.[2] To increase pressure on Pakistan, the Afghan government had also begun clandestine support of the Pakistani Taliban.[3]

Karzai was not interested in formal talks between his government and the insurgency, which he believed gave the latter political legitimacy they did not deserve. He rejected participation in reconciliation conferences in which his government was one of many "Afghan parties" along with the Taliban, political opposition figures, and civil society groups.[4] Many analysts in the US intelligence community did not believe Karzai's approach was realistic. The Taliban had political cohesion, their ties with the Pakistani government were weakening, and there was little chance of eradicating Taliban sanctuaries that resided in densely populated Afghan refugee camps and city sectors.[5] By 2014 only one estranged former Taliban senior leader, Agha Jan Mutasim, had defected. And he did that only after he was shot and left for dead by his erstwhile compatriots.[6]

Karzai's approach reflected the limited political space in Afghanistan for reconciliation. Elites from the former Northern Alliance, warlords, and civil society actors were adamantly opposed to negotiations with the Taliban.[7] Such talks might reopen the issue of accountability for war crimes during the Afghan civil war (anti-Taliban warlords had orchestrated an amnesty bill in 2007 that forgave them of such crimes).[8] They also feared that a deal with the Taliban would come at their political and economic cost. Civil society actors highlighted the potential for backsliding on human rights. A peace deal or process would likely result in discussions about political reform or changes to the constitution. Karzai's co-option model would not. India, meanwhile, was not keen on any reconciliation effort that diminished the influence of its supporters or increased Pakistani sway in Afghanistan.[9]

To build support for his approach, Karzai held a Consultative Peace Jirga in June 2010. The jirga appointed a High Peace Council under the leadership of former President Burhanuddin Rabbani.[10] Rabbani was the head of the Badakhshan-based Jamiat-e-Islami Afghanistan Party during the Soviet war and had served as the president of the Islamic State of Afghanistan from 1992 to 1996, when the Taliban ousted his government. The High Peace Council led by the fiercely anti-Taliban Rabbani could provide Karzai with the political cover he needed. As Indo-Afghan ties strengthened due to Pakistan's malign activity, so did Karzai's caution on reconciliation.

The view from the Taliban was quite different. Although many US and international officials regarded Secretary Clinton's February 2011 speech as a breakthrough toward peace, the Taliban viewed the red lines as preconditions.[11] To lay down arms and cut ties with their allies, even ones as problematic to them as al Qaeda, before entering talks seemed like surrender. Moreover, to forfeit so much leverage before negotiations would be foolish and potentially suicidal. The Taliban were under substantial military pressure, to be sure, but their sanctuaries remained intact, and they had no trouble recruiting in Afghanistan or funding the insurgency.[12] They were not on the brink of surrender or defeat, and many senior leaders remained confident in their prospects for eventual success. Besides, acceding to the constitution meant accepting the legitimacy of the Bonn process that excluded them and the provision that equated resistance to the Taliban with resistance to the Soviets.[13] Some current and former Taliban senior leaders considered the rule hypocritical, noting to me that Afghan officials and warlords do not themselves abide by the constitution.[14]

The Taliban developed their dual-track approach.[15] They were keenly interested in discussions with the United States, but not in negotiations to end the conflict. Their approach to talks was not cynical; it reflected strategic calculation.[16] They would continue the military campaign but use diplomacy to build legitimacy in the eyes of Afghans and the international community.[17] If the Taliban overthrew the government, they would need foreign assistance to survive.[18] As Tayyab Agha explained to me in September 2010, "We were the government once, but we were isolated from the international community. . . . When we return to government, we need to have good relations with the world, especially the United States."[19] If the Taliban could not win outright and the war came to a stalemate, their leverage in negotiations would be far higher after international forces left. The gradual drawdown of foreign troops increased the Taliban's bargaining leverage while reducing that of the United States.

The Taliban had audience costs to consider, too. They had to build support carefully within their diverse and highly decentralized movement, just like the United States and the Afghan government would have to bring along their constituencies. After vilifying the Karzai government as puppets and the international forces as infidel occupiers, a sudden move toward peace negotiations risked splintering the insurgency.[20] The dual-track approach sustained the military campaign while building international credibility that would be needed if the Taliban returned to power.[21]

Confirmation bias and political frictions undermined the US government's ability to exploit differences between the Taliban and al Qaeda. Many senior US officials and members of Congress equated the Taliban and al Qaeda and wanted no negotiations with terrorists.[22] Of course, such a view was overly simplistic. The relations between the Taliban and al Qaeda were always rocky and had atrophied.[23] As discussed in chapter 4, bin Laden was invited to Afghanistan in 1996 by Abdul Rasul Sayyaf with Rabbani's permission. Tayyab Agha commented several times during the talks that the Taliban had "inherited al Qaeda." They accepted their support and provided them sanctuary, too.

Nonetheless, it is interesting to examine some of the documents from the bin Laden raid that have been released recently by the Department of National Intelligence. In one, al Qaeda senior leader Abu Yahya felt compelled to issue a guidance letter about proper behavior to presumably Gulf Arab *Wahabbis* wanting to join the fight in Afghanistan among Afghan *Deobandis*.[24] In a letter to bin Laden, an al Qaeda operative describes their support as only moral and symbolic: "We are participating in the work in Afghanistan, and we have to do that, but praise be to God, Taliban almost does not need us."[25]

After bin Laden questioned Mullah Omar on the wisdom of talking with the international community, Tayyab Agha issued a strongly worded reply explaining the Taliban's rationale on religious and practical grounds and rejecting bin Laden's description of talks as appeasement.[26] The Taliban also seem to have asked bin Laden not to return to Afghanistan or to appoint an al Qaeda representative for Afghanistan.[27] The Taliban, in short, could potentially offer the United States what it wanted most (cutting ties with al Qaeda) in exchange for what the Taliban wanted most: withdrawal of international forces so the insurgency could fight an unaided Afghan government.[28]

To support their diplomatic efforts, the Taliban revamped their strategic communications and political program. Before 2008, their *Eid al-Fitr* and *Eid al-Adha* statements attributed to Mullah Omar were pedantic vitriol. By 2009, these statements became far more sophisticated and aimed at both Afghan and international audiences. The new narratives repeatedly noted the strictly national aspirations of the Taliban and that they posed no threat to other countries. The September 2009 Eid al-Fitr message explained:

> The Islamic Emirate of Afghanistan wants to maintain good and positive relations with all neighbors based on mutual respect and open a new chapter of good neighborliness of mutual cooperation and economic development.
>
> We consider the whole region as a common home against colonialism and want to play our role in peace and stability of the region. We assure all countries that the Islamic Emirate of Afghanistan, as a responsible force, will not extend its hand to cause jeopardy to others as it itself does not allow others to jeopardize us.[29]

The Taliban crafted the last line to distance themselves from international terrorism. Although western governments missed the nuance, the statement caused a stir within the jihadi community, who felt the Taliban were betraying them.[30] Despite the controversy, the Taliban doubled down on the sentiment in an October 14, 2009, open letter to the Shanghai Cooperation

Organization Conference.[31] The Taliban also began discussing issues such as good governance, civilian protection, and human rights in terms that were not dissimilar to Afghan government positions.[32] Some compared quite favorably to US allies in the Gulf.

One potential explanation for the change in tone and substance was *taqiyya*—the statements were propaganda designed to deceive Afghans and the international community about the Taliban's real agenda. Their increased use of suicide bombings and victim-operated IEDs in 2009, for instance, certainly undermined elements of the Taliban narrative. On the other hand, the Taliban may have surmised that they were getting little strategic benefit from al Qaeda but incurring considerable legitimacy costs. They may also have determined a need to focus on governance and to adapt to Afghan expectations on various political and social issues.[33] The "progressive" statements were likely more reflective of the views of the political commission rather than the rank and file. Nonetheless, the new narrative caused no apparent ripple within the Taliban movement.[34] Sincerity, of course, would need to be tested.

The Taliban senior leadership authorized their political commission in 2009 to begin outreach with the international community, including the United States. To support these efforts, the Taliban established a private office in Doha, Qatar, which would give those not on the UN sanctions lists, such as Tayyab Agha, access to the outside world. Having an office in Doha also helped the Taliban political commission operate independently of Pakistan.[35] Diplomatic efforts by the Taliban required freedom of movement and the ability to meet interlocutors, but an office in Pakistan could be subject to considerable pressure from the Inter-Services Intelligence (ISI). Because Qatar hosted a US military base, the Taliban believed that the Americans might be more willing to engage them in Doha. They likewise surmised that the Qataris could play an intermediary role. Qatar, aspiring to be a diplomatic force in the Gulf, accepted the opportunity.[36] The Taliban had reached out to over 20 countries by the end of 2011 and participated in unofficial conferences in Norway, Japan, and France.[37]

Pakistan, with its national security policy run by the army, was arguably more of a unitary actor than the others. Their spy service, ISI—whose external operations branch is responsible for monitoring, liaising, and supporting various "asymmetric groups"—is subordinate to the chief of army staff. Stephen Tankel argues that scholars have developed three broad terms to explain government relationships toward militant groups: collaboration (active support), enablement (passive support), and belligerence. Borrowing an economics word, Tankel adds a fourth relationship: coopetition, which denotes "frenemies" or groups that straddle more than one category.[38] The categorization has its limits but is sufficiently useful to describe the relationship between Pakistan and the Afghan Taliban, and its effects on reconciliation.

The ISI collaborates with Kashmir-oriented groups like *Lashkar e-Taiba.* Their support for Afghan insurgent groups during the Soviet war could be described as somewhere between collaborative and enabling—providing logistics, funding, and expertise, but not directing operations in Afghanistan.[39] In contrast, the ISI's relationship with *Tehrik-e-Taliban Pakistan* (TTP)—or the Pakistani Taliban—on the other hand, is belligerence.[40] The Pakistani relationship with the Afghan Taliban appears to be a hybrid. The Taliban have sanctuary in Pakistan, particularly within the densely populated refugee camps that have been in existence since the late 1970s, where they can recruit, train, strategize, and organize logistics. The Afghan Taliban, moreover, has been destabilizing the India-friendly Afghan government, something that works to Pakistan's interests. The latter may see little need to assert direct control—an attempt to do so would likely create animosity. The relationship, however, is distant; the Afghan Taliban needs and fears Pakistan.[41] They need the sanctuary; they fear being perceived as or coerced into being a puppet.[42] This dilemma helps explain why the Taliban sought a political office in Doha, a country outside Pakistan's orbit and pressure. The ISI's longstanding relationship with the Haqqani Taliban, on the other hand, is much closer to collaboration.[43] Occupying parts of North Waziristan, the Haqqanis (who reportedly have closer ties with al Qaeda) have been useful to the ISI as intermediaries with various TTP groups and with the Quetta-based Taliban leadership. In return, the Haqqanis receive direct support from the ISI, enabling them to carry out high-profile operations inside Afghanistan.

This view of the relationship is not universal. Many Afghans and some US officials and scholars view Pakistan's relationship with the Afghan Taliban as master and puppet—an extreme form of collaboration.[44] Pakistani officials, who have become open about the existence of Taliban sanctuaries in Pakistan, seem to corroborate these views.[45] ISI apprehended Mullah Omar's deputy Mullah Abdul Ghani Baradar in Karachi in early 2010, allegedly for participating in unauthorized talks.[46] A Pakistani security official told the *New York Times,* "We picked up Baradar . . . because [the Taliban] were trying to make a deal without us. We protect the Taliban. They are dependent on us. We are not going to allow them to make a deal with Karzai and the Indians."[47] Such statements seem proof positive of Pakistan's control of the Afghan Taliban. However, they are also consistent with the hybrid model, in which Pakistan acquiesces in Taliban presence and provides enabling support but will act against them when the Taliban threatens Pakistani interests. The hybrid model also better explains the Taliban's decision to put the political office in Doha and Pakistan's adverse reaction to it.[48]

The nature of the relationship is vital for reconciliation. Were Pakistan the puppet master, they could have delivered the Taliban to negotiations. These have been the demands of the Afghan and US governments. If, however, the relationship is a hybrid in which the Taliban are autonomous in their strategic decision-making but not entirely independent (due to the needs and constraints of the sanctuary), the prospects of Pakistan forcing the Taliban into negotiations are unrealistic. As Barnett Rubin and Ahmed Rashid conclude, "No state can be successfully pressured into acts it considers suicidal."[49] The idea of forcing the Afghan Taliban into the arms of the Karzai government is unrealistic, given Pakistan's historic tensions with Afghanistan and India. As such, they have opposed the capitulation and the defection models because they believe the result would be a stable, hostile, pro-India Afghanistan. Pakistan would prefer Afghanistan to be a client state but would settle for a less hostile or even neutral Afghan government.

These challenges and conflicting views among key actors would affect the trajectory and ultimate failure of the Obama administration's reconciliation efforts.

17

Exploratory Talks

Building and Damaging Confidence

Bureaucratic frictions can paralyze strategic decision-making. To manage national security affairs, the US government has developed powerful institutions. The National Security Council consists of the president of the United States, the vice president, the secretary of state, and the secretary of defense. They are supported by national security staff at the White House and by their departments. The Department of Defense manages the military, the Department of State runs diplomatic efforts, and the US Agency for International Development (technically part of the State Department) coordinates international aid and development. Various agencies provide intelligence (the Central Intelligence Agency, National Security Agency, and Defense Intelligence Agency, among others). A National Security Council meeting includes typically around twenty officials, fewer than half of whom have a voice. This national security structure, primarily constructed in 1947, was organized for conventional war. Diplomats aim to avoid conflict or build a coalition to fight the war.[1] Once a war is declared, military forces fight to win, lose, or draw. Diplomats return to the fore to negotiate peace. Then aid agencies move out to repair the damage.

The United States has amassed greater global reach and obligations since 1947, stretching this structure. The same small group of people manages nearly every national security crisis across the globe, in addition to persistent matters, such as space, cyber, climate change, nuclear weapons and materials, and the rise of China. They also have a domestic policy to manage and departments and agencies to run. The load is crushing. Even such incredibly capable people as US cabinet officials lack the bandwidth to attain expert knowledge on every issue, and their agencies may lack procedures for non-standard crises. In their landmark study of the 1962 Cuban Missile Crisis, Graham Allison and Philip Zelikow explain how organizational processes and bureaucratic politics can result in frictions that impede decision-making and execution.[2] The organizational process model explains how bureaucratic procedures can place limits on a policymaker's freedom of action.[3] Instead of examining a crisis holistically, governments tend to break down the problem along organizational lines. Each bureaucracy addresses its portion of the issue. Government agencies use their existing procedures to execute assigned tasks and peacetime routines. When applied during a crisis, they can result in delays, major oversights, or rigidity.

The bureaucratic politics model, on the other hand, explains decision-making as a product of politicking and negotiations among the government's top leaders. These leaders have varying levels of power based upon their charisma, relationship to the president, and their interpersonal and persuasive skills. The intense discussions, often filled with miscommunication and misunderstandings, can result in a consensus that differs significantly from individual preferences. Conversely, individuals may take actions that the group would not condone.[4]

The organizational process and bureaucratic politics models explain why the US government may develop suboptimal intervention strategies that overemphasize the military instrument and make implicit assumptions about decisive victory. They also illustrate the reasons why losing or ineffective approaches could be difficult to change—the same decision-makers who agreed on the strategy have a vested interest in making it work. Leaders prefer to tinker on the margins rather than to overturn the existing strategy. New governing coalitions are often necessary to change an approach. Interest misalignment with the host nation and moral hazard can paralyze decision-making.

Political and bureaucratic frictions within the US government, the lack of a body of expert knowledge for wartime negotiations, and resistance from Karzai undermined reconciliation even as exploratory talks began. In late 2010, US officials started meeting periodically with Tayyab Agha. After a few sessions, the discussions centered on confidence-building measures. The Taliban wanted several prisoners released, particularly five senior leaders detained in the US military prison at Guantanamo (GTMO). They also wanted UN sanctions on them lifted and recognition of their political office in Doha. In turn, the United States wanted the Taliban to release Sergeant Bowe Bergdahl, a US soldier they captured in 2009, and for the Taliban to denounce international terrorism, announce support for a peace process, and begin meeting with the Afghan government.[5] The Taliban said they would talk with the Afghan government at the end of the confidence-building process.[6] The talks continued throughout 2011 and into 2012. US diplomats informed Karzai of the main points after each meeting.[7]

Grossman and I joined the talks in mid-2011 as the confidence-building measures were gaining definition. The London Conference in early 2011 had solidified Karzai's call for December 2014 to be the end of the ISAF combat mission. Obama's speech in June 2011 specified that the transition to ANSF-led security was to be complete by the end of 2014.[8] The timeline announcements likely solidified the Taliban's negotiating strategy. Having withstood the surge, they had no incentive to negotiate an end to the conflict until they could take on the Afghan government after international forces had left. They also had little incentive to make compromises that might cause tension within their ranks, particularly any actions that might confer legitimacy on the Afghan government.

Grossman played the hand he was dealt as well as he could. He focused on confidence-building measures to gain the Taliban's consent to meet with the Afghan government. These steps included the opening of a Taliban political office in Doha and Taliban statements denouncing international terrorism and supporting a peace process, detainee releases (the GTMO-5 and Bergdahl), and a meeting with the Afghan government.[9] A couple of obstacles remained. First, the Taliban needed to agree to terms of reference for the detainees sent to Doha, which included limitations on activities and a travel ban until the end of 2014. Second, they needed to agree to rules for the political office. The most important restriction was that the office not appear as an embassy or use the name "Islamic Emirate of Afghanistan."

Tayyab Agha was soft-spoken, even-tempered, and highly pragmatic. His closely cropped hair and beard little resembled the 2001 pictures of him. He gained Mullah Omar's trust as his secretary and, in 2009, became the head of the political commission. He described the Taliban's views of the conflict in two dimensions: external and internal. The Taliban wanted to deal first with the external conflict (particularly with the United States), and then address the internal conflict.[10] The external conflict meant the withdrawal of *all* foreign forces—the international coalition as well as al Qaeda and other foreign militant organizations that were fighting the government. The lifting of international sanctions, the establishment of the political office, and the transfer of the GTMO-5 were the confidence builders he said the Taliban needed to agree to have the Afghan government in talks. GTMO was a symbol of injustice to them.[11] The five detainees they wanted to be released had either surrendered to the Northern Alliance or turned themselves in based on agreements that they would be free to live in peace. The GTMO transfers were their biggest test of American sincerity.

Some administration officials, including ones at State and Defense, viewed the Taliban participation in talks as insincere. They believed that if the Taliban wanted peace, they would first stop fighting. High-profile or large-scale attacks were cited as proof of Taliban deception, even as the United States continued night raids and operations against them. There was a mismatch in the view that the United States could fight and talk, but the Taliban could not. Some exchanges with very senior officials became intense over these issues. Such challenges illustrated the importance of taking the effort slowly and step-by-step to build political space on all sides. If US officials found the notion of talks with the Taliban distasteful, many Afghans had far stronger reactions. Skeptical officials demanded significant unilateral signals from the Taliban as proof of sincerity.

Intelligence officials and Taliban experts often described the Taliban as a decentralized, pragmatic, consensus-based organization.[12] If they have an ideology, it is unity and the prevention of discord. Although Mullah Omar was the iconic leader of the movement, he did not rule by diktat. The Taliban use councils (*shuras* and *jirgas*) to discuss issues, examine ideas, and come to a consensus.[13] In a traditional jirga system, decisions require unanimous approval. This approach lowers the risk of dispute but makes decision-making slow and conservative. The status quo bias in an organization like the Taliban tends to be quite high, so major unilateral concessions from the Taliban at that point were extremely unlikely.

Confidence-building measures consumed most of the US-Taliban meetings, which took place once every four to six weeks through the summer and fall of 2011. Unfortunately, no one took official minutes. Tayyab Agha and the Qatari intermediaries spoke excellent English, but it was clear to me from Agha's body language that he would sometimes miss parts of the conversation. I would note to Grossman when I detected this, and he reiterated vital points, but the absence of an agreed record heightened the risk of misunderstandings. Nonetheless, texts containing the terms of reference for the Taliban political office went back and forth. These were decided by November 2011, except for some ambiguity about the office's name. The Americans sent the terms of reference to the Afghan government in advance of the hoped-for office opening in early 2012.[14]

Tensions with Karzai over reconciliation were growing. The US officials provided Karzai the points of each meeting with the Taliban, but this was not creating ownership and buy-in.[15] Unlike the carefully coordinated security transition effort, the United States had not developed an agreed approach with the Afghan government on reconciliation.[16] Defense officials expressed concerns, but no changes occurred.[17] "Karzai is irrelevant," a senior SRAP official told me at the time. "His interests are so different from ours that there is no point in trying to discuss it."[18]

Karzai grew increasingly worried that the United States was attempting to make a deal with the Taliban as a cover for withdrawal, just as America did in Vietnam.[19] A *Foreign Affairs* article by former US diplomat Robert Blackwill argued for the soft partition of Afghanistan. Karzai viewed this article as reflecting an option being considered seriously by the Obama administration—ceding the south to the Taliban (and Pakistan) in exchange for a cease-fire and an end to the conflict.[20] He perceived the efforts by the United States on reconciliation to be dangerously naive and potentially catastrophic.[21] The lack of serious engagement and coordination with Karzai on an effort so central to the political order and future of Afghanistan would have grave consequences.

These problems exploded into controversy just before the beginning of the Bonn II Conference on December 5, 2011. The atmosphere in Kabul was tense, particularly after the High Peace Council chairman, Afghanistan's former President Rabbani, was killed in September by a suicide bomber posing as a Taliban representative. Karzai's political rival, Abdullah Abdullah, warned, "This is a lesson for all of us that we shouldn't fool ourselves that this group, who has carried out so many crimes against the people of Afghanistan, are willing to make peace."[22] A concession to the Taliban was likely to create a backlash in Kabul.

The way the Taliban office discussions were unfolding amplified these concerns. After receiving the unsigned agreement on the Taliban office, Karzai requested a meeting with Secretary Clinton. Participants with knowledge of the exchange recalled that Karzai was outraged at being blindsided with a fait accompli.[23] Knowing that the US military command and embassy in Kabul backed Karzai's views on reconciliation and opposed SRAP's probably gave Karzai the confidence to press the matter hard. He demanded suspension of the effort until he had the opportunity to review and comment on the document. He recalled his ambassador from Qatar a few days later.[24] The United States reluctantly acceded to his demands.

Frictions increased in Washington as well. Sensitive to how negotiating with the Taliban would be perceived by Congress, the White House arranged meetings with House and Senate leaders in the late fall of 2011 and January 2012.[25] An interagency team supported the briefings. Grossman outlined the different measures under consideration and explained that the entire focus of the effort was to arrange a meeting between the Afghan government and the Taliban. He noted that the chances of success were low and surmised that the assassination of Rabbani and the September 13, 2011, attack on the US embassy in Kabul might have reflected the Taliban's true intentions about peace.[26]

His explanation came across to House and Senate leaders as a high-risk, low-reward proposition. The administration, they perceived, was having Grossman offer significant concessions to get the Taliban to agree to a meeting with the Afghan government. Why, they questioned, was the administration even considering such a bad deal? These concessions, they argued, would improve the Taliban's legitimacy and capabilities while placing American soldiers at higher risk for no meaningful return. Those present for the briefings were many of the same congressional leaders that received testimony from Defense and State officials that everything was on track in Afghanistan and that the risks were manageable.[27] The discussion was leaked to the press immediately. Members voiced strong opposition to the talks.[28]

Still, the administration sought to lay the foundations for a potential detainee exchange. Some observers mistakenly believe the Defense Department opposed the effort and tried to derail reconciliation talks.[29] Defense officials did want to take smaller steps first because transferring detainees from GTMO had become highly politicized. Congress passed legislation in 2009 that the Secretary of Defense had to certify in writing to Congress that he had taken all measures necessary to ensure the transferred individual would no longer pose a national security threat. Congress maintained these requirements through 2014.[30] In short, the Secretary of Defense would be personally accountable if a transferred detainee returned to the battlefield. The Obama administration often complained that such provisions made GTMO transfers virtually impossible.[31] Prematurely moving forward on such a potentially explosive issue could create more obstacles, heighten suspicion and cynicism, and potentially undermine the entire effort.

The Principals Committee determined that the United States needed assurances from the Qatari government that the detainees would be monitored, not allowed to engage in acts against the United States or its allies, and not permitted to leave the country. A senior Defense official and I worked closely with the Qatari attorney general on the provisions, capturing them in writing over a series of meetings. After a few months, we agreed on the terms of reference and gained the approvals of Secretary of Defense Leon Panetta and Secretary of State Hillary Clinton. The Qataris, however, would enforce the provisions only if the Taliban agreed to them. They understandably wanted the Taliban to share the blame for any transgression. The Taliban, however, refused to accept the travel ban. When the Principals Committee discussed the matter in late 2011, their decision was unanimous and unequivocal—the transfers could not take place until the Taliban agreed to the travel ban.[32]

The lack of clear, authoritative language about negotiated outcomes, strategic incoherence, and poor coordination within the US government and with the Afghan government sorely undermined confidence in reconciliation. The damage was about to expand.

18

Coming Off the Rails

As these frictions undermined the prospects of reconciliation, the drawdown of international forces was eroding American leverage. With the Taliban intransigent on the travel ban and Karzai objecting to the office, the process was stuck. Discussions about the political office leaked to the press.[1] The report caused a major stir within the insurgency, noted former intelligence officials and Taliban experts. Taliban commanders wondered if the senior leaders were trying to cut a deal with Karzai.[2]

Perhaps wanting to keep the effort secret until they had an accord with the United States, the Taliban's leadership had not fully discussed the political office and its purpose with its membership. They moved to control the damage. "In this regard," explained Taliban spokesman Zabiullah Mujahid, "we have started preliminary talks, and we have reached a preliminary understanding with relevant sides, including the government of Qatar, to have a political office for negotiations with the international community." The office, he emphasized, was not going to talk immediately with the Karzai government (a condition Karzai said was unacceptable). Reflecting how the Taliban leadership considered the conflict to have external and internal dimensions, he continued: "There are two essential sides in the current situation in the country that has been ongoing for the past ten years. One is the Islamic Emirate of Afghanistan, and the other side is the United States of America and their foreign allies."[3] He mentioned, in particular, that the Taliban sought the release of GTMO detainees.[4]

This episode reveals how the Taliban leadership responds to the rank and file. The *Eid* statements attributed to Mullah Omar from 2009 onward discussed a variety of issues regarding relationships with the international community, distancing from terrorism, and relatively progressive comments on governance and human rights. None of these statements created any stir in the Taliban ranks. The announcement about the office did. From early January until mid-February 2012, the Taliban issued no fewer than seven statements concerning the office and ongoing negotiations.[5] The winter months are when Taliban leaders tend to gather in Pakistan for annual discussions about their upcoming military campaign, so the Doha office was likely a subject of debate. The Taliban leadership managed to build the necessary consensus about the office during that time but would need to proceed more carefully in the future.[6]

The Obama administration wanted to get the Taliban office moving forward again. Defense officials continued to express concerns about the lack of coordination and an agreed framework with the Afghan government. Grossman met with Karzai in January 2012 before heading to Doha for another round of talks with the Taliban. Karzai made three demands for the office: Qatari government representatives needed to come to Kabul to explain the office to him, the Afghan government would rewrite the rules as the grantor of the office to the Taliban, and the Taliban had to meet in advance with the Afghan government.[7] These were three well-crafted poison pills designed to derail what Karzai perceived to be a highly dangerous process.

Grossman flew to Doha and outlined Karzai's three demands. The Taliban and Qatari representatives reacted in shock.[8] Karzai had insulted the Qataris a month earlier by recalling his ambassador; the Gulf nation believed that they had done nothing wrong and were trying to help resolve the conflict. The Americans had said that Karzai was on board. For the Qataris to come penitently to Kabul to seek Karzai's forgiveness and blessing was tough to swallow. Second, the Taliban were not willing for the office to be a gift from the Afghan government; that would be recognizing the government's legitimacy before they were ready to do so and placing the Taliban in a supplicant position. Third, the meeting with the Afghan government was agreed to come at the *end* of the US-Taliban confidence-building process. Particularly given the internal strife over the office earlier that month and their explanations that they were not talking with Karzai, the Taliban could not accept a meeting with the Afghan government as a *precondition.*[9]

Tayyab Agha replied that he would consult with "the leadership" about the demands and reiterated the Taliban's nonconcurrence on the travel ban for the GTMO-5. I asked him to consider whether the five were better off in GTMO or spending time with their families in Doha. Two months later, the Taliban issued a statement suspending the talks, suggesting that the United States was negotiating in bad faith.[10] They had always considered Karzai a puppet. If the Americans were serious about the effort, they believed, they would have forced him to accept the office. The United States had a long history of ignoring or overriding Karzai's wishes and concerns on matters like civilian casualties, detentions, night raids, the 2009 surge, parallel governance structures, anticorruption measures, and reconciliation.

Reconciliation remained in limbo for the rest of 2012. On the first anniversary of the successful bin Laden raid, Obama announced during a speech at Bagram Air Base that he had signed a Strategic Partnership Agreement with the Afghan government. He also provided an update on the five lines of effort and the pace of the drawdown of US forces. "In coordination with the Afghan government," he explained, "my administration has been in direct discussions with the Taliban. We've made it clear that they can be a part of this future if they break with al Qaeda, renounce violence, and abide by Afghan laws. . . . The path to peace is now set before them. Those who refuse to walk it will face strong Afghan security forces, backed by the United States and our allies."[11] The Taliban were unmoved.

By continuing the drawdown, the United States was losing negotiating leverage. The military pressure kept declining. The Taliban had every reason to believe they could wait out the American presence and that their advantage would be higher when fighting with the Afghan government unaided by over 100,000 international troops. They would be very unlikely to risk dissension in the ranks to make such large concessions right away. If Obama viewed that working toward a peace process was more important than completing the drawdown, he could amass plenty of reasons for extending the troop presence—the September 2011 attack on the US embassy, the assassination of Rabbani, Pakistani intransigence, the Taliban's suspension of talks. His administration continued to testify that the war was on track.

In early 2013 the Qataris attempted to restart discussions on the Taliban office. The Taliban indicated that they were willing to make another go at it, and US officials put pressure on the Afghan government to move forward. Karzai acceded but insisted that the office was for negotiations between the High Peace Council and the Taliban. The joint statement by Obama and Karzai in January 2013 summarized the agreed points: "The Leaders said that they would support an office in Doha for the purpose of negotiations between the High Peace Council and the authorized representatives of the Taliban. In this context, the Leaders called on the armed opposition to join a political process, including by taking those steps necessary to open a Taliban office. They urged the Government of Qatar to facilitate this effort." They also agreed to negotiate a bilateral security agreement (BSA). "The scope and nature of any possible post-2014 U.S. presence, legal protections for U.S. forces, and security cooperation between the two countries is to be specified in the Bilateral Security Agreement."[12] Obama made it quite clear that the United States would not maintain troops in Afghanistan after 2014 without a BSA. His withdrawal from Iraq after the Status of Forces Agreement negotiations failed underscored the point.

US officials worked over the next few months to get an agreement to open the office. A team of us met with Afghan national security advisor Dadfar Spanta in May 2013. He said that the Afghan government could support a US-Qatar-Taliban agreement on the office if the Afghan government could review and approve the document and if the Qataris would agree to a strategic partnership with the Afghan government first. The Qataris demurred. Spanta then suggested that there be no written agreements at all. If there could be no written agreement between the two governments, there should be no written agreement between the Qataris and the Taliban. All parties, including the Taliban, agreed to this idea.[13] Seeking assurances, Karzai asked for a letter from President Obama that the office would be for negotiations with the High Peace Council and not refer to itself as "the Islamic Emirate of Afghanistan" or look or act as an embassy. Obama provided the letter in June.[14] One potentially fatal flaw, however, remained. Without a written agreement about the rules of the office, the Taliban could do whatever they wanted.

The date for the official opening was June 18, 2013, the same day as the transition ceremony marking the transfer of lead security responsibilities across the country from ISAF to the ANSF. The latter was the signature event of the civil-military campaign since 2009.[15] When I asked if the United States should offset the dates, a senior White House official brushed aside the concerns and said that having both on the same day would be a statement of progress. Indicators that the Taliban were going to make a spectacle of the office opening prompted Defense and Kabul embassy officials to suggest that the United States double-check that the Taliban were meeting US requirements.[16]

The military transition ceremony took place in the morning to moderate media coverage. A couple of hours later, the Qatari-based *al Jazeera* televised worldwide the opening ceremony for the Taliban office. A senior official from the Qatari foreign ministry stood at a podium next to a Taliban representative amid Taliban flags. A large banner behind the speakers declared the opening of the "Political Office of the *Islamic Emirate of Afghanistan*" (emphasis added). The office was in a large enclosed compound in the same area of

Doha as other embassies. "Political Office of the Islamic Emirate of Afghanistan" read a plaque on the outside wall. A Taliban flag flew in the courtyard.[17] A US official reportedly involved in the process explained that the event "very much reflects this whole process, which began with a series of Loya Jirgas that Karzai held in 2010 and 2011. It includes the Karzai visit here to Washington in January [2013]. And this is an Afghan initiative, and it's a perfect representation of what we mean by Afghan-led, Afghan-owned. So, if the Afghan delegation makes this a priority in their engagements with the Taliban, then that's completely in keeping with Afghan ownership."[18]

Karzai and Afghan officials were apoplectic. The United States asked Qatar to suspend the office and blamed the Qataris for this debacle.[19] But a substantial burden falls on the Obama administration for poor coordination, ignoring warning signs, and abysmal communication.[20] The next morning Karzai suspended talks on the BSA because of "inconsistent statements and actions in regard to the peace process."[21] The Taliban office opening had violated nearly every assurance given by Obama. Protests erupted in Kabul.[22]

The new SRAP, Ambassador Jim Dobbins, and an interagency team went to Kabul to try to assuage Karzai. The latter believed the fiasco was deliberate. "Unfortunately, the manner in which the office was announced, including the title given to the office and the imagery on display, were all in breach of the written assurances we received from the U.S. government," a senior Afghan official explained to the *Washington Post*.[23] "The bizarre turn of events following the opening of the Taleban office in Doha," Afghan analysts Borhan Osman and Kate Clark reflected, "has led many [Afghans] to wonder whether the affair could have been deliberately sabotaged. Was it possible it had just been badly handled?"[24]

The fiasco had adverse consequences on the civil-military campaign, the transition, the BSA, and regional diplomacy—not to mention American credibility. "The reconciliation effort became too limited," former Deputy SRAP Singh observed, "and was not fully connected to the strategy."[25]

19

Fallout

BSA, Bergdahl, and the 2014 Elections

The ever-decreasing American leverage added to the strategic uncertainty as the 2014 elections approached with a peace process nowhere in sight, the Taliban resilient, and the Afghan government continually plagued with corruption and predation. Russia, China, and Central Asian governments began hedging their bets by increasing their contacts with the Taliban. These factors eroded confidence in Kabul and persuaded Karzai to take an increasingly aggressive anti-American policy.

Karzai eventually permitted BSA negotiations to resume. While American and Afghan officials worked out an agreement, Karzai called for a Loya Jirga to examine and recommend approval or disapproval of the text. The deal was then to be sent to parliament for approval before being signed by Karzai. The timing coincided with the release of William Dalrymple's book *Return of a King,* which discussed the sad reign of Afghan king Shah Shuja—the last Popalzai ruler before Karzai—who was considered a British puppet and was overthrown shortly after the disastrous British retreat from Kabul in 1842.[1] Already sensitive to being considered an American puppet, Karzai wanted to distribute the political risk of signing a BSA as widely as possible.

The support from the assembled delegates in favor of the BSA was overwhelming. Karzai, however, took to the podium and urged members against it.[2] He waved the assurance letter Obama sent him regarding the BSA, recalling the earlier assurance letter about the Taliban office, and explained how US promises could not be trusted.[3] Karzai ultimately refused to sign the agreement—he had more to gain by standing up to the United States. The document would not be approved until October 2014, after the new Afghan administration had come to power, and just two months before the United States was to end its combat mission. Neither Ashraf Ghani (the new president) nor Abdullah Abdullah (in the newly created office of chief executive), signed the agreement. They delegated that task to an unelected official, National Security Advisor Hanif Atmar.

As the end of the US combat mission loomed closer, discussions within the Obama administration began to center on ways to try to recover the US Army's Sergeant Bowe Bergdahl—the only American soldier in Taliban captivity. With the ongoing counterterrorism missions in Afghanistan and Pakistan, the United States was unwilling to refocus more resources to find him. I supported the idea of the detainee exchange, provided the Taliban agreed to the travel ban and that the deal was linked to a broader peace process.

The Israelis, hardly fussy when it comes to dealing with militant groups, traded over 1,000 prisoners for Gilad Shalit in 2011, a soldier held by Hamas since 2006.[4] The five Taliban detainees, after more than a decade in GTMO, would probably prefer to live with their families in Doha rather than try to assume a battlefield leadership role.[5] Besides, the Taliban had capable people in those roles already and probably did not want these five competing for senior leader jobs.

The Taliban wanted them out of GTMO, which was leverage the Obama administration could use for a higher return. The Qataris valued their diplomatic role in the Gulf and wanted good relations with the United States.[6] They had every reason to uphold their end of the agreement, which we had negotiated in 2011 and 2012.

By May 2014, the Taliban finally caved on the travel ban, enabling the Qataris to sign the agreement and opening the door for the detainee transfers. Secretary of Defense Chuck Hagel had the political courage to approve the transfers and sign the certification letter.

Bergdahl was repatriated in an impressive battlefield link-up between the Haqqani network of the Taliban and US special operations forces on June 4, 2014. Some congressional members and pundits cried foul, saying that the effort was damaging to US interests.[7] As of December 2020, all five exchangees had remained quietly in Doha, following the rules we negotiated with the Qataris. The exchange, however, was not linked to any peace process.[8]

Had US and Afghan officials examined the issue dispassionately, they could have arrived at some conclusions about the Taliban. First, the exchange showed that the political commission in Doha represents the Taliban leadership. The transaction had to be communicated from the Qataris to the political commission, then to the Taliban senior leadership in Pakistan, and then to the Haqqani network who was holding Bergdahl. The link-up of hostile forces in a combat zone to conduct a prisoner handover like this one was complex and high risk, and the Taliban political commission made it possible. Second, it helped answer a lingering question about the relationship between the Haqqanis and the Taliban senior leadership. Despite public statements from Sirajuddin Haqqani that he was Taliban, some US intelligence agencies and commands have persisted in believing they were separate.[9] The exchange showed that the Haqqanis were responsive to Taliban senior leader orders. Although several of their top leaders were in custody, including a son of Jalaluddin Haqqani, the Haqqanis did not demand that their detainees be part of the exchange.[10] Third, it showed the political commission could make commitments on behalf of the senior leadership and follow through. These points suggest the political commission should be considered a legitimate conduit for talks. The US officials ignored or dismissed those facts until August 2018, when senior American military leaders pressed again for exploratory discussions.

The uncertainty over the bilateral security agreement, meanwhile, likely prompted more aggressive hedging strategies by key actors in Afghanistan and the region. Businesses were even more reluctant to invest in Afghanistan. The capital flight continued to be a problem. Real estate prices plummeted. Kleptocratic behavior became more aggressive as officials sought to extract all they could and tuck it away overseas in case things fell apart.[11] The worth of the Afghan currency plummeted. On August 1, 2013, the *afghani* had traded at 53 to the dollar. Two years later, it was trading at 64 to the dollar—a 21 percent decline. Figure 7 shows the post-2013 decline in optimism among Afghan respondents.

The situation may have created even greater incentives for electoral fraud in both the Ghani and Abdullah camps.[12] Ghani's vote tallied higher than Abdullah's. The latter refused to accept the results and tensions mounted. The impasse over the election results heightened the risk of political violence as Abdullah supporters, such as the powerful Balkh governor Atta Mohammad Noor, threatened civil war.[13] Carter Malkasian, an Afghanistan expert and a former senior political advisor to ISAF commander General Joseph Dunford (2013–2014), notes that these problems were occurring anyway due to the drawdown of international forces.[14] Nonetheless, the anxieties surrounding the BSA intensified the toxic environment of 2014 and increased American frustration and fatigue.

After months of wrangling, the disputed 2014 presidential election to replace Hamid Karzai resulted in a US-brokered Government of National Unity consisting of two rival camps. The new government was led by President Ashraf Ghani and by Abdullah Abdullah, who accepted the newly created position of

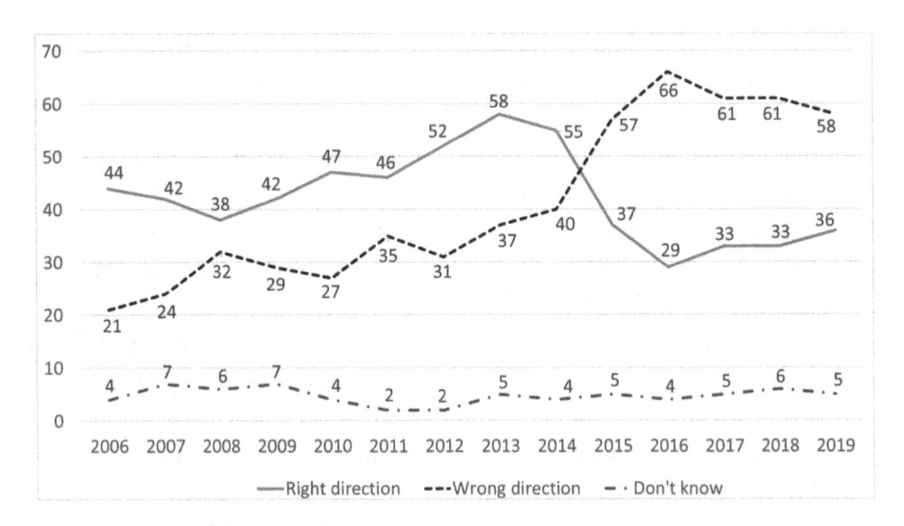

Figure 7. National Mood in Afghanistan

Question: "Overall, based on your own experience, do you think things in Afghanistan today are going in the right direction, or do you think they are going in the wrong direction?"

Source: Afghanistan in 2019: A Survey of the Afghan People, The Asia Foundation, 2019, 17; reprinted with permission.

chief executive. The provisions of the deal included, within two years, electoral reform before parliamentary elections scheduled for 2015, district council elections, and a constitutional Loya Jirga to consider, among other issues, whether to institutionalize a prime minister position.[15] The lingering animosities and competing visions of government guaranteed gridlock and made progress on these sensitive issues virtually impossible. Having rebuffed the BSA, Karzai positioned himself as a political figure of significant stature and has been agitating against the new government since he stepped down from power.[16]

The Afghan government continued its refusal to engage the Taliban political office, insisting instead that the Pakistanis deliver the Taliban to the table. Afghan president Ashraf Ghani, in the face of stiff political opposition, began efforts toward a rapprochement with Pakistan, which even included agreements on intelligence sharing.[17] To show their bona fides, the Pakistanis delivered three unempowered Taliban interlocutors to talks in Muree, Pakistan, in 2015.[18] The Taliban political commission, however, dismissed the talks and noted that they were the only Taliban entity authorized to participate in them.[19] The announcement of Mullah Omar's death two years earlier scuttled the meeting anyway. The Quadrilateral Coordination Group Process (Afghanistan, Pakistan, US, China) atrophied out of existence by 2016.[20]

Professor Paolo Cotta-Ramusino, the secretary general of Pugwash, an international organization, had maintained contact with the Taliban political commission and believed that private talks with them and former US officials could help restart a peace process. He recruited Ambassador Robin Raphel and me to join him in Doha for an informal conversation. After several meetings in which the Taliban members expressed a desire for talks so that Afghanistan could avoid becoming "another Syria," the group took the extraordinary step of writing an open letter to the American people asking for discussions to find ways to end the conflict. We conveyed our follow-up discussions to senior American and Afghan officials. President Ghani and the top NATO commander in Afghanistan, General John "Mick" Nicholson, bravely offered a cease-fire during the 2018 Eid al-Fitr holiday.

Top American military leaders and former Defense senior leaders encouraged President Trump to appoint Zalmay Khalilzad as US special envoy to resume talks with the Taliban. After nearly 18 months of negotiations, the United States and Taliban concluded an agreement in which the former agreed to withdraw all foreign troops within 14 months (by April 30, 2021), and the latter agreed to prevent al Qaeda and other terrorist groups from threatening the United States from Afghan soil. Both parties agreed to make efforts toward a peace process that included the Afghan government, provided the latter releases up to 5,000 Taliban prisoners.

The agreement was probably the best the United States could get at that time, given the state of the conflict. It is far less than what the Taliban wanted in 2012 to begin talks with the Afghan government.

President Trump, eager to end so-called endless wars, put his support behind the talks (including an ill-advised invitation for the Taliban to come to Camp David, which the latter rejected). The United States and the Afghan government missed opportunities in 2001, 2003, 2004, and 2012 when their leverage was far higher, and had to settle for what they got in the February 2020 agreement.

According to Statista, a data harvesting website, 581 American service members died from 2012 to 2019 in support of Operation Enduring Freedom.[21]

Conclusion to Part IV

Bargaining asymmetries undermined the prospects for reconciliation in Afghanistan. President Obama's decision on December 1, 2009, to announce both a surge of forces and a timeline for withdrawal limited American leverage. The Taliban aimed to gain concessions that improved its legitimacy while coaxing the United States to complete its withdrawal, but the insurgents were not interested in negotiating an end to the conflict at that time. As the drawdown continued, American bargaining power declined further. By March 2012, the Taliban postponed talks with the United States. The persistent and increasingly specific withdrawal announcements likely doomed any hopes of a peace process before the withdrawal of American troops. The Pentagon continued to receive a level of White House scrutiny about military operations that diplomats never underwent regarding reconciliation, regional diplomacy, and the 2014 elections. The number of meetings about those matters were comparatively few and far between.[1]

Bureaucratic frictions, lack of vision, poor coordination, inadequate strategic empathy, and sloppy execution damaged reconciliation. The lack of agreed conceptual frameworks for war termination inhibited clear communication and consensus-building within the US government, making the status quo harder to change.[2] The five strategic lines of effort became unprioritized silos. Individual actions in one, like the abortive opening of the Taliban political office, created setbacks in others. In the end, reconciliation further poisoned the relationship with Karzai, undermined American credibility in Afghanistan and the region, and heightened political uncertainty and instability as Afghanistan approached the 2014 elections and the end of the international combat mission. None of these problems were inevitable, nor were they easily preventable.

The lack of a coordinated strategy with the Afghan government increased the risk that sensitive issues such as reconciliation would become friction points. These problems exacerbated the prisoner's dilemma—each party recognized the benefits of cooperation toward peace. Still, no one trusted the others enough to take any risk to bring it about. Risk aversion and mutual suspicions created further obstacles. These issues made a credible reconciliation strategy even more necessary. The US government, however, proved incapable of delivering one.

In Iraq, Bush changed strategy when advisers affirmed his beliefs in the importance of success, presented facts on how the current approach was failing, and offered an alternative path to achieve his goals. That never occurred for Obama on Afghanistan, in part because his definition of success centered on ending the US role in the war rather than achieving specific security and stability outcomes. Withdrawal timelines became rigid. Defense and State failed to offer alternatives that might have addressed Obama's concerns about open-ended commitments, even though the status quo risked leaving Afghanistan on its own to deal with an industrial-scale insurgency that was still allied with al Qaeda. After foregoing opportunities for a peace process in 2001, 2003–2004, and 2010–2012 when America and the Afghan government had very high negotiating leverage and the Taliban had few conditions, the United States found itself in February 2020 accepting the Taliban's demands for a 14-month withdrawal timeline in exchange for the latter's assurances that they would not allow terrorists to use Afghan soil to foment attacks.

Part V

Pursuit of Decisive Victory in Iraq

Overview

Since Iraq invaded Kuwait in 1990, the United States was so suspicious of Saddam Hussein that regime change in Iraq eventually became official US policy with the Iraq Liberation Act of 1998. In the aftermath of the 9/11 attacks and after years of unsuccessful negotiations with Saddam Hussein over Iraq's suspected weapons of mass destruction (WMD) program, the United States ended talks and, on March 20, launched Operation Iraqi Freedom.[1] The Hussein regime fell on April 9, and the United States established a transitional government called the Coalition Provisional Authority (CPA).

Failing to prepare adequately for the occupation, the US military and civilian authorities faced an increasingly chaotic situation as the Iraqi military collapse turned into an insurgency.[2] Policies such as de-Ba'athification (which removed and banned members of Hussein's Ba'ath Party from the public sector) and disbanding the Iraqi Army, combined with heavy-handed military operations in the Sunni Triangle, stoked Sunni Arab fears of marginalization and repression by dominant Shi'a parties.[3] Senior Iraqi leaders often manipulated unwitting American forces to advance personal and sectarian agendas.[4] Sunni-Shi'a violence spiraled with the February 2006 bombing of the Shi'a mosque in Samarra.[5] Senior US officials focused on drawing down the US presence and handing over security responsibilities to the fledgling and sectarian Iraqi security forces.[6] They rebuffed recommendations by junior American officials to reach out to Sunni leaders.[7]

This intervention was to replace an existing regime with one more amicable to US interests. Failure to adequately consider war termination resulted in a strategy that fixated on the use of military force, wished away post-Saddam risks, and underappreciated the requirements for a favorable and durable outcome. The critical factors model outlined in chapter 2 suggests that success hinged upon, first, the establishment of a government that earned legitimacy across the political, ethnic, and sectarian spectra, and second, preventing armed resistance from becoming a sustainable insurgency. Decisive victory could have been possible had those two conditions been achieved. However, by 2006 the United States was backing a predatory sectarian Iraqi government that was fighting against a sustainable Sunni Arab insurgency, and Iran-supported Sadrist militias battled coalition forces while participating in the burgeoning civil war.

What Went Wrong?

The existing scholarship is broadly in agreement about why the Iraq War spiraled quickly from an overwhelming military success into a grinding civil war. The Bush administration failed to deploy enough troops to secure the country after the fall of the regime.[8] Compounding this error was the Department of Defense's failure to plan for the so-called Phase IV: post-war reconstruction.[9] The administration then stubbornly refused to deploy more troops as the situation grew worse. The US military, meanwhile, had deliberately unlearned counterinsurgency after Vietnam and employed counterproductive tactics that exacerbated rather than diminished the insurgency.[10] Within this broader explanation, some view Coalition Provisional Authority chief L. Paul Bremer's decisions on de-Ba'athification and the disbanding of the Iraqi Army to have made a lousy situation unrecoverable.[11] A RAND study notes, however, that de-Ba'athification affected far fewer Iraqis than did de-Nazification for Germans after World War II, and that the Iraqi Army had already disbanded itself.[12] These decisions, moreover, were briefed to the Bush administration and approved by Secretary of Defense Donald Rumsfeld.

Ambassador James Dobbins, perhaps the most prominent American scholar and practitioner on post-war reconstruction, largely exonerates the CPA from responsibility for the descent into civil war. After all, he notes, the CPA's progress ranks quite high in comparison to more than 20 other post-conflict missions.[13] The CPA succeeded in areas in which it had the lead. It was not, however, responsible for security.[14] Dobbins lays principal blame with the Department of Defense for mismanagement. "Experience in these and many other cases," Dobbins argues, "has dictated a prioritization of post-war tasks: beginning with security, then restoring basic public services, stabilizing the economy, and finally reforming the political system."[15] Secondarily, he criticizes the White House for giving the post-conflict mission to Defense and failing to supervise the planning and execution adequately. "By doing so, the President took himself and his staff out of the daily decision loop."[16] Had Defense done the appropriate planning, anticipated the scale of effort, and allocated enough forces for security after Saddam's fall, these arguments suggest, they might have prevented an insurgency or quickly defeated it.

The argument assumes that better planning, additional resources, and more efficient execution would have prevented disaster. The war termination framework developed in chapter 1 and the critical factors in chapter 2 provide ways to build upon the existing scholarship. The Bush administration assumed decisive military victory but failed to develop a strategy to gain the most favorable and durable outcome possible at the least cost in blood, treasure, and time. A consequence of this assumption was the failure to evaluate the risks of inadequate government legitimacy and a sustainable insurgency. As Part VI will show, having better plans and more American "boots on the ground" was not sufficient for success. The Bush administration's decisions and US actions in-country amplified rather than reduced the risks in at least three interrelated ways. First, US officials in Washington, D.C., and Baghdad lacked the strategic empathy to understand and manage the intense and often violent scrimmage for political power. Instead, senior American civilian and military officials super-empowered favored Iraqi elites, who used such backing for narrow personal and political advantage.

In many cases, they managed to dupe US officials into enriching their cronies and targeting their rivals. This problem damaged the foundations of legitimacy and gave Sunni Arabs a cause to fight. Second, US officials envisioned post-war security and reconstruction as an engineering task: break the problems down into components, arrange them into linear milestones over fixed timelines, and apply the necessary resources to achieve them. American officials did not appreciate how Iraqi elites could manipulate these milestones in ways damaging to government legitimacy. Third, civilian and military efforts worked in silos that were ably exploited by the adaptive Iraqi networks on both pro- and anti-government sides. As the United States focused on the efficient execution of its plans and congratulated itself on achieving bureaucratic milestones, the government was losing legitimacy and the insurgency was becoming sustainable. The Bush administration and its officials in Baghdad had plenty of detailed plans but no strategy adequate to address the dynamic complexity of the emerging conflict.

20

Operation Iraqi Freedom

Plans without a Strategy

President George W. Bush outlined his core goals for Iraq during a February 26 speech at the Hilton Hotel in Washington, D.C. The United States, by force, if necessary, would defend the American people and allies by removing the dual threat posed by Saddam Hussein's alleged weapons of mass destruction and sponsorship of international terrorism.[1] The invasion commenced after the Iraqi leader failed to comply with the US ultimatum. Secretary of Defense Donald Rumsfeld outlined eight supporting objectives:

- end the regime of Saddam Hussein by striking with force on a scope and scale that makes clear to Iraqis that he and his regime are finished.
- to identify, isolate, and eventually eliminate Iraq's weapons of mass destruction, their delivery systems, production capabilities, and distribution networks.
- search for, capture, drive out terrorists who have found safe harbor in Iraq.
- collect such intelligence as we can find related to terrorist networks in Iraq and beyond.
- collect such intelligence as we can find related to the global network of illicit weapons of mass destruction activity.
- to end sanctions and to immediately deliver humanitarian relief, food, and medicine to the displaced and the many needy Iraqi citizens.
- secure Iraq's oil fields and resources, which belong to the Iraqi people, and which they will need to develop their country after decades of neglect by the Iraqi regime.

- to help the Iraqi people create the conditions for a rapid transition to a representative self-government that is not a threat to its neighbors and is committed to ensuring the territorial integrity of that country.[2]

Colonel Kevin Benson, the senior campaign planner for the US Third Army (also known as Combined Forces Land Component Command [CFLCC]) and the architect of Operation Cobra II (the name of the military campaign), recalled the military objectives as follows: "Destabilize, isolate, and overthrow the Iraqi regime and provide support to a new, broad-based government; destroy Iraqi WMD capability and infrastructure; protect allies and supporters from Iraqi threats and attacks; destroy terrorist networks in Iraq; gather intelligence on global terrorism; detain terrorists and war criminals and free individuals unjustly detained under the Iraqi regime; and support international efforts to set conditions for long-term stability in Iraq and the region."[3]

War planning had been ongoing well before the public articulation of goals. On November 27, Secretary of Defense Donald Rumsfeld directed General Tommy Franks, the commander of the US Central Command (CENTCOM), to begin operational planning for the removal of Saddam Hussein from power.[4] Franks was no stranger to military planning for conflict in Iraq and was well versed in the governing contingency operation, code-named 1003-98.[5] As a former Third Army commander, he would have been responsible for ground operations in the event of war in the Middle East. Franks gave his first brief to Rumsfeld on December 7, and three weeks later he delivered his concept to President Bush.[6] The plan for Operation Iraqi Freedom (OIF), then called OPLAN 1003V, envisioned four major phases: (I) establishing international support and preparing for deployment; (II) shaping the battlespace; (III) major combat operations; (IV) postcombat operations.[7]

The concept aimed for a decisive victory. Phase III identified two primary goals: "regime forces defeated or capitulated" and "regime leaders dead, apprehended, or marginalized."[8] Franks and Rumsfeld had a series of meetings with Bush to discuss Phase III.[9] The five-pronged attack included a ground assault from bases in Kuwait and Turkey to defeat Saddam's fielded forces and special operations forces to neutralize SCUD missiles in western Iraq. Air and missile strikes in and around Baghdad would disrupt command

and control and attrite the Republican Guard formations, while psychological operations would erode Iraqi will.[10] Franks continued refining his plan, reducing troop numbers, and comparing force generation models (Generated Start and Running Start).[11] Benson notes that the goals and objectives never changed. The means, however, were under constant revision.[12] In the final version, Phase III was to take only 90 days and use a fraction of the forces planned initially.

Phase IV would probably require years, Franks argued. The objectives for Phase IV were "the establishment of a representative form of government, a country capable of defending its territorial borders and maintaining its internal security, without any weapons of mass destruction."[13] The plan envisaged that forces would continue to flow into Iraq until roughly 250,000 were supporting the occupation.[14] Franks was adamant that security and "civic action" were inextricably linked—a nod to the importance of Phase IV.[15] The plan assumed that the Iraqi military would remain relatively intact and be available to provide stability and support to reconstruction efforts.[16] Iraqi leaders, meanwhile, presumably would work together with US and international officials to establish a new government. This model, he believed, had worked recently in Afghanistan and could succeed in Iraq.[17]

The United Kingdom's prime minister, Tony Blair, tried to shape US decision-making. He aimed to convince Bush on the need for the inspectors to have sufficient time to do their jobs, investigating whether Saddam Hussein was fully compliant with UN Security Council Resolution 1441.[18] The Chilcot Report details the points Blair made to Bush during their January 31 meeting.[19] Blair wanted time to build public support in the United Kingdom and argued that they should assemble a broad international coalition for legitimacy if war became necessary. He was aware of Bush's optimism about a decisive military victory. The report shows no evidence that Blair challenged this view or raised the question of war termination.[20] French president Jacques Chirac told Blair that France opposed going to war in Iraq unless Saddam Hussein did something unacceptable. UK foreign minister Jack Straw advised Dominque de Villepin on January 29 that it was in neither country's interest for the United States to act unilaterally. "That would mean the international community losing influence over U.S. actions."[21] The United Kingdom's Joint Intelligence Council assessed that the Iraqi people would "acquiesce in Coalition military action to topple the regime, as long as civilian casualties are limited."[22] Blair did raise the importance of post-war planning, and Bush reportedly assured him that the plan was progressing well.[23] The Chilcot Report suggests that the United Kingdom's concerns about post-Saddam instability were not a deal-breaker. Instead, the United Kingdom aimed to get its portion of the reconstruction effort right.[24]

Phase III captured the attention of the Department of Defense and the US government. CENTCOM commander General Tommy Franks recalls, "While we at CENTCOM were executing the war plan, Washington should focus on policy-level issues. . . . I knew the President and Don Rumsfeld would back me up, so I felt free to pass the message along to the bureaucracy beneath them: You pay attention to the day after, and I'll pay attention to the day of."[25] A RAND study that examined planning for post-Saddam Iraq argues that Franks's mindset "reinforced an understandable tendency at CENTCOM to focus planning on major combat as an end in itself rather than as a part of a broader effort to create a stable, reasonably democratic Iraq. The result, arguably, was a military operation that made the latter, larger goal more difficult to achieve."[26]

As part of the standard military planning process, Franks and his staff, as well his subordinate commands, would conduct war games and rehearsals to test the feasibility of the military campaign against a competitive and uncooperative enemy. They would expose flaws and identify likely contingencies. If necessary, the military adds "branches" (deviations from the base plan to address threats and opportunities) and "sequels" (follow-on efforts) to the operation. In November 2002, the Third Army conducted an exercise called Lucky Warrior that exposed some problems in the CENTCOM plan. Lieutenant General David McKiernan, the commander, was not impressed by the likelihood of a quick collapse and believed that his troops would need to fight to Baghdad. He was concerned about the small size of the invasion force and the efficacy of the so-called Running Start concept. He outlined his concerns to Franks and drew up an alternative called Cobra II.[27]

As the invasion grew closer, CENTCOM commenced a "Rock Drill" on December 7–8, 2002, to rehearse the campaign with the subordinate commands.[28] The rehearsal would set conditions for Internal Look, which was to take place a few days later. Internal Look was a fully computerized war game designed to test the plan and the command and control systems against an adversary playing Saddam Hussein and the Iraqi military.[29] It identified issues like the ones McKiernan flagged earlier and opened the opportunity for him to brief Rumsfeld on Cobra II. The secretary approved Cobra II at the end of December but would retain tight control over troop levels.[30] The military commands continued rehearsing and refining the plans until the start of the war. Senators Joseph Biden and Chuck Hagel, who were on a fact-finding trip to the region, visited Internal Look. Biden noted his concerns about the lack of clarity on the post-war plan. "Phase IV worries America," he reportedly told the participants.[31]

Phase IV, however, received only a tiny fraction of Pentagon attention, compared to other phases. No analog of the deliberate planning and preparation process occurred for post-combat operations.[32] "The majority of activities required for Phase IV were perceived by the Department of Defense to be the responsibility of civilian agencies and departments," summarized a RAND report.[33] Phase IV rehearsals and war games that required participation by other departments and agencies within the US government and perhaps some international and nongovernmental organizations could have highlighted potential problems. Nonetheless, RAND notes that "military planners believed such collaboration would not be necessary for stability, reconstruction, and transition activities to succeed."[34] Defense even neglected to assess the troop levels needed. It was "hard to conceive that it would take more forces to provide stability in a post-Saddam Iraq," Deputy Secretary of Defense Paul D. Wolfowitz explained, "than it would take to conduct the war itself and to secure the surrender of Saddam's security forces and his army—hard to imagine."[35]

The US government never articulated an explicit theory of success for the war, but it is possible to piece one together.[36] They presumed that the military would defeat the Iraqi forces and depose Saddam Hussein. The vanquished Iraqi Army would remain intact and, with local police, would support the security and stability of Iraq backed-up by remaining international military forces. International civilian efforts would help Iraqi exiles and internal non-Ba'athist leaders establish a new democratic government that would become a partner in the war on terror. Reconstruction assistance would allow the Iraqi economy to recover. Iraq's oil wealth would enable the country to become self-sustainable, and Iraqis would take over international aid efforts, permitting foreign civilians and the military to withdraw. The exit strategy, Franks emphasized to Bush, must be based on effective Iraqi governance, not a fixed timeline.[37]

The campaign plan relied on three implicit assumptions. First, the Iraqi military and police had to remain intact and be willing to provide security with limited assistance from international forces.[38] Next, the Sunni Arabs had to accept a much lower share of power, acquiesce in the new government, and not fight back.[39] Third, and most importantly, a political leadership based mostly around formerly exiled elites had to earn the legitimacy to rule Iraq.[40] All three would have to be true for the US plan to succeed. All three turned out to be wrong, and the United States had no strategy to address them.

General Franks relied on the Afghanistan example to justify his belief that Phase IV would be relatively peaceful and focused on achieving political and economic milestones.[41] Did Rumsfeld and Franks have reason to believe the Afghanistan experience was representative of other post-conflict situations, and if not, did Afghanistan and Iraq share unique characteristics amenable to low-footprint, short-duration approaches?

For the first question, the Pentagon and the Bush administration had the benefit of four recent examples: Somalia (1992–1994), Haiti (1994–1996), Bosnia (1995–present), and Kosovo (1999–present). Peacekeeping missions were a lightning rod with the new Bush administration. They believed that the US military was too valuable to be wasted on tasks that other armies could do perfectly well.[42] Rumsfeld and other military senior leaders favored small footprint, short-duration post-conflict missions that could be handed off quickly to local or other international forces.[43]

A 2003 RAND study led by Ambassador James Dobbins examined force levels associated with other US nation-building efforts.[44] This study emerged from a conference in May 2003, so it would not have been available to Rumsfeld, Franks, and their staffs during the invasion planning. The information, however, was readily accessible. Bosnia and Kosovo, both relatively successful peacekeeping missions, had force-to-population ratios of 18.6 and 20 soldiers per 1,000 people, respectively. In other successful examples, force ratios of roughly 20 per 1,000 were used by the British in Malaya and Northern Ireland.[45] These were large-footprint, long-duration missions not favored by Rumsfeld. For an Iraqi population of roughly 26 million people, a 20 per 1,000 ratio would have amounted to 520,000 troops.

How did small-footprint, short-duration missions fare? Not well is the short answer. Somalia had 5 peacekeepers per 1,000 inhabitants and failed.[46] Similarly, Haiti had 3.5 American troops per 1,000 inhabitants. The United States managed to restore the elected leadership but left before durable political institutions could form. The country has been politically unstable ever since.[47]

The force-to-population ratio in Afghanistan, perceived to be going successfully as of 2003, was 5 per 1,000 in Kabul (mostly non-US forces), but only 0.46 countrywide.[48] Iraq was to have a 6.6 ratio overall, with 2.4 per 1,000 in Baghdad. Deputy Defense Secretary Wolfowitz chastised US Army Chief of Staff Eric Shinseki for expressing his estimate that the United States and its partners needed 400,000 troops for stability operations in Iraq.[49] In Shinseki's defense, the examples with roughly 20 per 1,000 ratios were consistently successful. The low ratio examples, Somalia and Haiti, showed poor results. In short, plenty of empirical evidence suggested that small-footprint, short-duration post-conflict missions had far lower rates of success than large-footprint, long-duration ones. The troubles brewing in Afghanistan should have raised doubts that powerful Iraqi elites could set aside deep-seated rivalries and personal aspirations to work together harmoniously.

The actual invasion force for Operation Iraqi Freedom totaled just over 200,000 troops, with roughly 140,000 on the ground in Iraq by April 2003.[50] The RAND study about Phase IV planning notes a CENTCOM and Third Army belief that follow-on forces would continue to flow into Iraq, based on the 1003V projection of a 250,000-troop requirement.[51] Whether Rumsfeld or Franks believed those numbers were necessary is doubtful. Franks canceled the deployment of the 1st Cavalry Division as US forces entered Baghdad.[52]

Perhaps Rumsfeld and Franks had other reasons to believe that Iraq would be fundamentally different than Somalia and Haiti. After all, like Afghanistan and unlike the other two examples, the United States was to defeat the Iraqi military, overthrow the regime, and work with a combination of exiles and internal opposition figures to form a new government. Should they have had any reason to doubt their prospects for success?

In June 1999, then CENTCOM commander Anthony Zinni conducted a classified exercise called Desert Crossing to examine potential courses of action if the Saddam Hussein regime collapsed and CENTCOM had to stabilize the country. The experts compared two approaches: "inside-out" envisioned Iraqis seizing power; "outside-in" imagined a US-imposed administration. The exercise determined that issues such as internal looting, sectarian strife, regional interference, and violent struggles for power were likely. "A change in regime does not guarantee stability," noted the after action report.[53] As war with Iraq became likely by 2002, Zinni attempted to meet with Franks and discuss the potential challenges of dealing with a failed state. The Pentagon reportedly blocked the trip. The Bush administration never studied Desert Crossing.[54]

During the lead-up to the Iraq War, the State Department developed the Future of Iraq Project, which would identify many of the post-invasion problems the United States encountered.[55] The effort had no authority to write a post-war plan.[56] It consisted of 17 working groups that amassed over 2,000 pages organized into 13 volumes.[57] It got Iraqi exiles, various American officials, and representatives from international and nongovernment organizations to discuss the future of the country.[58] Under Secretary of Defense Douglas Feith reportedly dismissed the project as a "bunch of concept papers."[59] The Pentagon ignored it and suspected that State was not on board with the war.[60]

Warnings about the potential for post-conflict violence surfaced from a wide variety of sources. Early in the planning, Rumsfeld identified a 29-point "Parade of Horribles" that were risks to success with the military operation.[61] Most of these centered on weapons of mass destruction and international reactions. Only one identified the likelihood of ethnic or sectarian strife. None mentioned the legitimacy of an interim Iraqi government. Other agencies pointed out the risks of post-invasion instability. The National Intelligence Council issued a January 2003 report forewarning that "a post-Saddam authority would face a deeply divided society with a significant chance that domestic groups would engage in violent conflict with each other unless an occupying force prevented them from doing so. . . . Score-settling," it noted, "would occur throughout Iraq."[62] A persistent debate within the Bush administration centered on whether "internals" (Iraqis from Iraq) or "externals" (Iraqi expatriates) should lead the interim Iraqi government.[63] The Office of the Secretary of Defense preferred externals because they could be pre-vetted, so the interim government could stand up more quickly and speed the transition and withdrawal. State and the Central Intelligence Agency, however, worried that externals would have no domestic legitimacy.[64]

Brent Scowcroft, a former national security advisor to Presidents Gerald R. Ford and George H. W. Bush, was so troubled by the notion of war with Iraq that he penned an op-ed in the *Wall Street Journal.* "An attack on Iraq at this time would seriously jeopardize, if not destroy, the global counterterrorist campaign we have undertaken." Moreover, he wrote, "if we are to achieve our strategic objectives in Iraq, a military campaign very likely would have to be followed by a large-scale, long-term military occupation."[65] Iraqi exiles also offered cautionary notes. "On many occasions, I told the Americans that from the very moment the regime fell, if an alternative government was not ready there would be a power vacuum, and there would be chaos and looting," claimed Massoud Barzani, leader of the Kurdistan Democratic Party (KDP). "Given our history, it is very obvious this would occur."[66]

After Internal Look in December 2002, barely four months before the invasion, the Joint Staff directed the US Joint Forces Command to create Task Force IV to work on Phase IV planning. The task force began to assemble in January 2003. Defense disbanded the task force in March 2003 and supplanted it with the Office of Reconstruction and Humanitarian Assistance (ORHA), led by Lieutenant General (Ret.) Jay Garner.[67] National Security Presidential Directive (NSPD) 24 authorized ORHA scarcely two months before the invasion. NSPD-24 gave Defense the responsibility for post-war planning and directed the formation of an office to execute it. ORHA planning focused mainly on potential humanitarian crises—most of which never materialized. "The problem," RAND's Nora Bensahel summarizes, "was not that no one in the US government thought about the challenges of post-Saddam Iraq. Rather, it was the failure to coordinate and integrate these various thoughts into a coherent, actionable plan."[68]

21

A Complicated Approach to a Complex Situation

The ORHA staff arrived in Iraq on April 21. Three days later, Rumsfeld informed Lieutenant General Garner that Ambassador L. Paul Bremer would be coming to Iraq as the head of the Coalition Provisional Authority (CPA). The Bush administration had not settled in advance on whether to turn administration over to Iraqis immediately or to govern as an occupying authority and transition more slowly to Iraqi control.[1] "The President's goal," Rumsfeld recorded in an October 14 memo, "is to stabilize Iraq and then turn it over to the Iraqis."[2] Indeed, this was the advice of figures such as Ahmed Chalabi.[3] As the Saddam regime disintegrated and Iraqi exiles began to bicker, the administration changed course.[4] The CPA would govern Iraq under United Nations authority with the support of an Iraqi Governing Council (IGC) until handing it over to an Iraqi government.[5] Bremer arrived in Baghdad on May 12. He made two controversial decisions: de-Ba'athifaction and disbanding the Iraqi Army.

Bremer issued CPA Order Number 1, known as de-Ba'athification, on May 16, his first official act. The order banned members of Hussein's Ba'ath Party from holding public office. Since Sunnis held most of the power in the Ba'ath Party, the order disproportionately affected that community.[6] Later that day, Bremer delayed the transfer of authority to Iraqi officials.[7] Hamid Bayati, from the Shi'a party Supreme Council for the Islamic Revolution in Iraq (SCIRI), reportedly warned CPA leaders, "the longer Iraqis are not in control of their political life, the more problems would arise." Chalabi and others registered concerns about a US broken promise to turn over power to Iraqis within weeks.[8] Bremer nonetheless appointed the 25 members of the IGC, which consisted of 13 exiles and 12 local Iraqis.[9] Twelve of the 25 were Shi'a, and five were Kurds. Five Sunni Arabs were included in the body (20 percent), three of whom were local.[10] Most Iraqis were reportedly unfamiliar with the IGC and its members.[11]

Bremer gave the IGC the responsibility for implementing de-Ba'athification. This decision gave those wanting to consolidate their power a potent tool to eliminate the competition. The IGC wasted no time in pressing for an expansion of the program while preventing former Ba'athists who had committed no crimes from returning to government.[12] Ahmed Chalabi, for instance, used these aggressive de-Ba'athification authorities to undercut support for his political rival Ayad Allawi, a secular Shi'a who aimed for greater Sunni inclusion.[13]

The decision to disband the army, CPA Order Number 2, was issued on May 23. Since Saddam's army was also led mostly by Sunni Arabs, the order had a disproportionate effect on them and risked alienating some 385,000 armed and trained men.[14] It prompted angry reactions.[15] Demonstrations occurred for weeks in Baghdad. Violent protests in Mosul, where Major General David Petraeus was trying to gain local support, wounded 16 American soldiers. One senior military official noted that "the insurgency went crazy. . . . One Iraqi who saved my life in an ambush said to me, 'I can't be your friend anymore.'"[16] On June 18, an estimated 2,000 former Iraqi soldiers protested outside the Green Zone. "We will not let the Americans rule us in such a humiliating way," declared one speaker.[17] American soldiers reportedly fired into the crowd, killing two.[18]

The military plan relied on the defeated Iraqi Army to provide security and reconstruction assistance.[19] Disbanding the army meant there was no local Iraqi force and far too few international soldiers to fill the security vacuum. Chaos and looting were rampant in Iraqi cities. Meanwhile, officials at the CPA and in the military command disapproved efforts by US military and intelligence officials to reach out to Sunni military leaders. One Sunni leader reportedly told his American counterpart, "All right, my friend, this is the last time we will speak, and I wish you luck in the hard times to come."[20]

The CPA's first two official orders, de-Ba'athification and disbanding the army, prompted Sunni Arab resistance.[21] Dobbins notes that de-Ba'athification and disbanding the Iraqi army—both orders which were cleared by the Department of Defense and the White House—could have benefitted from further review. But he downplays their significance. After all, de-Ba'athification affected 0.1 percent of the Iraqi population—25 times less than the de-Nazification policy in post-war Germany.[22] The Iraqi Army had already dissolved itself, he argues, so there was little reason to issue the order. Bremer instead could have put them on inactive status, sustained their pay, and recalled individuals and units selectively.[23] These orders did, as Dobbins observes, antagonize the Sunni community from which the insurgency grew.[24]

These acts sent a statement, whether intended or not, about who was welcome and who was not. Each order taken individually might not have been as problematic as some critics suggest. Still, together with other actions, they suggested that Sunni leaders had no place in the new order.[25] "I remember one prominent US National Security Council official telling me more than once that the answer for Iraq was the '80-percent solution,'" *Washington Post* columnist David Ignatius recalls. "Kurds and Shiites would build the new state regardless of opposition from the 20 percent of the population that was Sunni. This view was recklessness dressed up as *realpolitik*."[26] "The resistance developed," reflected a Sunni Arab political leader, "once it became clear to the Sunni community that they were being excluded from the political process."[27]

Military actions increased resentment. In April 2003, soldiers from the 82nd Airborne Division in Fallujah fired into a protesting crowd after allegedly taking fire from some militants, killing 17 and wounding more than 70.[28] Within weeks they reportedly shot and killed some of the local policemen after mistaking them for insurgents.[29] Similar problems were occurring around Tikrit, another part of the Sunni triangle. The 4th Infantry Division was using tactics like the ones seen in Anbar: large-scale sweeps, liberal use of firepower in populated areas, indiscriminate night raids that hauled in high volumes of detainees, many of whom were innocent.[30] Insurgent attacks increased nearly four-fold countrywide, from roughly 15 per day in June 2003 to 60 by November.[31]

These orders and military actions took place as a new communications revolution was getting underway. Without the benefit of mobile phones and digital technology, information had traveled slowly in previous post-war environments. Individuals and groups had fewer opportunities to share ideas and communicate problems. Officials had more reaction time.

Iraq in 2003 was vastly different. Mobile phones and the Internet speeded communication among families, leaders, and social networks. Outrage echoed across aggrieved Sunni Arab communities, amplifying indignation into resistance at a pace that far exceeded the ability of US officials to react. A prime beneficiary was a Jordanian named Abu Musab al-Zarqawi and his group, al Qaeda in Iraq.[32]

In addition to Sunni insurgent and terrorist groups, Muqtada al-Sadr's Iran-backed Shi'a militia, Jaish al-Mahdi (Mahdi Army), began agitating in

Baghdad and Najaf.[33] Sadr was challenging Ayatollah Ali Sistani, the senior Shi'a cleric in Iraq, for influence and took the competition into the streets. His militia allegedly murdered Ayatollah Abd al-Majid al-Khoei in April 2003. They also began a series of sectarian murders designed to purge Sunni Arabs from areas of Baghdad.[34] The killings prompted a debate between the CPA (who wanted Sadr arrested) and the military commander, Lieutenant General Ricardo Sánchez, the Pentagon, and the CIA (who did not).[35] The US decided not to detain him, and Sadr continued operating with impunity, taking over the Samir Hotel in Najaf on October 15 and naming it his Ministry of Defense.[36]

Despite the start-up challenges and an increasingly difficult security situation, the CPA did manage to put together an action plan by July 2003: "A durable peace for a unified and stable, democratic Iraq that provides effective and representative government for the Iraqi people; is underpinned by new and protected freedoms and a growing market economy; is able to defend itself but no longer poses a threat to its neighbors or international security."[37] The plan focused on four core foundations: security, governance, economy, and essential services. The CPA later added strategic communications to this list.[38] It listed benchmarks and timelines for each foundation.[39]

Meanwhile, Shi'a and Kurdish leaders were working the CPA process to press their advantage. A redistribution of power among the major groups in Iraq toward proportional representation was bound to work against the Sunnis but did not have to alienate them so aggressively. The IGC selected a 25-member cabinet, with only three positions going to Sunni Arabs (plus an interior minister from April to June 2004).[40] The cabinet's appointed status did not sit well with many Iraqi leaders. Sistani issued a fatwa, or religious ruling, on June 28 calling for Iraqis to elect a body that would write a new constitution.[41] Sistani and other Iraqi leaders (including some Sunnis) demanded elections as soon as possible.[42] With the Sunni political community fragmented and many leaders joining the insurgency, elections would heavily favor organized Shi'a political parties and movements.[43] To press their advantage, Shi'a leaders in the IGC called for more expansive de-Ba'athification.[44] Militants registered their protests by targeting IGC members as well as international forces.[45] Abu Musab al-Zarqawi's network Tawhid wal-Jihad (the precursor to al Qaeda in Iraq [AQI]) bombed the UN compound on August 19, killing Special Envoy Sergio Vieira de Mello and prompting the United Nations to pull out of the country.[46] Some IGC leaders advocated the use of militias, such as Peshmerga and the Badr Corps, to help stem the violence, but the CPA rebuffed them.[47]

The efforts to gain international legitimacy were running into problems, too. The Arab League announced in July 2003 that it would not recognize the IGC.[48] They relented after heavy lobbying by the Bush administration and the United Nations.[49] On September 23, the IGC banned the Qatar-based *al Jazeera* news channel along with *al Arabiya,* the Saudi-owned, Dubai-based network, on suspicion of encouraging violence and provoking sectarian strife.[50] Both are stations sponsored by Sunni Arab states.

Bremer refined his plan. In a September 8 op-ed in the *Washington Post,* reportedly without clearance from the Pentagon, he outlined his seven-step plan to Iraqi sovereignty. These steps included: (1) creation of a broadly representative Iraqi Governing Council, (2) IGC to name a constitutional preparatory committee (CPC) to develop a way forward in developing a new Iraqi constitution, (3) appointment by the IGC of 25 ministers to run the Iraqi government, (4) writing a new Iraqi constitution, (5) popular ratification of the constitution, (6) election of a new government, and (7) dissolving the CPA.[51] The CPA estimated that this seven-step process would require 540 days to complete.[52] Leaders in Washington and Iraq reportedly were stunned.

The Bush administration and the Pentagon were aiming for a near-term handover to Iraqi authorities. Bremer's plan would take far longer than the administration desired. National Security Adviser Condoleezza Rice grew increasingly concerned and asked Robert Blackwill, a former ambassador to India, to assess the situation. Blackwill characterized Bremer's plan as unrealistic and done without Iraqi support.[53] Ultimately, the Americans would hand over authority to an Interim Iraqi Government (IIG) in mid-2004 with Ayad Allawi as prime minister. National elections would be held six months later, in January 2005.[54]

While the Shi'a-dominated IGC showed little interest in winning over Sunni Arab support, US actions continued exacerbating the latter's sense of alienation. Derek Harvey, the chief intelligence analyst for the military command in Iraq, wrote a classified assessment in February 2004 of the burgeoning insurgency called "Sunni Arab Resistance: Politics of the Gun."[55] The report detailed how Saddam Hussein had expanded the Special Republican Guards (*Fedayeen Saddam*), the Iraqi Intelligence Service, and the Ba'ath Party militia as a hedge against internal rebellion. These forces had placed arms caches and explosive materials in safe houses throughout the country.

Harvey estimated that 65,000 to 95,000 of these men, a ready-made insurgent cadre, were in and around Baghdad.[56] As the political process excluded them, Sunni Arabs perceived that US military operations aimed to repress them, which created a mutually reinforcing cycle that inflamed the insurgency. The detentions system became a fertile recruiting ground.[57] Most detainees were Sunni Arabs. Military officials reportedly estimated that 70 to 90 percent of them were innocent. The Abu Ghraib torture and prisoner abuse scandal, combined with widespread perceptions of injustice, added fuel to the insurgency.

Harvey and others believed that outreach to Sunni Arabs was essential if the United States hoped to reduce the levels of violence. When American and British intelligence officials developed a plan to reach out to Sunni tribes, however, CPA officials demurred.[58] "I was struck by the desperation of Iraq's Sunni sheikhs, who feared and in many cases despised the brutal Zarqawi," writes David Ignatius, "But couldn't get tone-deaf U.S. officials in the international Green Zone to take their problems seriously."[59]

In Anbar, the 1st Marine Division took over from the 82nd Airborne in March 2004. Their commander, Major General Jim Mattis, issued instructions to limit civilian casualties.[60] Nonetheless, the marines found patrolling into Fallujah tough going. Abu Musab al-Zarqawi, who arrived in Iraq in August 2002 expecting a US invasion, was growing Tawhid wal-Jihad into a formidable terrorist network and angling to inflame a nascent sectarian civil war.[61] Large numbers of Anbaris, as well as foreign jihadis, flocked to join his ranks.[62] On March 31, they ambushed Blackwater security guards and hung their charred and beaten bodies on the city's main bridge. Fallujah had become violently anti-coalition.[63] Sánchez ordered an immediate offensive. Mattis argued that the timing was wrong and that the mission was ill-advised until they could improve popular support, but Sánchez insisted on moving forward.[64] Marines kicked off Operation Vigilant Resolve on April 6 to deal with the growing threat in Fallujah. Most Iraqi security forces who were to participate deserted immediately.[65] The offensive, the first large-scale assault on a major city since the end of Saddam, outraged Iraqis and played into Zarqawi's hands. A heated exchange took place as Bremer demanded that Sánchez call off the operation.[66] Sánchez relented.

The military shifted to the Sadr threat. Sánchez designed the plan and briefed Bremer, who sought White House permission. The latter disapproved of the plan to arrest Sadr, believing that such an operation could enrage Shi'as and disrupt the June transfer of power.[67] Washington failed to appreciate the internal Shi'a rivalry. The coalition had increased military pressure on the Sadrist Party-cum-militia Jaish al-Mahdi (JAM), which created a backlash in some Shi'a communities but bloodied the Mahdi Army. Sadr ordered his forces to stand down in the face of substantial losses.[68] He remained at large and his popularity grew, however, while Sunni Arabs seethed.[69]

22

From Decisive Victory to Transition

With decisive victory clearly out of reach, the United States' approach by 2004 drifted toward transition-and-withdraw. It was not a change of strategy; the Bush administration was lowering the bar of success as they sought a way out.[1] Bremer passed control to the IIG's Prime Minister Ayad Allawi on June 28; the CPA disbanded, and so did Combined Joint Task Force-7.

A new civil-military country team provided an opportunity to look at the situation with a fresh set of eyes. John Negroponte became the chief of mission at the new American embassy, and General George Casey took command of Multi-National Force—Iraq. They signed a Joint Mission Statement to reduce the civil-military rift between Bremer and Sánchez. They also established a team to assess the conflict and suggest a way forward.[2]

The preeminent threat, the new plan stated, was Sunni insurgents and members of the former regime—"Sunni Arab Rejectionists" and "Former Regime Elements." The goal for the coalition was not to defeat the insurgents, but to "reduce the insurgency to levels that can be contained by ISF [Iraqi security forces], and that progressively allow Iraqis to take charge of their own security."[3] President Bush described it succinctly in a June 28 speech at Fort Bragg, "As the Iraqis stand up, we will stand down."[4] The essence of the plan was to build and train Iraqi security forces, hand over control to them, and draw down the US presence.[5]

The rules for the January 30 election, however, entrenched Sunni resistance. In the run-up, Iraqi and American officials considered whether to divide Iraq into multiple voting districts or to treat it as a single national district. Sunni Arab candidates would win seats in predominantly Sunni Arab areas such as Anbar province.[6] In the single-district alternative, the Shi'a and Kurdish parties would have a significant advantage due to their sheer numbers and the probability that insecurity and insurgency would suppress the Sunni vote.

The decision to treat Iraq as a single national district was part of a UN-brokered compromise to appease Sistani's demands for elections in 2004. This poll was to elect the Iraqi National Assembly, which would form the Iraqi Transitional Government (ITG). Former CPA official Meghan O'Sullivan argues that treating Iraq as a single voting district was the only realistic way to pull off elections within this timeline.[7] The decision reinforced Sunni Arab marginalization and undermined the election's legitimacy.

Many Sunni Arab leaders called for a boycott of the election; others were probably too scared to vote due to threats made by Zarqawi and other insurgents.[8] In Anbar, the turnout was only 2 percent.[9] Violence on polling day was high, but 58 percent of eligible voters reportedly participated.[10] Allegations of rigging included ballot stuffing, voter intimidation, fraudulent voter registration, vote buying, and importation of non-Iraqi Kurds to cast votes for Kurdish parties.[11] The United Iraqi Alliance, a coalition of non-Sadrist Shi'a religious parties, came in first. The Kurdish parties were second; Allawi's Shi'a-Sunni coalition secured only 25 seats. For the 275-member parliament, Sunnis tallied only 8 percent of the representation—less than half their proportion of the population.[12]

After post-election jockeying by the parties to form a government, Ibrahim al-Jaafari finally secured the prime minister post. He was not an influential national figure backed by a militia, so he posed little threat to Iraq's Shi'a strongmen. Bayan Jabr from SCIRI gained the powerful Interior Ministry. He acted swiftly to intensify de-Ba'athification and remove Sunnis from leadership positions.[13] Sectarian violence, including atrocities, rose.[14] Iran, meanwhile, continued its influence campaign with the new Iraqi government (reportedly giving millions of dollars to Jaafari in advance of the election) while enhancing its lethal support to Shi'a militias that were fighting the coalition and engaging in sectarian cleansing.[15]

Shi'a and Kurdish advantages in parliament gave them control in drafting the constitution, which would be approved or rejected in an October 2005 referendum. In advance of the vote, Sistani explained to US officials his preference for votes to be counted by province instead of treating the whole country as a single district. The former, he argued, would reduce fraud from inflating the number of Kurdish seats in parliament and could increase Sunni representation.[16] A new electoral law created electoral districts instead of treating Iraq as a single national district.[17] This time the Sunni Arab leaders encouraged participation, because two-thirds of voters in three of Iraq's 18

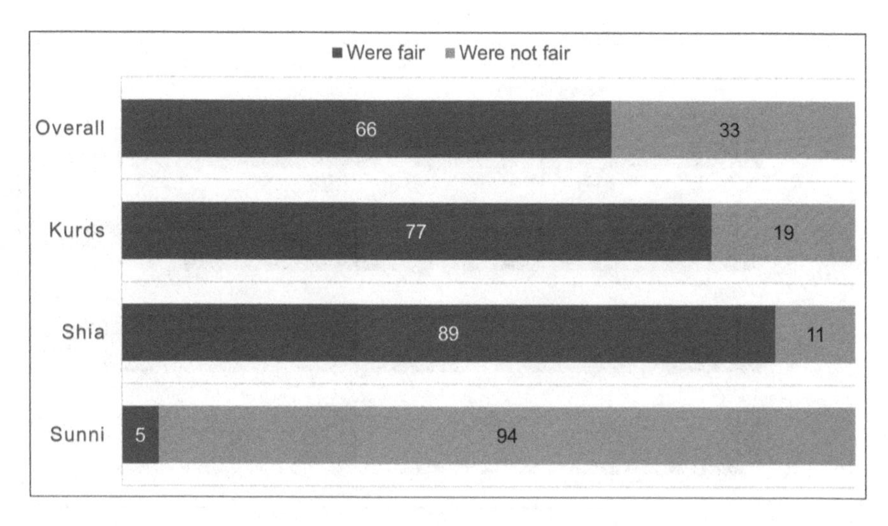

Figure 8. Perceived Fairness of 2005 Elections

Question: "Do you think that recent parliamentary elections were or were not free and fair?" Chart shows percentages.

Source: Based on data from WorldPublicOpinion.org, "What the Iraqi Public Wants," a WorldPublicOpinion.org Poll conducted by the Program on International Policy Attitudes, January 31, 2006, 7.

provinces could defeat the measure.[18] Voters approved the constitution, although Sunni Arabs alleged that election rigging repressed Sunni turnout and inflated Shi'a and Kurdish votes in Diyala, Salah ad-Din, and Ninewa.[19]

The final milestone was the parliamentary elections set for December 2005 to form a permanent Iraqi government. Concerns about electoral fraud, however, increased as the polling approached, including allegations of ballot stuffing, intimidation, fraudulent registration, and vote buying.[20] The de-Ba'athification commission removed 90 Sunni Arab candidates from eligibility.[21] After the polling, Sunni Arab and secular party leaders protested the massive fraud by the government and Shi'a and Kurdish parties.[22] Sunni Arab leaders complained to Khalilzad that their constituency was "under siege" from the terrorists, the government, and the coalition.[23] Violence reportedly increased, particularly around Baghdad.[24] Reflecting on the historic 2005 elections series, Allawi complained that the polls undercut rather than advanced democracy in Iraq.[25] Survey data draws a similar picture.

WorldPublicOpinion.org, together with the Program on International Policy Attitudes, conducted opinion surveys in Iraq entitled "What the Iraqi

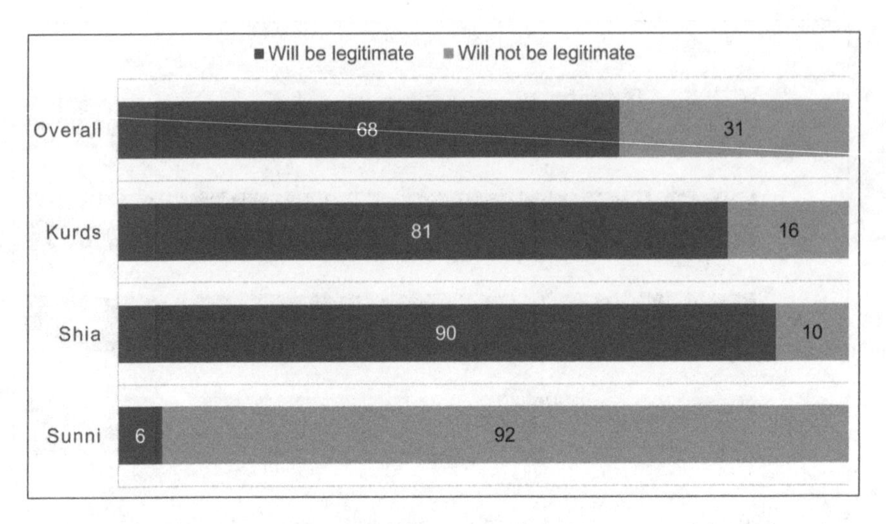

Figure 9. Future Government Legitimacy

Question: "Do you think that the government to be established by the newly-elected parliament will or will not be the legitimate representative of the Iraq people?" Chart shows percentages.

Source: Based on data from WorldPublicOpinion.org, "What the Iraqi Public Wants," a WorldPublicOpinion.org poll conducted by the Program on International Policy Attitudes, January 31, 2006, 7.

Public Wants." Their January 2006 edition asked about the fairness of the December 2005 elections (see figure 8) and the legitimacy of the new government (see figure 9). Both showed a clear sectarian divide. Eighty-four percent of Shi'a and 76 percent of Kurds considered the country to be going in the right direction, but only 6 percent of Sunnis agreed.[26]

Nouri Kamal al-Maliki, a relatively low-profile figure with a deep sectarian past, finally emerged as the prime minister.[27] Maliki used Iraqi forces and JAM to continue Sunni Arab repression.[28] A Sunni Arab leader told US officials that "The extrajudicial practices of Shi'a police in Anbar province fueled the insurgency."[29] When US officials presented Maliki with evidence of sectarian atrocities at the hands of these forces, he reportedly replied that things were worse under Saddam.[30] Saudi Arabia's King Abdullah complained to the US ambassador to Iraq, Zalmay Khalilzad, that the United States had given Iraq to Iran as a "gift on a golden platter."[31]

Polling data collected by the Defense Intelligence Agency showed that confidence in Iraqi security forces was dwindling among Sunni Arabs while support for armed resistance was increasing.[32] Attacks rose from 200 per week in January 2004 to more than 1,000 per week in July 2006. They would nearly double again to 1,800 per week in July 2007.[33] A sectarian civil war was well underway, with atrocities on both sides of the Sunni-Shi'a divide.[34] The Sunni insurgency had become sustainable, enjoying local popular and external support. The Maliki government exploited American forces and Jaish al-Mahdi to consolidate Shi'a power and exact retribution for decades of abuse at the hands of Saddam. Despite these shocking levels of violence, General Casey and Secretary of Defense Rumsfeld began canceling deployment orders for units scheduled to go to Iraq. They remained convinced that the United States had to draw down to win.[35] As US officials measured progress using milestones, they ignored evidence that the overall situation was spinning out of control. With a growing Sunni insurgency and Shi'a militancy and a government of dubious legitimacy in the eyes of many Iraqis, the conditions were emerging for a counterinsurgency loss.

Conclusion to Part V

Why did this toxic combination arise of an insurgency with tangible support and a host nation government unable to win the battle of legitimacy in contested and insurgent-controlled areas? Existing explanations focus on the lack of Phase IV planning, inadequate troop presence to maintain security, and poor politico-military integration.[1] These problems contributed to a more significant cause: a failure to develop a strategy resilient to the competitive, interactive, and often violent nature of politics in post-conflict societies. America deployed a well-trained and sophisticated military force, highly experienced diplomats, and extensive civilian expertise. By 2006, US forces were losing to a poorly trained, equipped, and resourced insurgency, as Iraqi elites manipulated American officials, thus creating a downward spiral into more violence and damaged legitimacy.

The Bush administration failed to consider war termination in its policy and strategy. They fixated on the military campaign, implicitly assumed a decisive military victory, ignored warnings about probable post-Saddam risks, and thus heightened the likelihood of quagmire. The administration failed to develop a strategy that integrated all elements of national power toward a favorable and durable outcome and that was resilient enough to adapt to emerging risks. The military campaign plan was sufficient to defeat a poorly trained and incompetently led Iraqi Army, but too rigid to adapt to the dynamic aftermath. Despite ample warnings from the intelligence community and a wide array of experts about the risks of post-regime instability, the Bush administration clung to the optimistic assumptions underpinning the campaign plan.[2] Wolfowitz's testimony on February 27, just weeks before the invasion, dismissing the need for stabilization forces suggests the Bush administration never seriously contemplated the risks.[3] The Department of Defense and the Bush administration fell victim to what Nobel Prize winner Daniel Kahneman called the planning fallacy—the tendency to underestimate the costs and overestimate the benefits of future actions. They were not alone. If State had grave concerns about the prospects for success, Powell could have recommended an interagency war game to Bush. As a former chairman of the Joint Chiefs, he was aware of the utility of this tool to expose problems. The failure to understand the difference between strategy and plans had significant consequences.

The CPA developed on the fly and implemented a milestone-centric plan that viewed post-war security and reconstruction as an engineering task. The approach, so deeply ingrained in US civilian and military officials, was unquestioned. Iraqi elites exploited each milestone. The Sunni Arab sense of alienation grew, as the coalition military and Iraqi government were perceived to be adding repression to political exclusion. As more Sunni Arabs fought back, Shi'a leaders pressed their advantage. Higher troop levels in 2003 might have delayed the onset of the insurgency. Still, they could not have ensured government legitimacy or convinced Sunni Arabs to accept what they perceived to be a state of oppression.

Rebalancing of political power to reflect Iraqi demographics was bound to reduce Sunni Arab influence and enhance Shi'a and Kurdish influence, so the Bush administration, CPA, and military should have taken extra care to address Sunni Arab concerns. The United States, however, was unprepared for the intense scrimmage for power among Iraqi elites.[4] De-Ba'athification and disbanding the Iraqi Army had disproportionately large effects on Sunni Arabs. Excessive military efforts in the Sunni Triangle, torture, and abuse fueled resentment. Together, they stoked armed resistance and a belief that there was no place in the new Iraq for Sunni Arabs.[5] These policies and actions were encouraged and amplified by Shi'a leaders eager to consolidate power and avenge years of humiliation and abuse. American officials lacked the strategic empathy to discern these traps. A strategy that was adaptable to the likely scrimmage for power, score-settling, and potential for sectarianism would have put the United States in a much better position to succeed. "Bush, Cheney, Rumsfeld, and Tommy Franks spent most of their time and energy on the least demanding task—defeating Saddam's weakened conventional forces—and the least amount of time on the most demanding—rehabilitation and security for the new Iraq."[6]

The administration's governing strategy document assumed that the political, military, and economic elements of its strategy were "integrated and mutually reinforcing,"[7] but each track operated on an internal logic and trajectory. Military and civilian officials worked hard in their silos, doing their best in a dynamic environment. The main problems of security, governance, development, and political inclusion, however, were interconnected and spiraled downward.[8]

No one in Iraq had the authority, responsibility, and accountability, however, to ensure these efforts worked in concert toward achieving a successful outcome. The White House and National Security Council were in no position to run a limited war full time when they had responsibilities across the globe, domestic policy to manage, and departments to run. In the absence of full-time governance of American and coalition efforts on the ground, political and military activities unwittingly self-synchronized to empower a highly sectarian Iraqi government and heighten Sunni resistance. The result by 2006 was a sustainable insurgency fighting against the coalition and an increasingly predatory Iraqi government that showed little interest in earning Sunni Arab support.

Part VI

Staying the Course in Iraq

Overview

From 2003 to the end of 2006, American military and political leaders were slow to recognize the character of the conflict: a Sunni Arab insurgency fighting a predatory sectarian Iraqi government and its coalition backers. Shi'a militias contributed to the sectarian violence and attacks against international forces. Despite a declining security and political situation, confirmation bias became evident as US leaders fixated on examples of progress and suppressed or dismissed contrary assessments.[1] Calls for alternative approaches became a groundswell in Washington, D.C., by early 2006. Supported by retired General Jack Keane, critics such as intelligence official Derek Harvey and scholar Frederick Kagan convinced the Bush administration that the situation in Iraq was nearing disaster and that a surge of military forces and capabilities under a new strategy was needed.[2] President Bush approved the plan in January 2007 as General David H. Petraeus and Ryan Crocker, the new commander and US ambassador, respectively, developed a US-led approach to the conflict, which included reconciliation with disaffected Sunni tribes.[3] By late 2008, the security situation in Iraq had stabilized. Coalition troops and their new Sunni Arab allies decimated al Qaeda in Iraq.[4] But the American public had tired of the conflict. They elected Barack Obama as president in 2008. He had made ending the war in Iraq a part of his campaign.

Why did the United States cling to a losing approach for three years? The existing scholarship focuses mainly on the reasons why the war went poorly and then on the relative merits of the 2007 surge. Bob Woodward's *State of*

Denial illustrates how the persistent narrative of optimism within the Bush administration and by officials in Baghdad impeded objective assessments and strategic decision-making.[5] Gordon and Trainor's authoritative volume *The Endgame* describes the interconnected problems of an expanding insurgency and growing sectarianism in the Iraqi government and security forces, civilian and military persistence with the campaign plan, and the process that led to the surge decision. Part VI examines, instead, why the paralysis occurred and how the Bush administration overcame it.[6]

The Bush administration had three mutually reinforcing problems. First, it fell victim to *confirmation bias,* in which officials described even strikingly unfavorable information as evidence of success. Second, severe *patron-client problems* trapped the United States in a downward spiral. Third was *loss aversion.* In comparing the status quo with the Democrat alternative to cut losses and withdraw, the administration was unwilling to pay the penalty of admitting defeat. Change became possible when an alternative approach provided the administration a chance to reverse a probable loss. Clinging stubbornly to a losing strategy, however, cost public support. Even though a more effective campaign plan was showing promise, the American people had had enough.

23

Achieving Milestones While Losing the War

The Bush administration and its civilian and military leaders in Iraq worked hard to understand the situation and make smart decisions. The narrative of progress reinforced a stubborn refusal to change.

The administration had plenty of data to support its narrative. Iraq was hitting all the major benchmarks. Politically, the elections were taking place. Iraqis drafted and voted approval for a new constitution; they established successive governments and enacted legislation. The Iraqi government achieved every benchmark outlined in UN Resolutions 1546 (2004) and 1723 (2006).[1] Iraqi security forces were being trained, equipped, and fielded on time.[2] Coalition special operations forces captured Saddam and killed AQI leader Abu Musab al-Zarqawi. The Sunni Arab and Sadrist insurgencies were suffering tremendous casualties. Economically, international aid agencies awarded major developmental contracts to Iraqi firms, and the oil industry was beginning to recover.[3] In short, the administration was meeting its standards for success in its political, military, and economic lines of effort. It broke down the requirements into parts and organized resources to achieve each one, a classic attempt to solve the complexity of post-Saddam Iraq with an assembly line.

The Bush administration measured progress by individual milestones and silos. By October 2004, for instance, they had differentiated the enemy in Iraq into three tiers: Sunni Arab rejectionists, former regime elements, and international terrorists.[4] Sunni Arab rejectionists were those "who have not embraced the shift from Saddam Hussein's Iraq to a democratically governed state," while Saddamists and former regime elements "harbor dreams of establishing a Ba'athist dictatorship."[5] American officials estimated that the first two tiers had some 3,500 fighters and 12,000–20,000 supporters. Foreign terrorists totaled roughly 1,000.

Nonetheless, the intelligence community, even in late 2005, had a tough time understanding the nature of the enemy, its relationship to the population,

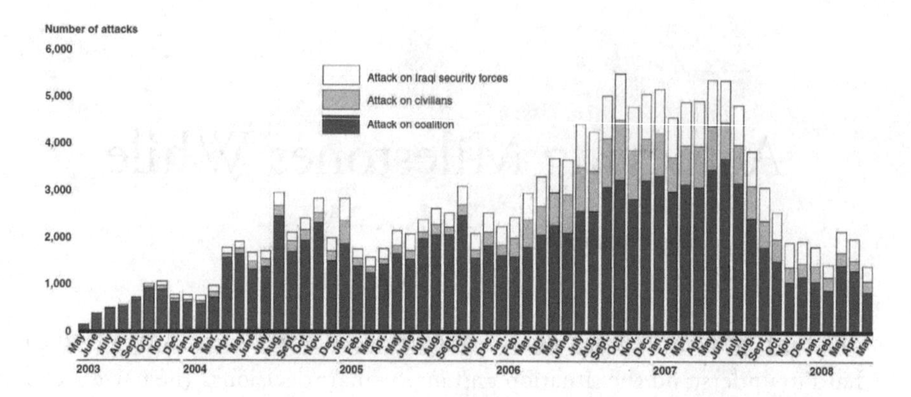

Figure 10. Annotated Enemy-Initiated Attacks, Iraq, 2003–2008
Source: US Department of Defense, *Measuring Stability and Security in Iraq,* Report to Congress, March 2008, 18 (events and trend lines added). For more on the challenges of assessing insurgencies, see Thomas C. Mayer, *War without Fronts: The American Experience in Vietnam* (Boulder, Colo.: Westview Press, 1985). On analytical measures, see James G. Roche and Barry D. Watts, "Choosing Analytic Measures," in *Journal of Strategic Studies* 14, no. 2 (1991): 165–209.

and the implications for US strategy.[6] The difference between Sunni Arab Rejectionists (deemed reconcilable) and Former Regime Elements (considered irreconcilable) was hard to determine. Moreover, these distinctions treated the insurgency as a collection of individuals rather than as groups with political constituencies that had goals and objectives.

The Department of Defense graph in figure 10 charts enemy-initiated attacks (EIAs).[7] These attacks were an imprecise but consistent indicator of insurgent strength levels. Weekly attacks increased roughly three-fold to 600 per week from January 2004 to July 2005. The upward trend continued to 800 per week when Maliki came to power in May 2006. Sectarian violence, as noted in the previous chapter, intensified. Curiously, the Bush administration did not include violence levels in the security assessment.[8]

Polling data revealed a more complicated picture. Surveys can be problematic in combat zones, especially if people believe they might be at risk of harm based on how they answer.[9] Nonetheless, a significant percentage of Iraqis polled across ethnic and sectarian lines in March/April 2004 supported the idea of parliamentary democracy (54 percent), believed they would be better off five years from now (63 percent), and wanted laws guaranteeing freedom of speech (94 percent), assembly (77 percent), and religion

(73 percent). Most surveyed believed US forces behaved badly (58 percent) and viewed them as occupiers (71 percent). The same poll showed widespread support for Iraqi police and the new Iraqi Army.[10] A September 2004 poll conducted by the State Department, however, showed that Iraqi perceptions of security continued to decline.[11] A significant percentage of Sunni Arabs (88 percent by January 2006) supported the insurgency or believed attacks on coalition and government forces were justified.[12] "The Sunni Arab insurgency is gaining strength and increasing capacity," a May 24, 2006, intelligence assessment surmised, "despite political progress and security forces development."[13]

The interpretations of the data reveal the dominant mental model within the Bush administration and among US officials in Baghdad. Based on the progress in meeting political, military, and economic benchmarks, Vice President Cheney concluded on May 30, 2005, that the insurgency was "in its last throes."[14] CENTCOM commander General John Abizaid explained that the number of Iraqis in the rebellion was only 0.1 percent of the population.[15] Secretary of Defense Donald Rumsfeld, even as late as August 2005, described the opposition as "dead-enders" and remnants of the former regime while downplaying the risk of civil war.[16] The October 2005 Department of Defense report to Congress boasted, "One noteworthy strategic indicator of progress in the security environment is the continued inability of insurgents to derail the political process and timelines."[17] When questioned about the rising number of attacks, Rumsfeld explained that the military's data collection was improving and "we're categorizing more things as attacks."[18]

The increasing violence remained an acute cause for concern, but accepted explanations for it conformed to the administration's assumptions and way forward. Foreign occupation, the military command assessed, was a core motivator of violence. American officials believed that most Sunni Arabs supported the government but were prevented from showing it due to coercion and intimidation.[19] General Abizaid explained that coalition troops generated "anti-bodies" within Iraqi society that were determined to throw out the foreigners.[20] Once foreign forces withdrew, the insurgency would die out.

Notably, the administration resisted describing the armed resistance as an insurgency. Defense Secretary Donald Rumsfeld, who had been using the term, decided in November 2005 to reject "insurgency" as an accurate description of those fighting the coalition and Iraqi government. "I think that you can have a legitimate insurgency in a country that has popular support and has a cohesiveness and has a legitimate gripe," he noted. "These people don't have a legitimate gripe." He preferred the label "Enemies of the legitimate Iraqi government."[21] Embedded in this odd terminology dispute is Rumsfeld's belief that Sunni Arabs had no legitimate reasons for fighting.

The US government's tendency to divide the war into parts within bureaucratic silos reinforced the tendencies toward confirmation bias. Assessments simply added up tangible achievements within each department and agency silo rather than viewing the situation as a whole. Lieutenant General James Dubik offered an illustrative example of this mentality occurring as late as 2007. One of his general officers told Dubik that he had trained the requisite numbers of Iraqi soldiers, "so his task was complete," and he was ready to go home.[22]

In short, officials in Washington, D.C., and Iraq assumed all the individualized progress was adding up to a successful outcome—reinforcing a belief that the war was on track. Confirmation bias masked how Iraqi elites were manipulating coalition efforts in those silos and how the silos began to self-synchronize in damaging ways.

The administration convinced itself that Iraqi public support would increase as the government achieved its political, military, and economic milestones, and thus reduce the insurgency.[23] General Casey, for example, viewed the January 2005 election as a success despite the Sunni Arab boycott. The inability of al Qaeda to block the vote, as he saw it, was an indicator that the insurgency had little popular support.[24] The focus on progress made along the political, economic, and security benchmarks would continue to dominate the quarterly reports submitted by the Department of Defense through August 2006.[25] The Bush administration interpreted the relevant data and concluded that the war was on track. In this view, the best course of action was to follow the Casey operational plan and begin to hand over security to fledgling Iraqi forces.

Our theories, Albert Einstein reportedly observed, are what we measure. They also inform how officials make sense of the mountains of data bombarding them in a dynamic, interconnected world. Smart and experienced US officials interpreted these data points—including the massive increases in violence—as evidence that the war was on track. Alternatively, the data could suggest that Sunni Arabs were fighting against the new Shi'a-dominated order as well as foreign forces. If this latter interpretation were accurate, the civil war would continue even if American troops departed, but this was not the view within the Bush administration. Their stubborn insistence that the strategy was working prompted Senator Chuck Hagel to conclude, "The White House is completely disconnected from reality."[26]

The American political left's relentless calls for immediate disengagement—an approach which would have probably led to defeat—may have unwittingly reinforced the administration's views. Bush resisted these calls, countering that withdrawal should be "conditions-based" along the standards outlined in the strategy.[27] As will be discussed later, Rumsfeld and Casey dug in their heels until the very end against a proposed troop increase and change in the plan because they did not see the situation or their strategy as failing.

24

Trapped by Partners in a Losing Strategy

Just as in Afghanistan, the situation in Iraq suffered from client-patron problems involving interest misalignment, information asymmetry, and moral hazard. The US government (the patron) and the Iraqi government (the client) never developed a common strategy for the war.[1] These problems, coupled with US officials' poor strategic empathy, masked significant interest misalignment. Subtle divergences and outright conflicts over objectives such as political inclusion, Iraqi security forces development, and US military operations and troop dispositions were undermining the prospects of success. These problems, which often occurred along the seams of US diplomatic and military silos, were masked by information asymmetries that Iraqi officials repeatedly used to their advantage. US officials found themselves unable to hold Iraqi officials accountable for actions that were intensifying the conflict. Manipulation by Iraqi officials reinforced Bush administration perceptions that the Casey transition-and-withdraw plan was the right one.

The problem of Sunni Arab marginalization was not unknown to US officials. Sunni Arab acceptance of the new government was essential to achieve the stability conditions that would enable America to withdraw, and cables from Baghdad show US officials encouraging political inclusion and their Iraqi counterparts responding in agreement.[2] As early as September 2005, the US embassy in Baghdad formulated an action plan to address "deep-seated anxieties among the Sunni Arab population.[3] The Americans aimed to assist the Iraqi government in reaching out to Sunni Arab Rejectionists, who were considered reconcilable, and in defeating the Former Regime Elements (Ba'athists), who were not.[4] In 2006, the Bush administration listed "security and national reconciliation" first in their three-track approach with the Maliki government.[5]

Such backing must have seemed successful. Because the Sunni Arab boycott of the January 2005 elections left them little representation in the Iraqi Transitional Government that was to write the constitution, Prime

Minister Ibrahim al-Jaafari expanded the constitutional committee to include them.[6] Maliki developed a plan for reconciliation soon after taking office in 2006.[7] The US government does not appear to have devoted any resources or conditionality to addressing the "deep-seated anxieties" of the Sunni Arab community or measuring whether reconciliation efforts were having any effect. US civilian and military leaders viewed reconciliation as a matter for the Iraqi government. When some commanders in Anbar experimented successfully with outreach to Sunni tribes in 2005, Casey applauded their efforts but denied requests for additional support.[8] Casey assured Iraqi officials that US commanders were not "negotiating with insurgents."[9] The October 2005 US "civ-mil action plan" focused on capacity-building and did not address issues such as sectarianism or corruption.[10] US officials would often encourage Iraqi officials to "do more," but they stopped short of conditionality.[11]

The Jaafari and Maliki governments, however, maintained a dual-track policy of convincing the Americans of their politically inclusive bona fides while relentlessly pursuing the Shi'a consolidation of power. Iraqi security forces had been engaging in systematic human rights abuses, mainly targeted at Sunni Arabs. The Jaafari and Maliki administrations each defined Sunni Arab insurgents as Ba'athists, which promoted repression.[12] Within its first 90 days in Iraq in 2005, the US 3rd Infantry Division alone reported 57 allegations of detainee abuse by Iraqi officials.[13] US officials received many reports about these and other problems, conveyed their concerns to Iraqi officials, but took little to no action to address them (other than creating more capacity-building efforts).[14]

For their part, Iraqi officials were adept at assuaging US concerns. Jaafari's notoriously sectarian Minister of Interior Bayan Jabr repeatedly assured American officials that human rights reform was at the top of his reform agenda.[15] He dismissed reports that one of his units, the so-called Wolf brigade, was engaged in human rights abuses, noting that they were his "most effective" special police forces.[16] Iraqi investigations into reports of torture and abuse of 168 detainees at the so-called Bunker facility (a Ministry of Interior prison) were slow-rolled and largely forgotten.[17] Despite such extensive reports, US officials visiting from Washington, D.C., would take efforts to support their Iraqi counterparts. One senior Pentagon delegation told Jabr that they had been "impressed by what they had seen" at police training sites.[18] The official does not appear to have raised human rights concerns.

The American plan to stand-up Iraqi security forces, turn over battlespace, and withdraw from the cities reinforced the Iraqi government's sectarian strategy. Jaafari told the *Washington Post* on June 24, 2005, "We strongly prefer an increase in quality of Iraqi forces, increase in number, increase in efficiency, increase in the effectiveness of tactics they use, as well as increase in equipment . . . anything that will raise efficiency of Iraqi forces is something that will be very welcomed because it will allow other forces, especially American forces, to withdraw."[19]

US military officials poured money into police training. General Casey declared 2006 to be the "Year of the Police." The American general officer in charge of Iraqi police development told the *New York Times,* "We're trying to develop the police capability to the point where by the end of 2006 we can begin the transfer to civil security."[20] With Sunni Arabs increasingly marginalized in the security ministries, the ISF provided the Iraqi government with powerful muscle for the struggle against their sectarian rivals.[21] Absent conditionality, capacity-building efforts made the ISF into more active predators.

As reported extensively in chapter 22, US military operations unwittingly played into Sunni Arab fears.[22] American forces targeted and detained Iraqi Islamic Party leader Dr. Muhsin Abdal Hamid during a May 30, 2005, operation. Placards in Sunni Arab parts of Baghdad denounced the raid as "terrorism American-style" and vowed to throw out the occupier. US officials apologized after learning of the incident, and then congratulated themselves on their "damage control" efforts.[23] Despite US military operations against Sadrist militia Jaish al-Mahdi, perceptions of bias persisted. Jaafari's Minister of Defense Saadoun al-Dulaimi, a Sunni Arab, warned US officials that Sunni Arabs believe that Iraqi and American forces are "blatantly anti-Sunni Arab."[24] Sectarian reprisals soared. In Baghdad alone, one eyewitness reported in early 2006, roughly 50 bodies were found daily. "The Sunnis usually beheaded their victims, while the Shiites drilled holes in their heads."[25]

Sunni Arab participation in the December 2005 election did not reduce the violence. It got worse. Even before becoming prime minister, Maliki's sectarianism was well-known.[26] He was a compromise candidate put forward by US ambassador Zalmay Khalilzad when neither Jaafari nor Allawi could form a government.[27] Maliki's position was fragile. He faced an internal Shi'a threat from the Sadrists, who had opposed him in favor of Jaafari. Any move to accommodate Sunni Arabs risked fracturing his governing coalition and ousting him from power.[28] Eliminating Ba'athists, however, was a unifying theme. A Dawa Party official (Maliki's party) confided that "the Shi'a's deep-seated fear of the Ba'athists' return to power drives all Shi'a political decisions."[29] Maliki told US officials that Ba'athists were the "primary threat" to Iraq's security.[30] The divergent views between US and Iraqi officials of the Sunni Arab insurgency had consequences. An assessment of sectarian violence in Baghdad concluded that American operations had unwittingly trapped Sunnis in their neighborhoods, allowing Shi'a militias to go after and eliminate them. The Shi'a-dominated National Police "were using us to cleanse areas of Sunni presence," reported an American battalion commander serving in Baghdad, "and we essentially have no option because we're supposed to partner with these guys."[31]

Although most reporting focused on the police, the Iraqi Army had sectarian challenges as well. Jaafari's Minister of Defense Dulaimi complained of pressures to put more Shi'a officials into the ministry.[32] Frictions between the Ministries of Interior and Defense sometimes resulted in violence.[33] Maliki's minister of defense, Abdul Qader Obeidi, assured US officials that the Iraqi Army was nonsectarian and loyal to the Iraqi state.[34] Iraqi officers, however, reported that Shi'a militia influence was growing in the ranks.[35]

Maliki and his officials largely dismissed mounting evidence of sectarian atrocities given to them by American officials, with few consequences.[36] Jawad al-Bolani, Maliki's interior minister, assured US officials that promoting human rights was among his top priorities.[37] US officials issued a demarche after hearing more allegations of police abuse.[38]

In response to a congressional query regarding widespread human rights violations, US officials emphasized "vetting arrangements" with the Ministry of Interior, human rights, and rule of law training for Iraqi police, their efforts to "press senior Iraqi officials" on the matter, and demarches.[39] A US congressional delegation urged Bolani to "purge" sectarian elements from the police.[40] It seems that at no time before 2007 did US military officials, diplomats, or Congress penalize Iraqi officials or the government for perpetrating such abuses. The American general in charge of the police training mission insisted that reports of corruption, infiltration by sectarian militias, and dysfunction within the Ministry of Interior were inaccurate and unfair.[41] One cable, while criticizing him for lackluster reform efforts, gave Bolani "good marks" for positive changes in human rights.[42]

Assurances from the Iraqi government that they would take the lead on Sunni Arab political inclusion and conflict resolution would be proven cynical by such rampant sectarianism.[43] These promises kept the Americans out of the reconciliation business until 2007. For Jaafari and Maliki especially, Sunni Arab political inclusion threatened their consolidation of power and their ability to prevent a return of an authoritarian Sunni Arab regime that could once again persecute Shi'a Iraqis.[44] Even Iraqi president Jalal Talabani, a Kurdish leader, rejected the Iraq Study Group's recommendation that the government reconcile with former Ba'athists.[45] SCIRI leader Abd al-Aziz al-Hakim flatly stated that Iraq already had a government of national unity, so there was no need for a reconciliation program. "We will never reconcile with the Saddamists. They were killing us for the last thirty-five years, and now we are paying them back."[46]

The flaccid response by US officials toward the systematic sectarianism is puzzling. These actions were alienating Sunni Arabs and creating support for insurgent groups that were killing American soldiers. Inducements—encouragement and better training—seemed to be the preferred approach. Sovereignty concerns probably provided some rationale—US officials wanted to avoid the appearance of meddling in Iraqi political affairs (despite Khalilzad's activist approach in promoting Maliki for prime minister). US officials do not appear to have considered the consequences to American credibility of acquiescing in such sectarianism.

25

Mirror Imaging Civil-Military Relations

In the aftermath of World War II, as the United States created its first large-scale standing army to keep the post-war peace, Samuel Huntington developed the intellectual underpinnings of the right relationship between political authorities and military officials. The key for America was to have a professional military that was under the control of elected political leaders and would not threaten a coup or play a political role. In *The Soldier and the State,* Huntington explained the difference between subjective and objective control of the military.[1]

Subjective control occurs when the political leadership co-opts the military, often through some combination of personal enrichment or political influence balanced by pitting military officials against one another to prevent one from getting too powerful. The advantage of this system is that the political authority can use the military as an instrument of political control without fear of a coup. The downside is that the incentives toward personal power and enrichment tend to be more potent than the ones for military readiness and professionalism. In subjective control systems, military officials tend to buy their positions from the political leaders for a significant amount of money. To make good on the cost of the job and to turn a profit, military leaders become participants in the system. They use their positions to extract bribes, extort customs or business revenues, participate in illicit economies, seize land and lucrative contracts, sell military positions in their commands, and participate in other income- or asset-producing activities. In some cases, the military is allowed or even expected to use military force to suppress internal dissent and rival groups. These activities come at the cost of military readiness.

Objective control is different. In this model, the military enjoys professional autonomy in exchange for being apolitical. The military focuses on training and readiness against external threats while getting sufficient pay and support. In a force-on-force engagement, a professional army that spends its time training will defeat a comparably sized subjective control army that spends its time looting. Objectively controlled military leaders stay far away from politics. Military education indoctrinates the force on loyalty to the constitution or system of government—the state—rather than allegiance to a particular individual or group. This principle of civilian control of the military is unquestioned among senior officials in the United States and NATO countries. Most western military officers have never been exposed to subjective control and do not understand its incentives.

The professional military culture that makes objective control second nature to western officials may also heighten their risk of mirror imaging when working with subjective control militaries. Mirror imaging is a cognitive bias that occurs when group A views group B through their own lenses and experiences rather than trying to understand group B's perspective. In this case, military officials raised under objective control will perceive their counterparts as equally interested in professionalism and being apolitical. Subjective control officials, meanwhile, may play into these perceptions to maintain a positive relationship and mask self-enrichment efforts, or might mirror image that US officials played the same game.

Mirror imaging often masks patron-client problems. US military and civilian officials never questioned whether the Iraqi government practiced *objective control*; Maliki, however, wanted *subjective control* because he feared a coup.[2] Much of the Iraqi security forces were under the operational control of the US military. Although American officials in Iraq did not contemplate encouraging a coup, the United States had a history of supporting them during the Cold War.[3] A much stronger fear for Maliki was the growing influence of Sadrist and Sunni militias within the army and police.[4] "I am afraid to clash with militias and tribes," Maliki confided in late October 2006 to Khalilzad, "*because I am afraid the army or police might commit treason.*"[5]

The most effective way for Maliki to reduce his vulnerability to an American-sponsored coup was to weaken the US military's grip on the ISF. Maliki repeatedly expressed concerns about the weaknesses of the ISF, the need for better weapons and equipment, and his desire to accelerate the transition and withdrawal of American troops.[6] Maliki told a visiting congressional delegation in October 2006 that significant numbers of American forces could leave within a year.[7] He badgered Casey that coalition military operations were damaging his political reconciliation efforts.[8] In early November, Maliki told Bush's national security advisor, Stephen Hadley, that he wanted greater control over Iraq's security. Hadley replied that President Bush supported the objective.[9] Maliki pressed the issue with Casey in early February 2007, demanding full control of Iraqi special operations forces.[10] To address the threat of a coup, Maliki told Casey and Khalilzad that "reliability" should be a key consideration for ISF senior leaders.[11] As Maliki gained more control over the ISF, he purged key officials who opposed his political agenda and advanced those who supported it.[12] Maliki's March 2008 Charge of the Knights operation in Basra and subsequent operations in Sadr City served as tests of loyalty as the prime minister took on the Sadrist threat to his regime.[13]

For Bush, Iraq was too important to fail, and Iraqi leaders used that leverage to their advantage to maintain the status quo—sustaining lucrative capacity-building efforts and easing international forces out of the cities and the country, all while resisting suggestions about reconciliation or troop surges. Even though the Americans would insist on outreach to the Sunnis, the Bush administration believed until late 2006 that preventing failure overall meant uncritically supporting Maliki despite his sectarianism and the risks this posed to success. For his part, Maliki had far greater incentive to keep the Shi'a coalition on his side and to consolidate his grip on power than to risk fracturing his base by addressing Sunni Arab inclusion—especially while he had Americans fighting the Sunnis for him. Iraqi government incentives aligned with those of American leaders who advocated staying the course. The result was the perpetuation of a status quo that was damaging to US interests.

26

To Surge or Not to Surge

A Possible Win Beats a Certain Loss

Most people feel losses more intensely than gains, argues Nobel Prize winner Daniel Kahneman; they would rather not lose than gain the same value.[1] Of course, some people are far more tolerant of loss—such as professional risk-takers in the financial markets—but, within certain bounds, most seek to avoid certain losses and to hedge against uncertain ones. People even tend to prefer a sure thing that is of reasonably less expected value than a gamble that could have a little higher payoff.[2] When people buy insurance, for example, they are paying a premium against uncertainty. Kahneman and Amos Tversky have referred to this phenomenon as loss aversion.[3] This tendency, however, is not absolute. Kahneman argues the following:

> In a mixed gamble, where both a gain and a loss are possible, loss aversion causes extremely risk-averse choices.
>
> In bad choices, when comparing a sure loss to a more substantial loss that is merely probable, diminishing sensitivity causes risk-seeking.[4]

These are the central insights of prospect theory, for which Kahneman and Tversky won a Nobel Prize in economics. An individual's reference point plays a critical role in decision-making.

People tend to be risk-averse when they are in a gain frame of reference. They do not want to forfeit their gains, and future increases tend to be less relevant to them than previous ones. They also tend to place a much higher value on what they have, merely because they have it (also known as the endowment effect).[5] The perception of military gains made during the intervention, and the reluctance to forfeit them and other achievements, can amplify loss aversion.[6] To demonstrate that their strategy is working, leaders can get trapped in the rhetoric of progress.

In a loss frame of reference, however, people tend to be risk-seeking. They prefer to take risks to avoid or to recover losses.[7] Loss aversion, in part, accounts for why leaders could gamble for resurrection rather than accept losing, even if a lost gamble could be irrecoverable.[8]

In short, loss aversion, combined with confirmation bias, can create situations in which leaders believe a strategy is working even when there are compelling indications that it is not. Leaders may also recognize that an approach is not succeeding, but prefer to stick with it if they think the available alternatives could forfeit gains or undermine more critical interests.[9] Finally, leaders may double-down or escalate a conflict in hopes of reversing a probable loss, even if the odds of succeeding are low and the likely costs are much higher.

Confirmation bias, progress in political and security milestones, and encouragement by Iraqi officials to maintain the transition-and-withdraw plan reinforced the status quo against Democratic Party calls to end the war. Loss aversion created the desire to preserve perceived gains and resist changes in strategy that could place them at risk—even as the cost of the current policy was high, and its prospects of success were low. Trapped between an approach that they believed might work and an alternative that assured a loss, the Bush administration dug in. The administration reframed its reference point from perceiving to be winning to acknowledging it was losing and found a new approach that promised to reverse a failing situation.

The removal of the Saddam Hussein regime and eventual replacement with an elected government was an achievement for the Bush administration. Iraq's success in meeting benchmarks reinforced perceptions of progress. They described the risks of forfeiting these gains in stark terms in their 2005 *National Strategy for Victory in Iraq.* The war on terror was the "defining challenge of our generation," like struggles against communism and fascism before. "The terrorists regard Iraq as the central front in their war against humanity," the administration explained, "And we must recognize Iraq as the central front in our war on terror." Failure in Iraq would create a new terrorist safe haven from which al Qaeda could launch more September 11-style attacks. "Ceding ground to terrorists in one of the world's most strategic regions will threaten the world's economy and America's security, growth, and prosperity, for decades to come." The current strategy "will help Iraqis overcome remaining challenges but defeating the multi-headed enemy in Iraq—and *ensuring that it cannot threaten Iraq's democratic gains* once we leave—requires persistent effort across many fronts."[10]

Reports from the Department of Defense echoed the gains. Each Department of Defense quarterly report to Congress from July 2005 (when the reporting requirement began) to August 2006 marshaled evidence in the political, economic, and security lines of effort to show the strategy was working. Despite alarming increases in violence, the growth of sectarian atrocities, and the Iraqi government's transparent Sunni Arab marginalization, not a single report during that period raised questions about the strategy's viability.[11]

In a fall 2005 article in *Foreign Affairs,* Andrew Krepinevich argued for a revised strategy he called the "oil spot" approach that focused on creating and expanding secure enclaves.[12] These kinds of suggestions were opposed consistently by the military command and US embassy and gained no traction with the Bush administration.[13] Senator Joseph Biden provided the only alternative to the Democratic Party's demands to end the war on a specified timeline, calling for a soft partition of Iraq into Shi'a, Sunni Arab, and Kurdish enclaves and a troop withdrawal by 2008.[14] The White House was not yet prepared to take such radical steps.

While projecting confidence, doubts reportedly crept into President Bush's mind by mid-2006, but the lack of alternatives for success reinforced the need to stay the course.[15] Bush wanted to protect gains in Iraq, which he believed would be forfeited under the Democrats' "cut and run" concept and would be put at risk in Casey's "conditions-based" transition-and-withdrawal plan. He did not seem to question the underlying logic of a strategy that was leading toward disaster.[16] The nature of the problem eluded the Baker-Hamilton commission—the so-called Iraq Study Group—that was directed by the US Congress in 2006 to review the situation in Iraq and recommend a more productive way forward. The conclusions reached by the group largely confirmed the status quo—a gradual handover of security to the Iraqis and drawdown of American troops.[17] During a June 2006 National Security Council discussion on Iraq at Camp David, Casey reportedly pressed the logic of his plan. "This strategy is shaped by a central tenet: Enduring, strategic success in Iraq will be achieved by Iraqis," he argued. "Completion of political process [the recent formation of Maliki's government] and recent operations [which included killing Zarqawi] have positioned us for a decisive action over the next year."[18]

Retired general Jack Keane, the former vice chief of the army staff, began to challenge that logic. Championing ideas proposed by defense intellectual

Frederick Kagan, intelligence official Derek Harvey, and others, Keane argued that the American effort in Iraq was failing and would lose under the current strategy.[19] He and others noted that the Sunni Arab insurgency was directed both at coalition forces and the Iraqi government. Turning over security responsibility to Iraqi forces and withdrawing would play into the hands of those on both sides of the Sunni-Shi'a divide who were bent on civil war. Immediate withdrawal, as Democrats proposed, would have the same effect. Keane argued that a five-brigade troop surge using a classic counterinsurgency approach that emphasized working with and protecting the people on the ground in local communities could be successful. Outreach to and relationships with Sunni Arab leaders was critical, as was reducing the sectarian violence in mixed neighborhoods—particularly in Baghdad.[20] This approach was not dissimilar to the one Krepinevich advocated in 2005, but Keane's ability to show that the current plan would guarantee a loss and a new approach could result in a win at acceptable cost was critical. When questioned about whether surging five American Brigade Combat Teams risked breaking the army, Keane reportedly replied that such risk existed, but "the stress and strain that would come from having to live with a humiliating defeat would be quite staggering."[21]

Having heard the assessments and recommendations from Casey, the Iraq Study Group, the Democrats, and Keane, National Security Advisor Stephen Hadley decided to assess the situation. Arriving in Baghdad on October 29, 2006, Hadley became convinced that Keane was right. His memo to President Bush questioned whether the United States and Maliki shared the same vision and if the latter could rise above sectarianism. He noted that the actions of the Iraqi government suggested an explicit campaign to consolidate Shi'a power at the expense of the Sunni Arabs.[22]

After the 2006 midterm elections, which were a significant setback for the Bush administration, the National Security Council began to review their options in Iraq. Bush fired Rumsfeld—a clear signal that he wanted an overhaul to the strategy—and replaced him with former CIA Director Robert Gates.[23] Secretary of State Condoleezza Rice was reportedly skeptical that adding more troops would make any sustainable difference.[24] General Casey pushed back on the surge idea vigorously, arguing that any tactical gains made by the surge could damage the progress made in Iraqi ownership of security and governance.[25] In its December 2006 report to Congress, the Department of Defense finally accepted that the conditions were present for a civil war and that more effective reconciliation efforts by the Iraqi government were needed to arrest this trend.[26] The subtle change in sentiment neatly aligned with the Hadley memo. Still, the December 2006 report did not discuss at what point such risks required a shift in strategy. Maliki was reticent, too, about a surge of American forces.[27]

27

A New Plan on Shaky Foundations

Bush announced the new approach on January 10, 2007.[1] Similar to Obama's later decision to surge in Afghanistan, Bush's decision came at a time when US public support for the war was at an all-time low.[2] To execute the new approach, Bush changed his command team in Iraq, kicking Casey upstairs to be army chief of staff and selecting General David Petraeus to replace him in Iraq. Middle East expert Ambassador Ryan Crocker took over from Khalilzad as chief of mission in Baghdad.

The new team wasted no time implementing the new strategy. Several close to the changes noted that the new approach "reversed virtually all of the previous concepts."[3] Bush began to hold weekly video teleconferences with Petraeus and Crocker and with Maliki. "Iraq consumed 80 percent of the NSC's bandwidth in 2007 and 2008," Bush's former Deputy National Security Advisor Doug Lute recalled, to the neglect of other issues.[4]

The March 2007 Department of Defense report to Congress, authored mainly by Petraeus, was a radical departure from the earlier themes of optimism and progress: "The strategic goal of the United States for Iraq remains a unified, democratic, federal Iraq that can govern itself, defend itself, and sustain itself, and that is an ally in the war on terror. . . . To regain the initiative, the GOI is working with the United States and its Coalition partners, embarking on a new approach to restore the confidence of the Iraqi people in their government; to build strong security institutions capable of securing domestic peace and defending Iraq from outside aggression; and to gain support for Iraq among its neighbors, the region, and the international community." The report noted that its assessment of the situation should be "read as a baseline from which to measure future progress, and indications of success must be heavily caveated given the dynamic situation in Iraq."[5] In other words, Congress should discard the claims from previous reports.

The new approach was tactically successful. Weekly attacks spiked in the summer as the surge brigades arrived and began operations. An AQI-inspired civil war trapped Sunni Arab leaders in a war they could not win. The surge and the so-called Anbar Awakening—when Sunni leaders in Anbar province turned against AQI and worked with the Americans—became mutually reinforcing.[6] Sunni Arab leaders began turning to the American military for protection against AQI reprisals and from the Iraqi government and its supportive militias. Violence levels fell dramatically in late 2007 as the Awakening spread throughout Sunni Arab communities.[7]

Maliki's Charge of the Knights, a surprise Iraqi military offensive in Basra, ended an ulcerating Sadrist threat from Muqtada al-Sadr's Shi'a militia.[8] Senior US officials viewed this operation as proof that Maliki was a "post-sectarian" leader.[9] As discussed earlier, a better explanation is that Maliki needed to put an end to Sadr's threat to his governing coalition and to test the loyalty of the Iraqi security forces. US officials continued to underestimate these internal political challenges, reflected former senior White House and Defense officials.[10] Maliki's political strategy focused on consolidating his power among the Shi'a parties as the Americans were busy coopting Sunni Arab tribes. With his base secure, he could take on the perceived Ba'athist threat. Roughly 15 months after the surge began, violence levels had plummeted from a high of 1,800 per week to approximately 400 per week. They declined to under 200 per week in 2010.

Some fundamental problems remained that would undermine efforts to conclude the war in Iraq successfully. First was the continued lack of a coordinated strategy between the United States and Iraq.[11] Lute argued that this masked divergent interests and objectives, which gave Maliki more space for his sectarian agenda.[12] The Bush administration focused on milestones but underappreciated how Iraqi elites were manipulating them, a problem that heightened after Petraeus and Crocker left Iraq.

Second, while more assertive efforts by Petraeus and Crocker curbed some of Maliki's sectarian tendencies, they were unsuccessful in changing his political calculus and its effect on Iraqi institutions. An independent commission led by Marine Corps general James L. Jones sent to investigate the Iraqi security forces offered a scathing review of the Interior Ministry and recommended disbanding and rebuilding the Iraqi National Police.[13] Despite an influx of resources and a new approach to ISF development, the Iraqi security forces achieved only temporary gains in readiness and performance.[14] The United States never addressed the *subjective control* problem, so sectarianism resumed. "Maliki asserted his control over the security forces," Lieutenant

General (Ret.) James Dubik recalls. "He assumed the positions of Minister of Defense and Interior, used the Office of the Commander in Chief to sell positions to those he considered to be politically reliable . . . and he directed their operations" to advance his sectarian agenda.[15] The continued alienation of Sunni Arabs fostered the rise of the Islamic State of Iraq and Syria (ISIS).[16] General Petraeus lamented that by 2014 ISIS made "short work of the ISF which were led by individuals I'd insisted be fired back in 2007."[17]

Third, there was little thought about how to create a favorable and durable outcome. The surge was supposed to reverse a declining situation and did not address war termination.[18] Senior officials acknowledged the concerns about sectarianism and political inclusion but had no political strategy.[19] In the difficult days of 2007, it is understandable that the US senior leadership in Baghdad focused on the immense near-term challenges. Still, no one in Washington, D.C., appeared to focus on creating durable success.[20]

Conclusion to Part VI

Clausewitz argues that a balance of determination and ability to adapt to circumstances is essential to military genius.[1] Cognitive bias, entrapment by Iraqi leaders, and loss aversion reinforced the Bush administration's obstinacy and impeded their ability to learn and adapt. Although the Bush administration had developed metrics to assess progress, the most virulent, strategically damaging problems were intangible or difficult to measure. These included factors such as the political scrimmage for power, predatory sectarianism, and growing corruption. These frequently occurred along the seams of bureaucratic silos, so they were never accurately measured or considered in assessments of strategic risk. The absence of such considerations may have played a role in the willingness of the Bush administration to discount violence levels as strategically irrelevant.

Bush made a bold decision for the surge, but American public support for the war had deteriorated substantially from 2003 to 2007. Seventy-two percent of Americans surveyed by Pew in March 2003 believed that going to war in Iraq was the right decision. Twenty-two percent were opposed. By February 2005, opinions for and against were tied at 47 percent. By February 2008, despite positive results from the surge, opposition to the war grew to 54 percent. Only 38 percent remained supportive. Even though perceptions of how well the war was going improved from March 2007 to February 2008, a higher percentage of Americans still wanted to bring troops home as soon as possible rather than keep them in Iraq.[2]

War fatigue had set in among the American public and had become a divisive issue. During the 2008 presidential election, the vast majority of Democratic Party voters and 53 percent of Independents favored bringing troops home from Iraq, while large percentages of Republican voters wanted to keep them in. By contrast, 61 percent of Americans polled supported keeping forces in Afghanistan, including majorities in both parties and among Independents.[3] Tapping into such sentiments, Democratic Party nominee Barack Obama campaigned on a promise to wind down the war in Iraq within sixteen months and to refocus American energy on the war in Afghanistan.[4] Although the economy was by far the number one reason Americans voted as they did in 2008, the Iraq War was second.[5] Obama won the presidency and made good on his promise to end the war. The last convoy of American forces left Iraq on December 18, 2011.[6]

Part VII

Ending the War in Iraq

Overview

The United States rejected early bargaining opportunities with former Saddam regime officials in 2003 and unwittingly encouraged sectarian violence.[1] Until 2007, US officials, often at the urging of Shi'a leaders, ignored or blocked opportunities to negotiate with Sunni leaders willing to turn on al Qaeda in Iraq (AQI). As noted earlier, the US surge emphasized the importance of reconciliation—the increased troop levels combined with the Sons of Iraq movement (which spread from the Arab Awakening) resulted in steep reductions in violence.[2] US officials, however, may have misread Sunni intentions, believing that the latter supported the Iraqi government when actually they were allying with the Americans for survival against the two-fold threats of AQI and Maliki.[3] The United States missed opportunities to foster and incentivize genuine political inclusion.[4] The Iraqi government's sectarianism and corruption undermined the prospects of a successful transition.

The United States by this time was tired of the war. Maliki backed the timeline Obama outlined during the 2008 US presidential campaign.[5] The Obama administration's determination to withdraw may have led Maliki to consolidate Shi'a political dominance more aggressively—the small American presence envisioned by Obama would be a political liability without the benefit of protecting against a Sunni resurgence.[6] US forces left Iraq, as promised by Obama, at the end of 2011.

Maliki's renewed efforts to marginalize Sunnis led to the growth of the Islamic State of Iraq and Syria (ISIS).[7] The group emerged from the remnants of AQI and newly disaffected Sunni tribes.[8] ISIS took over the city of Mosul, and much of northern and western Iraq, in the summer of 2014 as the American-trained Iraqi security forces (ISF) fled, leaving large quantities of US military equipment, including tanks.[9] In 2015, President Obama ordered American advisors back into Iraq.[10]

28

The Surge Misunderstood

Why did the United States fail to achieve its desired outcomes in Iraq despite the apparent success of the surge? Most recent scholarship views the surge as a military success but a political failure.[1] The debate analyzes reasons for the significant reductions in violence, including US-centric reasons[2] (new troops, new doctrine, new strategy), Awakening-centric reasons,[3] and synergy between the two.[4] This debate matters, Stephen Biddle argues, because if policymakers adopt a US-centric view of the significant reduction in Iraq's violence, they are likely to apply the same methodology elsewhere. If, however, political conditions on the ground interacted synergistically with the surge, as Biddle suggests, then a US-centric template will be necessary but not sufficient in other conflicts.[5] These arguments would prove crucial during the Obama administration's decisions on Afghanistan in late 2009.

The substantial reduction of violence in Iraq, however, was not accompanied by greater political inclusion. Outlining the goals of the new Iraq strategy, President Bush argued, "A successful strategy for Iraq goes beyond military operations. Ordinary Iraqi citizens must see that military operations are accompanied by visible improvements in their neighborhoods and communities." A prosperous Iraq, he continued, would be "a functioning democracy that polices its territory, upholds the rule of law, respects fundamental human liberties, and answers to its people."[6] General David H. Petraeus saw reconciliation as critical for success: "Beyond securing the people by living with them, foremost among the elements of the new strategy was promoting reconciliation between disaffected Sunni Arabs and our forces—and then with the Shiite-dominated Iraqi government."[7] Understanding why political reconciliation failed is crucial in assessing war outcomes in Iraq.

The main explanations for the failure of political reconciliation focus on Maliki's sectarianism and Obama's inability to secure an agreement to extend US troop presence beyond 2011.[8] The first explanation largely exonerates the US government from any substantive role in promoting durable political inclusion. Why leave such a critical element of the war entirely in the hands of the Iraqi government? The second argument assumes that a lengthier military presence would have eventually led to political reform and reconciliation. Did either administration have a political strategy to advance reconciliation that could address the chronic patron-client problems outlined in chapter 24?

Despite the reduction in violence, the United States was unsuccessful in fostering a durable political outcome in Iraq for three interlocking reasons. First, the US desire to withdraw to zero or to what was perceived by Iraqi leaders as an ineffectual presence reduced American bargaining leverage. Unless the United States could provide sufficiently compelling incentives for reform and reconciliation, Iraqi leaders would be unwilling to make painful sacrifices only to see the Americans depart. Second, the reduction in violence led US officials to overestimate Maliki's inclusiveness, resulting in a significant shift in priorities under the Obama administration. Third, US leverage was further dissipated by civil-military tensions and strategic incoherence in theater, making coordination nearly impossible. As officials in Baghdad kicked problems to Washington, D.C., the National Security Council had to deal with highly complex issues that its officials had little bandwidth to navigate successfully. The results were often ham-fisted efforts that had unintended consequences.

29

The Absence of a Political Strategy Erodes US Leverage

Falling back from a decisive victory war termination outcome to transition can be appealing because it avoids the need to negotiate with the insurgency or its sponsor. Transition uses the crossover point premise: that the intervening power can alter the relative balance of power in the conflict by degrading or defeating the insurgency while building the capacity of the host nation's government and security forces. To paraphrase President George W. Bush, "As they stand up, we stand down."[1] The patron-client problems discussed earlier, however, can prevent that point from being reached.

The host nation's security forces might increase in size, but corruption and poor leadership may degrade readiness and battlefield performance. Predatory behavior can undermine the government's ability to win the battle of legitimacy and inspire disaffected groups to support the insurgency. Conversely, too slow a capacity-building effort and insufficient action against the insurgency could enable the latter to grow so strong that the host nation can never catch up. As the intervening power withdraws from the conflict, the incentives for reform may decrease if internal rivals remain a more immediate threat than the insurgency. If political leaders believe that the payoff is not worth the political cost, they may be unwilling to accept a small presence of foreign forces as trainers or advisors. Patron-client problems result in more substantial losses of influence than the intervening power might expect.[2] Transition thus might not be a low-risk alternative if decisive victory fails.[3] If the so-called crossover point remains elusive, the intervening country could face an unappealing choice between withdrawal (and hoping that the client survives) and an open-ended presence.

Iraq differs from Afghanistan (and Vietnam) in the absence of negotiations with the insurgency or its third-party sponsor. While reconciliation efforts during the Sons of Iraq movement entailed negotiation, those were primarily tactical and aimed at facilitating revolt against AQI in exchange for a cease-fire with US forces and protection from pro-government predation.[4]

The United States assumed that the Iraqis would win a political victory over AQI and the Sunni and Sadrist insurgencies, because violence levels reduced in the fall of 2007.[5] The surge and Awakening, however, largely failed to lead to national reconciliation and broader political inclusion.[6] While only Iraqis could make political inclusion work, the United States was not powerless to advance the prospects. Prime Minister Maliki's sectarianism was still evident, but the United States took an increasingly passive stance. The United States had three significant opportunities to apply leverage for political reform and reconciliation: the 2008 Strategic Framework Agreement (SFA) and Status of Forces Agreement (SOFA) negotiations, the 2010 parliamentary elections, and the 2011 SOFA negotiations. It failed to do so each time.

The surge in Iraq was a time-limited force uplift.[7] President Bush made a bold decision to double down on success in Iraq but had only a year left in office. Both competitors for the Democratic Party nomination, Senators Hillary Clinton and Barack Obama, campaigned to end the war in Iraq (the so-called war of choice) and reinvest in Afghanistan (the so-called war of necessity).[8] General David H. Petraeus, taking command of Multi-National Force—Iraq (MNF-I) in February 2007, had about seven months to make demonstrable progress before his congressional testimony in September of that year. He and Secretary of Defense Bob Gates believed that success by then could fend off demands for a rapid withdrawal timeline.

Meanwhile, the UN Security Council resolution that provided the international legal basis for American military presence in Iraq was due to expire at the end of 2008 and was not likely to be renewed.[9] Without a new framework in place, the US mission would be even more tenuous. The Bush administration wanted to keep American troops in Iraq, while Iraqi leaders sought an agreement that had greater respect for their sovereignty. They agreed to negotiate a Status of Forces Agreement (SOFA) to govern the US military presence and a Strategic Framework Agreement (SFA) to outline the bilateral diplomatic, economic, and cultural relations. The negotiations began in March 2008.

Maliki sought to increase his leverage as the withdrawal timelines became a political football during the 2008 US presidential election campaign. By the time presumptive Democrat nominee Barack Obama visited Petraeus in Iraq in July 2008, US forces had been steadily drawing down for the past eight months.[10] Obama campaigned for a 16-month timeline to withdraw troops from Iraq, which Iraq's Prime Minister Maliki endorsed during an interview with *Der Spiegel* earlier in the month.[11] That Maliki's remarks occurred during the SFA and SOFA negotiations was probably no accident.[12] Despite the steep reductions in violence, the American military presence remained deeply unpopular in Iraq.[13] With AQI decimated and the Sadrist militia defeated during the Charge of the Knights operation, Maliki's worries shifted to fears of a Ba'athist coup.[14] He was looking toward the 2010 parliamentary elections to solidify his grip on power.[15] The Americans were a good hedge against a feared Ba'athist resurgence until the Iraqi security forces could take on the task of internal security. Petraeus forecasted that this transition could occur by the end of 2011.[16]

Internationally, Maliki indicated to US officials that he suspected the Gulf Arab states were fomenting instability in Iraq. Iran, he believed, was helpful.[17] Iran also had leverage over the Sadrists, which could be useful in the elections.[18] Securing a withdrawal timeline from the Americans would bolster Maliki's chances of reelection, improve his status with Iran, and give him greater leverage over the Sunni Arabs and Kurds.[19] Stipulating the need for parliamentary approval of the SOFA would also likely win him support, improve his bargaining position regarding the withdrawal timeline, and distribute the risk as widely as possible.[20] Once Maliki had heard from Defense Secretary Gates that the United States had no "Plan B" to keep troops in Iraq without a SOFA, he could stand firm and resist unwelcome provisions.[21] He had everything to gain by securing the timeline and nothing to lose, as long as he left an opening in case things turned sour before the Americans departed.

The Bush administration, meanwhile, wanted a withdrawal framework based on the security conditions as the United States interpreted them, specifically the levels of violence and the capabilities of the Iraqi security forces.[22] President Bush had faced down the Democrats on that issue, having vetoed a 2007 war spending bill that included a withdrawal timeline.[23] Agreeing to one now would be a loss of face, but Bush was by nature less wary of audience costs and was moving toward the last year of his presidency. Giving way on the timeline to keep troops in Iraq until the end of 2011, when Petraeus forecasted the ISF would be ready to handle security, was deemed an acceptable outcome.[24] Notably, the conditions did not include political reform and reconciliation—which Bush had established as a vital goal of the surge.[25]

The US side began the effort in disarray, with departmental disagreements.[26] Without an interagency headquarters in Baghdad, the disputes and the negotiations had to be managed by the NSC. Deputy National Security Advisor Stephen Hadley initially allowed Defense to negotiate the SOFA separately, not wanting to create interagency friction over the matter. Still, US leverage was likely to be much higher if the SOFA and broader Strategic Framework Agreement (that the Iraqis keenly wanted) were tied together. The Iraqi government valued US economic and diplomatic support more than troop presence. Separating the two allowed the Iraqis to stick to their guns on the timeline without suffering economic or diplomatic penalties—or having to make painful political concessions. After three months of stalled negotiations by the Defense team, State took over the SOFA negotiations.[27] The Iraqis maintained their demands for a timeline.

As the negotiations dragged on, Maliki insisted US forces be out of Iraq's cities by early 2009 and completely out of Iraq by the end of 2010.[28] His demands aligned with Democratic presidential candidate Barack Obama's timeline. Administration officials interviewed by Gordon and Trainor claim that Crocker asked for guidance on July 29 from Bush. The latter wanted to avoid concrete deadlines. The president reportedly replied that he could only accept the end of 2011 as a withdrawal timeline and preferred an "as soon as possible" (as defined by the United States) wording on the withdrawal from Iraqi cities. In a video teleconference with Maliki the next day, Bush reportedly offered mid-2009 to be out of the cities and the end of 2011 as a "goal" for removing US troops. Maliki agreed but cautioned that his government could still reject the 2011 date.[29] For Maliki, getting American forces out of the cities showed progress on sovereignty and increased his freedom of action to deal with perceived internal security threats. A former senior White House official close to the negotiations recalled that Maliki even wanted to rename the section on troop withdrawal to "Retreat of Coalition forces from Iraq."[30]

Wanting to finalize the agreement, Secretary of State Condoleezza Rice went to Baghdad to try to nail down the remaining issues in a one-on-one discussion with Maliki. She believed they had come to a final agreement that included a residual force of 40,000 US troops to assist Iraqi forces with training and logistics after 2011.[31] That Maliki would have made such a concession seems unlikely. Besides, he could not make it unilaterally, and the Iraqi parliament would likely reject it.

When Bush met with Iraq's President Talabani on September 10 in the Oval Office, the latter was adamant that without a SOFA, US forces must leave. Bush agreed to being out of Iraqi cities by the end of June 2009 and to a December 31, 2011, withdrawal of all American troops from Iraq.[32] The United States caved on the deadline. Bush recognized that three years was the best he was going to get, confided a senior White House official close to the negotiations.[33] The Iraqi parliament approved the SOFA on November 27, 149 in favor, 35 opposed, 14 abstentions, and 77 not present.[34]

As important as political reconciliation supposedly was to American interests, it did not figure into any of the SOFA discussions.[35] The United States made no effort to trade timelines for progress on reconciliation and political reform.[36] The absence of a political strategy and a coordinated US-Iraq strategy undermined the ability of the Bush administration to set the conditions for durable success and to hold Maliki accountable for reforms. Maliki, meanwhile, was taking steps to consolidate power, using the ISF as a "political targeting force."[37]

The US leadership changed in 2009. Barack Obama was elected president of the United States, and General Ray Odierno replaced Petraeus as the US commander in Iraq. To bring the new administration up to speed, Odierno and Crocker prepared an assessment of the situation in Iraq. Notably, according to interviews by Gordon and Trainor, the US team in Baghdad cautioned against conditioning troop withdrawal to political progress. "While our military presence is key for the large issue of guaranteeing an environment for progress in the political process, it does not predetermine the outcome of that process."[38] US forces had defused several ethnic and sectarian flashpoints; losing that capacity entailed strategic risk. However, the United States had no strategy for advancing political reconciliation beyond crisis management, meetings, and encouraging pieces of legislation.[39]

Perhaps most significantly to diplomatic continuity, Ambassador Ryan Crocker departed. In one of his final cables, Crocker highlighted Maliki's use of Iraqi special operations forces as tools of repression. "Maliki has shown that he is either unwilling or unable to take the lead in the give-and-take needed to build broad consensus for the Government's policies among competing power blocks." A key question, Crocker posed, was whether Maliki was becoming "a nondemocratic dictator" or was "attempting to rebalance political and security authority back to the center. . . ." Hedging his answer, Crocker believed "the answer lies closer to the latter than to the former." He recommended that the United States "press the PM [Prime Minister Maliki] on institutional and political consensus-building as key to sustaining and advancing our relationship and support."[40] Crocker, one of the most talented and accomplished ambassadors in modern American history, highlighted the absence of a strategy to advance political reform and reconciliation. To an administration official reading the cable, the message was to stay the course and to keep encouraging Maliki to be more inclusive.

No doubt, this was the message conveyed by Crocker to his successor. Ambassador Chris Hill had a distinguished career and most recently had been the US negotiator with North Korea over their nuclear weapons program. Hill was reportedly determined to be the "un-Crocker."[41] He brought in his team, all with reliable diplomatic records but little to no experience in the region. Hill was also determined to redefine the civil-military relationship. Petraeus and Crocker were "joined at the hip," which experts and former officials believed was instrumental in preventing Maliki's worst tendencies.[42] Hill reportedly wanted the embassy to stop acting as an "adjunct to the military" and show that the diplomats were really in charge.[43] Whereas Crocker was skeptical of Maliki's intentions and would pester him about reform, Hill enthused to Vice President Biden of the PM's plan to build a broad-based cross-sectarian alliance.[44] Such efforts were blocked, he assessed, by Sunni and Kurdish rivals who refused to get on board with Maliki's inclusive agenda.[45] Hill argued in January 2010 that the Arab-Kurdish divide, not the sectarian conflict, was the "greatest remaining challenge for the US effort in Iraq."[46]

30

New Administration, Similar Challenges

The Obama administration embarked quickly on a strategy review for Iraq, determined to wind down the war and refocus attention on Afghanistan.[1] When the administration settled on a new policy, Secretary of State Hillary Clinton issued guidance to the embassy in Baghdad in an April 8, 2009, cable entitled "U.S. Policy on Political Engagement in Iraq."[2] It outlined six "critical" objectives: successful national elections; avoid violent Kurd-Arab confrontation; develop non-sectarian, politically neutral, and more capable security forces; avoid Sunni-GOI (Government of Iraq) breakdown; prevent government paralysis; maintain macroeconomic stability. After noting some less critical objectives, the cable articulated a "new way forward" based on a "grand process(es)" policy that "focuses on setting in motion and energizing productive processes, but not necessarily resolution, on the full range of critical and significant challenges." The United States "Will offer to play the role of honest broker and/or third-party guarantor of the Iraqi and U.N. reconciliation processes." This engagement would address five issues, focused primarily on the Kurdish-Arab frictions. It would support the UN Assistance Mission for Iraq (UNAMI) efforts to address disputed internal boundaries, support an election law specific to Kirkuk, promote the passage of a hydrocarbon law, and sustain an MNF-I coordinating committee with ISF and Peshmerga leaders. The fifth issue was Sunni Arab accommodation.

A February 2009 survey of the Iraqi population suggested a slightly more optimistic mood that could have been built upon to support the transition. Compared to pre-surge days, both Sunni and Shi'a were more satisfied with their lives (see figure 11) and with how things were going in Iraq (see figure 12). When asked about their satisfaction with specific areas, people attested an improvement across sectors, including security, their freedoms, and their economic situation.[3] At the same time, confidence in the government and the Iraqi Army was rising, while trust in local militia had decreased since 2007.

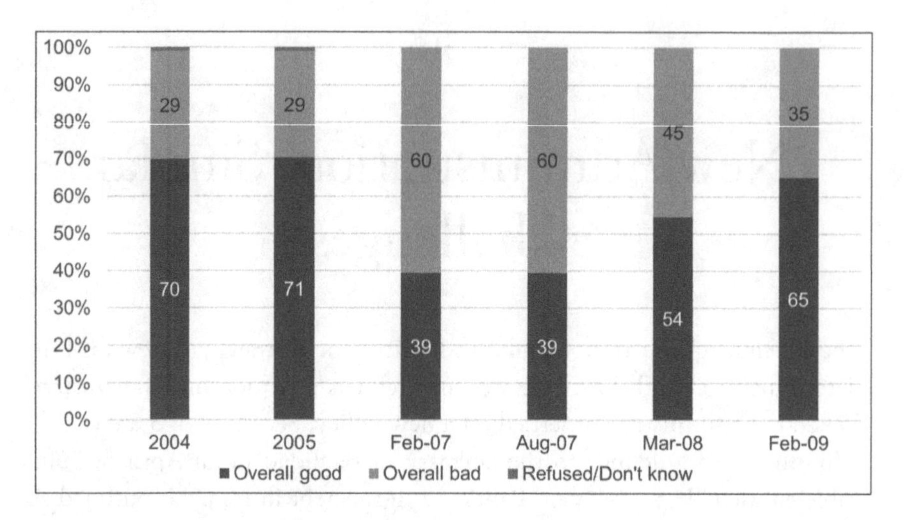

Figure 11. Iraqis' Satisfaction with Their Lives

Question: "Overall, how would you say things are going in your life these days? Would you say things are very good, quite good, quite bad, or very bad?"

Source: Based on data in D3 Systems and KA Research Ltd., "Iraq Poll February 2009," 1.

Support for a unified Iraq and a democratic political system was also as high as ever.

Assuming these figures were representative, this would have presented an opportunity for the United States (and a willing and able Iraqi government) to enhance political performance and inclusion in Iraq. However, bureaucratic and patron-client frictions complicated matters.

Due in part to the administration's shift in strategic priority to Afghanistan, Clinton's cable offered no resources or new authorities. "To carry out this policy, the Embassy and MNF-I are encouraged to reconfigure resources to support the five major processes (listed below) and to formalize an Embassy/MNF-I/UNAMI Working Group (including other parties as necessary) to coordinate these efforts." The embassy in Iraq was to advance US objectives through meetings, working groups, and support to UNAMI. With no authority over the military mission, the embassy could not direct MNF-I efforts or resources; the latter was free to implement US policy guidance as it saw fit. The cable did not direct the embassy to devise an implementation strategy or outline ways to use existing US leverage to advance its objectives. The guidance cable stated its desire not to limit or constrain the

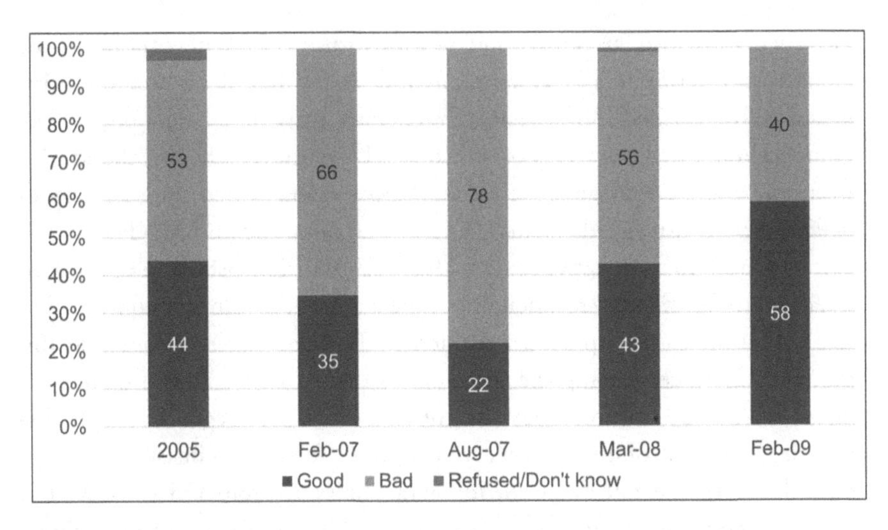

Figure 12. Iraqis' Satisfaction with Their Country

Question: "Now thinking about how things are going, not for you personally, but for Iraq as a whole, how would you say things are going in our country overall these days?"

Source: Based on data in D3 Systems and KA Research Ltd., "Iraq Poll February 2009," 3.

mission. "Indeed, creative and responsible initiatives from the field that effectively advance the stated policy are encouraged when appropriately proposed and approved." In other words, the embassy had no latitude. Ambassador Hill found that the easiest way to advance Iraqi agreement on the nonmilitary aspects of the policy was to pressure the Kurds and Sunni Arabs to accommodate Maliki's demands, rather than the reverse.[4]

The most likely explanation for the emphasis on the Arab-Kurd frictions was its recent intensity coupled with the belief that enough Sunni Arab reconciliation was occurring to lower the risk of instability.[5] The cable mentions areas of potential backsliding on the part of the Iraqi government on reconciliation, but orients on managing Sunni reactions: "Sunni political leadership is deeply fractured, rendering them more likely to advocate unhelpful, extreme stances, particularly during an election year. If mistrust grows, it could push Sunni Arabs out of the Iraqi national government and push more hardline Sunni Arabs towards a resumption of violence." There was no mention in the cable of exploring ways that the United States might capitalize on the elections or apply other leverage on the Iraqi government to advance reconciliation.[6]

There was also no suggestion of any future SOFA negotiations. The Departments of State and Defense, as well as the intelligence community, presumably provided inputs that were concurrent with the new policy. The guidance sent by the secretary of state to the US embassy reflected the administration's policy decision—the same decision upon which the secretary of defense would issue guidance to US Central Command and MNF-I.

Despite Hill's upbeat view of Maliki and the administration's decreased attention to Sunni Arab reconciliation, the Iraqi prime minister continued to weaken Sunni leadership.[7] With al Qaeda in Iraq decimated, violence declining, and US forces leaving the cities, Maliki had a freer hand to reduce the power of the 100,000-member Sons of Iraq (SoI).[8] He began to slowly dismantle the Sunni volunteers, after some initial efforts at US insistence to support them. In the fall of 2008, following pressure from Odierno, Maliki agreed to continue paying the salaries of 60,000 members and incorporate 40,000 into the ISF. The first tranche of volunteers went onto the government payrolls in October without incident.[9] When a budget dispute erupted, Maliki took money from the Ministry of Interior budget to pay the SoI groups. In March 2009, however, he started arresting Awakening leaders.[10] One such leader predicted to a US official that the government would "arrest all of us, one by one."[11] Then the Iraqi government stopped or substantially delayed paying member salaries. An Awakening member who wanted a job in the ISF had to navigate volumes of red tape, even for menial janitorial or servant positions. Sunnis in the mixed-sectarian Diyala province, east of Baghdad, were targeted disproportionately for alleged terrorist activities.[12]

Meanwhile, al Qaeda in Iraq was morphing into an underground organization, later to become the so-called Islamic State. "A few embers of Zarqawi's Islamic state remained, kept alive by flickering Sunni rage," writes David Ignatius. "The flame was nurtured at U.S.-organized Iraqi prisons such as Camp Bucca, where religious Sunni detainees mingled with former members of Saddam's Ba'ath Party, and the nucleus of a reborn movement took shape."[13] The Islamic State was also aiming at Awakening leaders. Former army officer and battalion commander in Iraq Craig Whiteside, now a professor at the Naval Post Graduate School, counts 1,345 Awakening members killed by the Islamic State between 2009 and 2013.[14] "Was anyone watching in Washington? Evidently not," Ignatius argues. "Officials in Baghdad, meanwhile, didn't seem to care; Maliki's government was probably as happy to see the killing of potentially powerful Sunnis as was ISIS."[15]

In Baghdad, Hill orchestrated Maliki's visit to Washington in July 2009. If political reform was essential, it should have featured prominently in the churn of cables and talking points that go into scripting head of state visits.[16] Maliki wanted to launch the Strategic Framework Agreement.[17] Hill suggested in his cable to Washington that the administration advance the economic discussion, help facilitate a resolution to decades-long frictions with neighboring Kuwait, and encourage Maliki to visit Arlington National Cemetery to honor American sacrifices in Iraq.

The cable was silent on reform and reconciliation, painting Maliki as a post-sectarian leader.[18] In the event, Maliki pressed the Iranian threat. If the United States could not get Sunni Arab states to stop fomenting unrest among Iraq's Sunnis, he reportedly cautioned, Iran was likely to intervene more aggressively in Iraqi politics. He also protested that representatives of the coalition military met with exiled Ba'athists in Turkey.[19] Political reform seemed to be absent from the discussions.

The next opportunity arose during the 2010 parliamentary elections. In the lead-up to the polls, Maliki aimed to outmaneuver his potential opponents.[20] By securing a withdrawal timeline, he showed his ability to advance Iraqi sovereignty and demonstrate political independence from the Americans.[21] He worked to marginalize Sadrists and other rival Shi'a parties while using his security forces to keep Sunnis off balance and in political disarray.[22] He also aimed to undermine their campaigns. Shi'a candidates sometimes deployed graphic anti-Ba'athist themes in efforts to increase Shi'a turnout.[23] Vice President Tariq al-Hashimi, a Sunni, meanwhile, threatened to veto the election law in hopes of gaining Sunni Arab seats from the Kurds. President Obama intervened successfully to pressure Kurdish president Massoud Barzani to accept the bill.[24] Despite the fractious political maneuvering, all seemed on track, as the US embassy expressed confidence in the Iraqi High Election Commission (IHEC).[25] That confidence shattered in January 2010 when the Accountability and Justice Commission (formerly the de-Ba'athification commission) barred roughly 500 candidates for parliament, mostly Sunni, over alleged Ba'athist ties.[26] Hill noted the subdued reaction of some Sunni communities to the ruling, and even suggested it may be the "first tangible example of cross-sectarian cooperation."[27]

Odierno reportedly took a dimmer view and raised a red flag to Washington. Hill was instructed by the Obama administration to address the issue with Maliki.[28] The prime minister was intransigent.[29] Obama next sent his point man on Iraq, Vice President Joe Biden, to Baghdad. According to an embassy reporting cable cleared by Biden's office, the vice president noted two concerns to UN Special Representative for Iraq Ad Melkert. "[T]hat the government had a serious responsibility to continue service delivery during the [post-election] transition, and that it was critical not to waste time during the period of government formation."[30] Advancing political reconciliation was absent from the list, and, according to a reporting cable, Biden did not raise the issue with Maliki. The prime minister assured Biden that he was not paranoid about Ba'athists but viewed them as a "malignant virus." Biden praised Maliki for democratic progress and political consensus, assuring US support while Iraq handled the de-Ba'athifcation issue according to its laws.[31] While the Americans were able to get many barred candidates reinstated, the bans had a disruptive effect. Maliki believed his State of Law list of candidates would win handily.[32]

The US country team was not so sure. Opinion polls suggested a tight race with Ayad Allawi's Iraqiya Party—a cross-sectarian Shi'a-Sunni coalition.[33] Maliki, however, had tools in place to shape the outcome and stacked the intelligence and security services with loyalists.[34] The Accountability and Justice Commission could disqualify winning candidates. Maliki controlled the judiciary and had the executive powers to declare a state of emergency. The US country team consulted with the White House, where, according to Gordon and Trainor's interviews with participants, Odierno pushed for guidance if this scenario came about. "We need a Maliki strategy," Hill reportedly said, "he is the only one with the tools to screw up democracy."[35] This admission by the US ambassador was stunning. If anyone should have developed and briefed a "Maliki strategy," it should have been Hill. Instead, he pushed the matter to Washington officials, who were in no position to create it from scratch. Unsurprisingly, the United States punted.[36] They were in the final stages of a contentious strategy review for Afghanistan, a higher priority for them than Iraq.

The elections provided yet another golden opportunity to exercise leverage. If the results were to be as close as the polls suggested, a disputed outcome would be likely. According to Iraq's election law, the party that wins the largest bloc of seats gets the first opportunity to form a government.[37] American support, one way or another, could tip the balance. Allawi's cross-sectarian coalition was more likely to promote political inclusion. Still, it would need US support in preventing Maliki from using extra-legal or even violent means to stay in power. Hill reportedly believed that Maliki would emerge from the election as the prime minister, whether legally or not.[38] Either way, the United States had a chance to advance political reconciliation.

The results on March 7 were as Odierno guessed. Allawi's secular Iraqiya list won 91 seats to Maliki's State of Law coalition's 89 and narrowly won the popular vote.[39] Iraqi National Alliance (INA), the competing Sadrist Shi'a bloc, took 70 seats. INA and State of Law had split because the former did not intend to endorse Maliki for another term. The Kurdish bloc won 57 seats. Maliki was incensed and moved forward aggressively to challenge the results.[40] He alleged that his opponents had cheated, and that the UN was complicit, so he reportedly sent a letter to the Americans demanding a recount in Baghdad and potentially in Mosul and Kirkuk.[41] He also aimed to use de-Ba'athification. The Accountability and Justice Commission conducted another review of candidates for alleged Ba'athist ties. Maliki asked for a ruling on the "largest bloc" language in Article 76 of the constitution. The pro-Maliki judiciary said the largest bloc could mean the party that won the most seats or a bloc assembled in parliament after the election.[42] These efforts gave Maliki the time and the opportunity he needed. If Maliki could gain INA support, he could claim the largest bloc.

Gordon and Trainor report that Odierno argued for the United States to get involved. He supported the Iraqiya case. Hill appeared to be less concerned about a Maliki victory and believed that the Saudis bankrolled Iraqiya and other Sunni parties.[43] With no in-country political strategy in place and the ambassador and commander in disagreement, the NSC had to handle the issue. While Washington considered the contrasting views from the field and deliberated whether and to what extent to get involved, Hill urged Iraq's politicians to start forming a government.[44]

Iran moved more quickly. They invited Iraq's Shi'a politicians to Tehran for Nowruz celebrations and urged them to come together. Although that was not yet agreed, the Shi'a parties on March 22 did support Talabani, who was also in attendance, to remain president.[45] By early May, Iran had convinced the State of Law and INA parties to merge into a single coalition: the National Alliance, with Maliki as the head. Together, they tallied 159 seats, just four fewer than the 163 needed to form a government. Iraqiya sought international support for its right to try to form a government, but nothing of substance was forthcoming.

Meanwhile, on April 26, a special judicial panel upheld a decision by the Accountability and Justice Commission to disqualify 52 candidates, one of whom was Iraqiya. Even if Allawi had the opportunity to form a government, he was highly unlikely to amass a majority. With the Shi'a mega-coalition created and the election outcome safely in Maliki's hands, an Iraqi appeals court overturned the earlier de-Ba'athification disqualifications, which removed an obstacle to certifying the election results.[46] The Obama administration rationalized that Allawi would have been unlikely to win enough support to form a government anyway, but Iraqiya never got the opportunity to try. Sunnis saw American acquiescence as a betrayal of the democratic process.[47] Maliki still needed the support of the Kurds to secure the 163-seat majority, and Barzani pushed hard to extract the best price for his help.[48]

At this point, the US team in Baghdad rotated. General Lloyd Austin replaced Odierno, and James Jeffrey (who had served previously in Iraq) took Hill's place as ambassador.[49] The Obama administration tried to bandage the election dispute by supporting a power-sharing arrangement among the rivals. James Steinberg, the deputy secretary of state, reportedly objected to the decision, because he believed it was more likely to produce gridlock and antagonism.[50] The administration persisted in the approach through the summer and into the fall of 2010. Maliki would remain as prime minister, but Allawi would head a newly created Strategic Policies Council. Allawi saw the powerlessness in the manufactured position and rejected it. The Americans then sought to promote Allawi as the president, which would mean Talabani would need to step down. That was a political blow to the Kurds and Talabani himself, but also a problem for his rival Barzani, who did not want him back in Kurdistan.[51] Various US officials attempted to persuade Talabani to give up the presidency. He refused.

On November 4, the administration took the extraordinary step of arranging a phone call between Obama and Talabani. The former pressed his fellow sitting president to step down. Talabani refused. The Kurdish leaders felt taken for granted.[52] By November 10, with Iran brokering an agreement between Maliki and the Kurds, Allawi likely recognized that the Strategic Policies Council was the best he would get. He reluctantly accepted, but the arrangement collapsed immediately. Allawi never joined the government. Most of Maliki's promises to the Kurds never materialized.[53]

Gates argues that the absence of sectarian violence between rival parties was a "mark of significant progress."[54] Such an indicator is misleading. Both leading candidates were Shi'a, so sectarian violence between them was unlikely. Maliki could manipulate the law to tilt the scales in his favor. Allawi would need American support if he hoped to form a government. Political violence by his party would have undermined any hopes of securing US backing. By focusing on putting the election crisis in the rearview mirror, the administration got nothing for the effort in terms of advancing reform and reconciliation.[55] Their efforts resulted in greater Sunni Arab and Kurdish resentment. Over the next 18 months, Maliki moved aggressively to crush Sunni leadership.[56]

A final opportunity for the United States to exercise its waning influence for reform and reconciliation came in 2011 as the administration attempted to renew the 2008 Status of Forces Agreement.[57] The United States had agreed in 2008 for all troops to leave the country by December 31, 2011—an outcome both Maliki and most Iraqis wanted. They were not alone. With the Americans out of the country and the governing coalition overwhelmingly Shi'a, the biggest winner was Iran. The Obama administration's enthusiasm for maintaining troops there was low. He had campaigned on a promise to get out of Iraq, and now his new country team was pushing to maintain a substantial presence. The Arab Spring was convulsing the Middle East and, along with Afghanistan, occupying the attention of the administration.[58]

As has been discussed in Part IV, Obama had dramatically escalated the war in Afghanistan, but the Taliban appeared no closer to collapse or to entering a peace process. The Karzai government was rife with corruption. Tensions between the two presidents were high. The US troop surge there was to begin receding in July 2011. Obama and his inner circle may have been sensitive to criticism about ignoring the advice of his commanders on the ground. They had accused the military of trying to "box in" the president regarding the troop surge in Afghanistan. They then fired General Stanley McChrystal after disparaging remarks by his staff were reported in *Rolling Stone*.[59] With the 2012 US elections just around the corner, another crisis with the military could be unhelpful. In short, both Maliki and Obama had incentives to let the SOFA expire while avoiding blame for doing so. The concessions needed to secure an agreement would need to be high enough for both to justify the political risk.[60]

Gordon and Trainor report that General Austin offered his estimate for the post-2011 force to cover the training, advising, and counterterrorism missions: 20,000 to 24,000 troops, which he assessed would still entail moderate risk. The Pentagon asked Austin to review the numbers again, which the latter revised downward with a preferred option of 19,000 troops, a middle option at 16,000, and a low choice of 10,000, which he deemed high risk.[61] The Pentagon must have massaged the numbers a bit more because an April 29, 2011, Principals Committee discussed options at 16,000, 10,000, and 8,000 troops.[62] Secretary of Defense Gates thought the lower two options could work.[63] Admiral Michael Mullen, the chairman of the Joint Chiefs of Staff, however, supported Austin's recommendation of 16,000. Both flag officers reportedly believed that having no troops at all in Iraq was better than having too few. Mullen exercised his legal right as chairman and the president's principal uniformed military advisor to express his concerns in a written memo to the national security advisor, Tom Donilon.[64] The arguments, however, remained fixated on troop numbers, without mention of political reform and reconciliation.[65]

In May 2011, Maliki hinted that he might support an American military presence if he could garner enough political support.[66] Obama, fresh off the successful Abbottabad raid that killed Osama bin Laden in his villa in Pakistan, approved a residual force in Iraq of up to 10,000 troops.[67] By early June, the Obama administration communicated four conditions that the Iraqis would have to meet for US forces to remain. First, the government of Iraq needed to make an official request. Second, Maliki would need to gain parliamentary approval for a SOFA continuing the same 2008 legal immunities for US soldiers in Iraq. Third, Maliki would need to fill the vacant Ministry of Defense position and other open positions within the security ministries. Finally, Maliki had to act against Iran-supported militants, which had been using EFPs (explosive force penetrators) and IRAMs (rocket-assisted mortars) against US troops.[68] Whether intended or not, the second requirement was a poison pill. The United States appeared to be dictating internal Iraqi government procedures. Even if Maliki wanted US troops to remain, the concessions he would likely have to make to gain approval would have been substantial—especially considering Iraqi public opinion on the matter.[69] Two former senior White House officials with knowledge of both the 2008 and 2011 negotiations note that Maliki got what he wanted in 2008; he would be highly unlikely to overturn the timeline without major US concessions.[70]

Nonetheless, this provision gave both Obama and Maliki a reasonable escape from potential blame.[71] In August, Obama further reduced the maximum presence to 5,000.[72] The political risk of meeting the conditions for so little gain was likely deemed by Maliki to be too high. In the event, Iraqi leaders supported US military trainers but ruled out immunities. Obama ended the negotiations on October 21.[73] A former senior White House official told *New York Times* correspondent Peter Baker, "We really didn't want to be there, and he really didn't want us there. . . . It was almost a mutual decision, not said directly to each other, but, in reality, that's what it became. And you had a president who was going to be running for re-election, and getting out of Iraq was going to be a big statement."[74]

As the US forces prepared to leave, Maliki's sectarianism was in full swing. Provinces with significant Sunni populations such as Diyala, Salah ad-Din, Ninewa, and Anbar began to demand autonomy under a provision in the Iraqi constitution.[75] Shi'as stormed the provincial council building in Baqubah (Diyala).[76] In a joint press conference on December 12, just weeks before the last US troops were to leave Iraq, Obama praised Maliki's efforts in leading Iraq's "most inclusive government yet."[77] Deputy Prime Minister Saleh al-Mutlaq told CNN he was shocked that Obama greeted Maliki as the "elected leader of a sovereign, self-reliant and democratic Iraq" in light of his continued aggressive targeting of Sunni Arab leaders.[78] Maliki reportedly told Obama that Iraq's Vice President Tariq al-Hashimi and other Sunnis in his government supported terrorism.[79] One week after being lauded by Obama for inclusiveness, Maliki sent troops to arrest Hashimi. The latter fled in time, but 13 of his bodyguards were tortured and sentenced to death.[80]

By mid-2014, less than three years later, a new Sunni Arab insurgency was flourishing.[81] Daesh, the so-called Islamic State, had taken Ramadi, Fallujah, Mosul, and Tikrit and established a proto-state along the Euphrates in Iraq and Syria by feeding on the alienation of Sunni Arabs and engaging in a sophisticated combination of coercion, selective violence, and local governance.[82] In September 2014, US Director of National Intelligence James Clapper confessed, "We underestimated ISIL [the Islamic State] and overestimated the fighting capability of the Iraqi Army. . . . I didn't see the collapse of the Iraqi security force in the north coming. . . . It boils down to predicting the will to fight, which is an imponderable."[83] That might be true, but the misjudgment was more significant. Winding down the Iraq War was a higher priority than taking the steps needed to bring about a favorable and durable outcome, which may have motivated US policymakers to rationalize the myriad signs of trouble that pointed to a potentially explosive political fragility. "U.S. policymakers and planners did not pro-actively consider the transformative nature of the withdrawal of U.S. military forces," argues a RAND study, "and the effects that transformation would have on strategic- and policy-level issues for both Iraq and the region."[84]

Conclusion to Part VII

The picture that emerges of war termination in Iraq is one of sophisticated military efforts and fragmented political activities that were powered by poor strategic empathy and untethered to an integrated strategy. As discussed in Part IV, the Bush administration assumed a decisive victory over Saddam Hussein's fielded forces would yield lasting success. Obsessed with military details, the US government failed to develop a strategy that brought together and managed the elements of national power to bring about a favorable and durable outcome. When decisive victory failed to materialize, the United States was left scrambling for a way forward. The failure to consider war termination led to a myopic strategy that fixated on the military campaign and ignored the aftermath, and set the stage for the super-empowerment of mostly Shi'a exiles and elites and decisions to launch a de-Ba'athification campaign and disband the Iraqi Army. Aggressive military efforts fed perceptions of Sunni Arab disenfranchisement. The latter fought back, igniting a fierce insurgency. This gap in strategy heightened the risk that the war would turn into a quagmire.

Part V showed how the strategy became intractable as the unprioritized elements of national power self-synchronized in unproductive ways. Confirmation bias, political and bureaucratic frictions, and patron-client problems impeded the Bush administration's ability to recognize and modify a failing strategy. Political development became an engineering project, but the milestone-centric approach failed to account for the aggressive and sometimes bloody scrimmage for power. Local elites manipulated the milestones and the gaps between US military, political, and economic silos. The fault lines between political and military silos damaged perceptions of legitimacy while amplifying violence. Sectarianism and insurgency fed on one another in a downward spiral. Meanwhile, US officials stubbornly refused to change strategy, citing examples of progress in achieving milestones while evidence of disaster mounted.

When offered a new approach to salvage the war, Bush boldly decided to surge in the face of opposition calls for withdrawal. The plan succeeded in diminishing the Sunni Arab insurgency but failed to advance reconciliation and substantive political inclusion. The continued absence of a political strategy and waning support in the United States and Iraq for the American troops' presence damaged US leverage, which undermined the prospects for a favorable and durable outcome. US military leaders and diplomats managed to curb some of the worst excesses and defuse multiple crises but could not change the underlying logic of Maliki's aggressive sectarianism. The Bush administration missed an opportunity to use the SOFA negotiations to advance what they considered to be an essential requirement for success. Maliki got the troop withdrawal dates he so eagerly sought in exchange for no political concessions.

The Obama administration, eager to end the Iraq War, attempted to apply low-leverage conditionality to offers of a continued troop presence that unwittingly played into Maliki's hands. Ambassador Hill and General Odierno had very different views on the way forward in Iraq. Neither one had the authority to manage the US efforts or the relationship with Maliki and the Iraqi government, so they kicked issues to the NSC. Instead of advancing the prospects of reconciliation, the Obama administration took the path of least resistance—pressuring the Sunni Arabs and Kurds to go along with Maliki. In the end, both leaders got what they wanted—a complete withdrawal of US forces.

Kenneth Pollack, a Middle East expert at the Brookings Institution, argues in the case of Iraq that "military success is not being matched with the commensurate political-economic efforts that will ultimately determine whether battlefield successes are translated into lasting achievements."[1] What the United States lacked was not a set of plans but a strategy to achieve a favorable and durable outcome that accounts for the competing and conflicting interests of others. In his landmark study on strategy, historian Lawrence Freedman described the ancient Greek concept of *metis* as a form of strategic intelligence. It "conveyed a sense of a capacity to think ahead . . . grasp how others think and behave . . . and stay focused on the ultimate goal even when caught in ambiguous and uncertain situations."[2] The combined challenges of cognitive bias, political frictions, patron-client problems, and bargaining asymmetries undermined the prospects of a favorable and durable outcome. Both US administrations relied on a transition method for war termination but failed to address the critical risks to success.

Part VIII

Implications

America's military, claim General David H. Petraeus and scholar Michael O'Hanlon, is the world's best.[1] If that is so, why have major post-9/11 military interventions in Iraq and Afghanistan turned into bloody and expensive quagmires? Even smaller-scale interventions in places such as Libya, Syria, and Yemen have had poor results. The Chilcot Report, the official inquiry into the United Kingdom's decision to go to war in Iraq, has important implications about following America's lead.[2] It also reveals the limited influence of American allies on US decision-making.

A consistent shortfall in policymaking for Iraq and Afghanistan was a lack of a war termination framework beyond decisive victory, which induced three significant problems that made quagmires more likely. First, the United States undertook these major interventions with a myopic strategy that assumed decisive military victory. This presumption led to an underappreciation for the importance of critical factors, such as host nation politics and insurgent sustainability. It promoted a tendency to ignore or dismiss early negotiating opportunities. Second, the US government was slow to modify a losing or ineffective strategy due to cognitive biases, political and bureaucratic frictions, and patron-client problems. Third, as the United States tired of each war and signaled a desire to withdraw, bargaining asymmetries prevented favorable and durable outcomes. Attempts at negotiations and transition consistently fell far short of expectations. These problems interacted with one another in unique ways to produce different trajectories into intractable conflicts.

This book uses war termination as a lens to clarify why these conflicts were unlikely to result in a zero-sum decisive victory. The critical factors framework serves as a guide that policymakers can use to assess the likelihood of decisive victory.[3] To date, assessments about the prospects of success have relied mostly on subjective judgments and gut instinct. Lyndon Johnson's Under Secretary of State George Ball, for instance, wrote on July 1, 1965, that the Vietnam War was unwinnable, arguing, "No one has demonstrated that a white ground force of whatever size can win a guerrilla war."[4] Those believing that the United States could be successful countered with other subjective arguments, which President Johnson accepted.

Regarding Iraq, James Dobbins wrote in 2005, "The beginning of wisdom is to recognize that the ongoing war in Iraq is not one that the United States can win. As a result of its initial miscalculations, misdirected planning, and inadequate preparation, Washington has lost the Iraqi people's confidence and consent, and it is unlikely to win them back."[5] The near-term success of the 2007 surge seemed to vindicate those who believed the war could still be a zero-sum decisive victory. Maliki's aggressive sectarianism, unconstrained due to the absence of American forces, brought about a Sunni Arab backlash in the form of the Islamic State. The pro and con arguments around Obama's Afghanistan surge, this book has shown, were similarly subjective and not informed by empirical studies.

Chapter 2, including the critical factors framework, provides a more empirical basis for arguments about the prospects for intervention. In the case of Vietnam, both critical factors pointed in the wrong direction: the insurgency had tangible internal and external support, and the host nation's government was unable to win the battle of legitimacy in contested and insurgent-controlled areas. The empirical analyses discussed in this book show that no counterinsurgency has been successful when the critical factors are negative.

Iraq and Afghanistan entailed regime changes. Success relied on preventing these two critical factors from materializing. Neglect of the war outcome issue during strategy development impeded the United States' ability to recognize these factors and take appropriate action. In both cases, the critical factors turned sour. Whether an external power from a different culture can have a successful regime change and prevent the critical factors from becoming contrary requires further research. The critical factors framework and the taxonomy of three ways to win in chapter 1 give policymakers and scholars a methodology to evaluate the prospects of a successful intervention against an insurgency.

The presumption of decisive victory and the belief that with enough commitment and goodwill the United States can solve any problem across the globe could also be part of American strategic culture.[6] Such exceptionalism could be motivating the United States to intervene in conflicts that it cannot possibly win or to pursue strategies that have little chance of success. Likewise, the notion of strategic distance could be a factor.[7] A longitudinal study could examine how often an intervening power from a very different culture succeeded. Impossible strategic distance could become a compelling explanation. To be sure, the quality of insurgent strategy and capabilities, as well as that of the host nation, is critical to the larger question of which side wins or loses. Those questions are beyond the scope of this book but should be part of the broader explanation for why the conflicts turned out as they did.

31

Iraq and Afghanistan Compared

Failure to consider war outcomes has heightened the risk of selecting myopic strategies that ignore or underestimate the critical strategic risk factors: an insurgency with sustainable local support and an external sanctuary and a host nation government that is unable to win the battle of legitimacy. In both Iraq and Afghanistan, military action resulted in the overthrow of an existing regime. Still, the United States had not thought through the risks to a favorable and durable outcome and the requirements to prevent or mitigate those risks. When those problems materialized, the United States was slow to recognize them and unable to address them adequately.

Exclusionary regimes took control in both countries and soon became predatory. The United States rebuffed efforts by both Sunni Arab leaders in Iraq and Taliban senior leaders in Afghanistan to negotiate some form of cessation of hostilities and political inclusion at the encouragement mainly of Shi'a elites on the one hand and Northern Alliance leaders on the other. Statistically, as Dobbins notes, politically inclusive governments are more apt to be successful when supported by peacekeeping troops than are exclusionary governments backed by a peace-enforcement mission. The data he used was available in 2001.[1]

In both cases, the risk factors became quickly entrenched. The United States, however, was slow to recognize the problems and to modify a losing or ineffective approach. The primary reasons had subtle differences in each conflict. Issues such as confirmation bias, inadequate strategic empathy, bureaucratic frictions, and patron-client problems were common. In neither case did the United States put someone on the ground in charge of coordinating and managing the full range of American efforts. Both administrations in both conflicts operated in bureaucratic silos. Strategic risks were emerging in the seams and fault lines between and among these silos. Local elites captured milestones and institutions. Predation and corruption, civilian harm, inadequate governance, and hollow capacity-building eroded legitimacy. Both the Bush and Obama administrations tried to treat the *complex* task of state-building as a *complicated* project. They placed political, diplomatic, military, and economic milestones along a timeline, and gave each to a department to handle. These silos were vulnerable to manipulation by local elites, even as they created a misleading impression of progress. Both administrations simply aggregated the in-silo metrics as evidence of success, while the overall situations deteriorated. These problems damaged the credibility of both administrations and public support for the missions. Loss aversion was more prevalent for Bush in Iraq than for Obama. Audience costs and risk aversion, however, seemed more salient to Obama's decision-making in Afghanistan than to Bush's for either conflict.

Bargaining asymmetries undermined prospects for a favorable and durable outcome in both conflicts but unfolded in different ways. Bush was unable to secure an open-ended or "conditions-based" American military presence in Iraq and accepted a strict timeline. Defense and State efforts to extend the timeframe during the Obama administration were unsuccessful. Political reconciliation efforts faltered and were probably doomed as US forces drew down, leaving Sunni Arabs exposed to Maliki's repression. Even the most intensive efforts during the Petraeus-Crocker time could only constrain Maliki's sectarianism, not promote reform. There was no negotiation with a senior insurgent leadership or external sponsor. However, discussions between US officials in Baghdad and Iranian Revolutionary Guards officials may have played a role in bringing about al-Sadr's cease-fire. US military officials did engage with tribal elders allied with or in charge of Sunni Arab resistance forces and convinced many to join with US forces against AQI. Such agreements also served to protect Sunni Arabs from Iraqi government predation.

The Iraqi government's sectarianism severely undermined the transition. It proved unable to win the battle of legitimacy among Sunni Arab populations. The first transition effort in 2005–2006 ended in near disaster. General Casey organized his military campaign plan around building Iraqi units and handing over security responsibility to them. Still, Iraqi security force readiness took a back seat to American military efforts to fight insurgents. The training mission was inadequately resourced and focused.

Meanwhile, little to no American efforts seem to have been applied toward good governance and addressing predatory sectarianism—problems that were undermining the legitimacy of the government and promoting resistance among Sunni Arabs. Even after the initial success of the 2007–2008 surge in lowering levels of violence and convincing Sunni Arabs to work with the US military against AQI, the United States failed to make political reconciliation a strategic priority. The Bush administration neglected to secure credible commitments on reconciliation in exchange for agreeing to a timeline. The Obama administration took the path of least resistance toward Maliki while staying focused on the withdrawal timeline.

ISIS was the biggest beneficiary, as alienated Sunni Arabs began to resist a new round of predation without the presence of US forces to keep sectarianism in check. Meanwhile, endemic corruption in the Iraqi security forces sapped the readiness of the army and police in Sunni Arab areas as Maliki sold positions and chose leaders based on personal loyalty and political reliability rather than performance. Their disastrous defeats at the hands of ISIS showed how even the best resourced and managed capacity-building efforts could have feet of clay. Obama deployed American forces back to Iraq to reverse the ISIS onslaught. Governance and political reconciliation remained back burner issues, even after ISIS's so-called caliphate disintegrated in 2018.

Patron-client problems undermined the transition in Afghanistan, too. The disconnect between the transition policy and the operations on the ground by US officials was striking. The military command focused mainly on defeating the Taliban through kinetic operations, while the ANSF development effort was, in McCaffrey's words, miserably under-resourced. The United States and NATO allies filled only a fraction of the required training teams before 2010. The ANSF was desperately short of critical equipment. Nearly six years after the overthrow of the Taliban, the Afghan Army was a paltry 34,000-strong force for a country of roughly 30 million people. The predatory kleptocracy formed beneath the noses of American civilian and military officials, who also failed to appreciate the impact of insurgent sanctuaries in Pakistan. The Obama administration corrected many of the problems with the ANSF development effort. Still, it remained unwilling to fully understand or address the predatory corruption and the damage it was doing to the transition. The military command made some efforts to deal with corruption, but Afghan elites simply outmaneuvered them. The Obama administration failed to find a realistic Pakistan policy, so Taliban sanctuaries stayed intact, and Pakistan's military continued getting American aid. Obama clung stubbornly to the timeline even though the so-called crossover point became impossible to reach. Figure 13 shows the trajectory of US troops in both conflicts.

The attempts to explore negotiations with the Taliban also failed. The announcement of withdrawal dates enabled the Taliban to play for time. The insurgency wanted exploratory talks to improve their international standing,

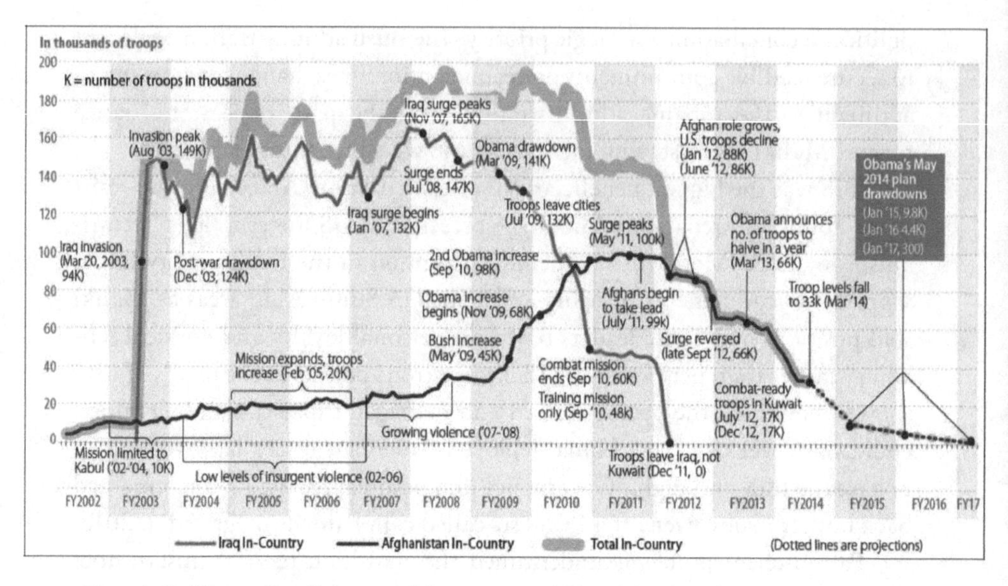

Figure 13. Troop Numbers in Afghanistan and Iraq Compared

Notes: Reflects US troops in-country; excludes troops providing in-theater support or conducting counterterror operations outside the region.

Source: Reprinted from Amy Belasco, "The Cost of Iraq, Afghanistan, and Other Global War on Terror Operations Since 9/11," Congressional Research Service, December 8, 2014, 9.

to gain concessions, and to speed American withdrawal, but not to end the conflict. Their leverage, they calculated, would be higher after US forces left. When the United States appeared to them as negotiating in bad faith, the Taliban walked away from talks. Failure to coordinate on reconciliation amplified Karzai's concerns about American intentions and credibility, which led him to undermine the effort. Karzai also refused American demands for reform. The Obama administration applied conditionality haphazardly, and Karzai calculated that the risks to internal stability far outweighed any benefits or penalties from the United States. He also figured that military presence in Afghanistan was a critical American interest that the United States would not risk. Karzai overestimated Afghanistan's centrality, but his assessment that the United States would not enforce conditions or impose penalties was prescient. He used this leverage to avoid signing the bilateral security agreement, which improved his domestic bona fides.

As the reconciliation effort met failure after failure, the kleptocratic nature of the Afghan government and its growing dependency on international advisors, support, and firepower undermined the transition. Positions in the government and security forces often went for sale at exorbitant prices. Corruption and poor leadership undermined ANSF readiness faster than international advisers could build it. By the end of 2014, the military command handed over responsibility for defeating a resilient insurgency to a government and security forces that were deeply corrupt and unable to win the battle of legitimacy in contested and Taliban-controlled areas. Although the ANSF, to date, have not experienced the same widespread collapses as the ISF, they have steadily lost ground to the insurgency. By 2015, Obama recognized that he had to slow and eventually stop the drawdown lest the Afghan government collapse. The latter has proven unable to win the battle of legitimacy in contested and Taliban-controlled areas.

Failure to include war outcomes into strategic considerations led to myopic strategies in both conflicts that presumed decisive military victory and ignored identified risks that were likely to materialize in the aftermath. These overly optimistic assumptions led Bush to conduct the invasion of Afghanistan with no plan for the result. His administration's focus shifted to Iraq and would remain there for the years to come. But even though the Iraq War preoccupied (especially) the Bush administration's minds and resources, they never questioned their strategic assumptions. More resources could not make up for a failing strategy.

As the critical risk factors put decisive victory out of reach, the United States was slow to modify the strategies in both conflicts due to cognitive obstacles, political and bureaucratic frictions, and patron-client problems. In neither case was the empathy-poor United States able to understand or address the inability of the host governments to win the battle of legitimacy in the contested and insurgent-controlled areas or the sustainability of the insurgencies.[2] The Obama administration did shift the focus to Afghanistan, and the surge had some success in driving back insurgency attacks. But here, too, higher budget and troop numbers were no panacea for the misconceptions and numerous problems that were troubling the war effort. When the United States tired of the wars and decided to withdraw, the erosion of leverage undermined efforts to negotiate and transition.

32

Implications for US Foreign Policy

Winning in war occurs when the concerned side, in this case the United States, achieves its policy aims. There are many paths to achieving policy aims, to paraphrase Clausewitz, and not all of them require zero-sum decisive military victory. The reflexive pursuit of the latter can undermine the prospects of success while increasing the costs in blood, treasure, and time. Large-scale US military interventions into irregular conflicts have consistently failed to achieve American goals in the expected amount of time and cost. If America wants to start winning again, the US national security establishment needs to address the chronic problems I have highlighted in this book.

A chronic problem exists when three elements are present: your actual results deviate from expected results; you do not know the cause of the deviation; and fixing the deviation is important.

I hold out hope that the US government and the American national security establishment care enough to fix the deviation. None of the presidential candidates in the 2020 election had a national security reform agenda. It seems that thought leaders are either wishing away the problem or (more likely) consumed by the four apocalyptic horses: the COVID pandemic, economic shutdowns, cultural upheaval, and a divisive 2020 presidential election.

In the expectation that the US government wants to avoid systemic errors, here are my observations and recommendations.

1. End the zero-sum decisive victory fixation. The United States created and persisted too long in following strategies that sought decisive victory in Afghanistan, Iraq, and Vietnam.

This book has uncovered systemic policy and strategy shortfalls by successive administrations. The United States has tended to produce unrealistic intervention strategies that presume decisive military victory and thus rely excessively on military force to deliver political aims. Many scholars have rightly pointed out that US aims can sometimes overreach. In the case of Afghanistan, US core aims focused on preventing the country from once again becoming a terror safe haven. That aim was not unrealistic, but the assumption that a decisive victory against the Taliban was necessary to achieve it was misguided.

Similarly, in Vietnam and Iraq, the core aims were not necessarily farfetched, only the presumption that decisive victory was necessary. Consideration of alternative war termination outcomes might have enabled US administrations to seize upon early negotiating opportunities and to take host nation legitimacy challenges more seriously.

Unfortunately, the policy and strategy processes neglected alternative war termination outcomes, resulting in strategic blind spots. Lieutenant General Terry A. Wolff, the Joint Staff's former plans and policy chief, noted that the United States "has no organized way of thinking about war termination."[1] Clausewitz's advice—do not take the first step before considering the last—should inform the development and evaluation of strategic options.

The critical factors framework outlined in this book could be a useful way to consider war termination outcomes, particularly by showing when decisive victory is impossible. Insurgencies that have durable internal support and external sanctuary tend not to suffer decisive defeats. Host nation governments that cannot retake and retain contested and insurgent-controlled areas tend not to win decisive military victories. Strategies that aim for decisive victory under one of these conditions have a high probability of failure and creating quagmires. "You need to decide what outcome is most realistic," McChrystal reflected, "not just what is most desirable."[2]

If the United States aims to rescue a troubled ally or partner, like in South Vietnam, it should have compelling reasons to believe that the intervention can reverse negative critical factors—and can succeed when interventions under similar conditions for different conflicts have failed. If there is a regime change, as in Afghanistan and Iraq, the strategy should focus on preventing the contrary critical factors from emerging. Options that include war termination outcomes can discipline the policy process, reduce groupthink, and improve alignment. America's recent experience in these interventions should also produce a greater degree of realism in affecting the internal dynamics of foreign countries.

If the critical factors suggest that outright military victory is unlikely, the United States is probably better off aiming for a negotiated settlement early, enforcing strict conditionality for issues such as governance, political inclusion, and anticorruption, or not intervening at all. As each case study has shown, the United States pays penalties in public support as interventions drag on. Once the United States decides to withdraw, the bargaining becomes asymmetric and undermines negotiations or transition. America's leverage is likely to be much higher (both on the insurgency and the host government) before intervention than once it is fully committed. Diplomatic-centric strategies and political-centric strategies that aim for negotiated outcomes that meet US interests could reduce the probability of quagmires.

2. Develop an interagency policy and strategy framework for waging war so that US officials can communicate clearly and reduce confusion.

First and foremost, this framework needs to help American political and military leaders understand the differences between waging and fighting a war. Waging war requires the integration of national power—political, diplomatic, economic, and military, among others—so that the whole is greater than the sum of its parts. In Colin Gray's formulation, there is no war without warfare—or warfighting—but there is more to war than combat.

By conflating war and warfare, the United States has undermined its ability to integrate other elements of national power. The military instrument predominates, and the others support. This practice has reduced the options available to each president and heightened the likelihood of quagmire.

To be sure, there are immense bureaucratic challenges toward such a framework. Diplomats may resist doctrine as constraining. The military has manuals covering tactical and operational levels of war, but nothing that qualifies as strategic doctrine. Still, a broadly framed national security policy that has presidential backing can provide a common language without being overly restrictive. This policy and strategy framework should address the following, at a minimum:

Authoritative terms and concepts for waging war, so officials can coordinate with precision and reduce the false cognates that impede communication and increase tensions. US military doctrine is not sufficient because its terms and concepts are mostly tactical. The US military's Joint Doctrine, for instance, lacks strategic definitions for words such as defeat, destroy, and degrade—terms which US presidents have used to describe their intentions.[3]

Different assumptions about defeat, the term McChrystal used when describing the ISAF mission to Obama, confused the president's civilian advisors. They saw it as an effort to coerce the president into approving a troop surge. Obama assigned the military the mission to "degrade" the Taliban, which is a term with no doctrinal definition. This ambiguity inhibited Obama's ability to govern the military and enabled the latter to do business-as-usual.

War termination outcomes. Any official national security language must include war termination outcomes. As this book has established, a strategy does not need to presume a decisive victory. In many cases, a decisive victory is unnecessary, and pursuing it could be counterproductive. Most wars do not end that way, either.[4] Providing for outcomes broader than decisive victory can help policy- and strategy-makers develop a more realistic array of options and greater clarity on their opportunities and pitfalls.

3. Right-size the military's role in the policy and strategy process, so the president receives integrated rather than military-centric options.

The president and national security advisor should demand integrated policy and strategy options rather than military-centric ones. A persistent problem for the United States has been the inadequate consideration of the political and diplomatic dimensions of national power. These elements tend to be more important than the military in creating a favorable and durable outcome. Turning first to the military for options wrong-foots the conversation and sets the policy on an unsupportable foundation. The president should, instead, direct the national security adviser to develop alternatives based on war termination outcomes.

The focus on military-centric options could stem from the assumption that military objectives deliver political results. As Clausewitz noted, "The political object—the original motive for the war—will thus determine both the military objective to be reached and the amount of effort required."[5] In this spirit, Colin S. Gray contended that "Strategy is, or should be, a purpose-built bridge linking military power to political goals."[6]

That view is too limited, which Gray would later recognize.[7] Military objectives are unlikely to deliver political results during irregular war interventions because the most critical issues tend to center on political legitimacy rather than force-on-force battles. Basing your policy and strategy on the presumption that the military effort will determine questions of political legitimacy is a fundamental flaw.

The US military's definition of strategy, "A prudent idea or set of ideas for employing the instruments of national power in a synchronized and integrated fashion to achieve theater, national, and/or multinational objectives," is a step in the right direction.[8] The rest of the government has not caught up. When *only* the military provides options, these more critical elements are left out because they have no authority or professional competencies to integrate them. The result is military-centric options that have a very low probability of success cheered on by nonmilitary efforts as supporting functions.

As seen in the Afghanistan case study, a former senior Obama administration official noted with frustration that the military presented options only in terms of troop and associated risk levels. Because the military, like every other agency, wants to remain in its (bureaucratic) lane, it has no authority for developing anything other than military options. For this problem to have escaped senior members of the Obama administration is disconcerting. The administration never asked for approaches in which diplomacy or strengthening of the Afghan state was the main effort. War, as Georges Clemenceau remarked, is too important to be left to the military.[9]

4. Reduce cognitive obstacles that impair decision-making.

The US government has experienced specific yet preventable errors that have undermined sound decision-making. Recognizing the persistence of these errors should enable the US government to take steps to reduce their influence. In particular, senior officials on the National Security Staff, such as the national security adviser, deputies, and senior directors, should be attuned to these decision-making errors and play the role of the honest broker to prevent the mistakes from undermining the decision-making processes. They should pay particular attention to these three cognitive biases:

Confirmation Bias has been the most consistent problem. With no interagency headquarters in the field responsible for coordinating US efforts, the NSC ends up running the war, and every agency grades its own homework. This combination has created incentives to place excessive emphasis on favorable data points while ignoring disconfirming evidence. In each case study, the US administrations damaged their credibility by claiming progress as the situation deteriorated. This problem undermined support for the wars and impeded the updating process needed to make sound policy and strategy decisions.

In-silo metrics create confusion because the intelligence and policy communities tend to assess the direction of the war using very different variables. The intelligence community considered enemy strength and capabilities (and sometimes host nation governance) as growing problems in Iraq and Afghanistan. The Defense and State departments, on the other hand, often used their metrics (numbers of enemy killed, civil servants trained, roads and schools built, etc.) to portray intelligence community assessments as "too negative." Such apples-to-oranges comparisons strengthen status quo bias.

Status Quo Bias and *Loss Aversion.* These two decision-making challenges may be more difficult to tackle. The US national security decision-making system is largely consensus-based. The scope and scale of problems the United States faces across the globe are overwhelming. It is only natural that different agencies have different points on view on a particular matter. Ideally, the NSC staff frames disagreements for discussion, debate, and decision among the principals. This process, however, often requires more bandwidth than most administrations can afford to devote to all but the most pressing issues. The NSC staff thus tends to seek consensus rather than debate. This approach empowers the most obstinate and favors the status quo.

Adding to the challenge is the human tendency toward loss aversion. As the administration touts the gains of its policies, it becomes more likely to dig in its heels against any changes that could put those gains, however great or small, at risk. Addressing these challenges probably would require giving the NSC staff psychological distance from running the wars so that they can more objectively assess the state of the conflict and make policy adjustments that increase the probability of success.

5. Address America's Bureaucratic Way of War by decentralizing to an in-theater, interagency command so that someone is accountable for results. Congress can hold meaningful hearings, and senior officials can stop misleading Americans with their claims of in-silo progress.

America's tendency to operate in bureaucratic silos undermines the prospects of success. No entity below the president has the authority and responsibility to direct and manage the full range of US government elements of national power deployed to a war zone. Bureaucratic silos impede coordination and induce NSC micromanagement. These practices erode the quality of assessments and decision-making, undermining America's ability to modify losing strategies. In both conflicts addressed in this book, the United States attempted reductionist, milestone-centric methods to address the complex challenges of state-building and reform.

Competing personalities, parties, and interests create feedback loops that cause actors to adapt.[10] US officials, focusing myopically on the milestones, failed to understand how super-empowered local elites were manipulating them for political advantage. These largely successful efforts damaged the legitimacy of the host governments in the eyes of the losers and those left out. The tendency to operate in bureaucratic silos created gaps that were repeatedly exploited by local elites and created situations in which efforts in one silo undermined efforts in others.

The troubling result was that US efforts tended to self-synchronize in damaging ways. Critical risks in Iraq and Afghanistan—predatory sectarianism and kleptocracy, bad governance, civilian harm, external support, and sanctuary, among others—materialized in these seams and fault lines. No US official had responsibility or was held accountable for addressing them.

Bridging silos will not solve significant structural problems in a war. Efforts to improve integration can indeed bring about different challenges, such as prolonging the policy process and creating disconnects between congressional appropriations and budget execution. Nonetheless, bridging bureaucratic silos can reduce policy and strategy errors and allow policymakers to address the structural challenges they face in war more effectively.

The United States can reduce chronic cognitive problems and silos by decentralizing authority. One way to do so is to replace senior military commands with a strategic headquarters led by a presidentially appointed and Senate-confirmed official.[11] That person should have the authority and responsibility to direct and manage all deployed elements of US national power and be held accountable for success. This organization will also help the US government differentiate between policy (which the NSC should manage) and strategy (which the strategic headquarters should oversee).

The Coalition Provisional Authority (CPA) in Iraq was an attempt to integrate America's nonlethal elements better, and it showed some success in that regard. It is important to emphasize that the military and intelligence were outside the CPA's authority. Still, the CPA made significant errors, and no one was accountable for success.

A senior leader who has the authority to direct and manage all US government entities in theater is no panacea. The president may wish to have that leader report through the Secretary of Defense. Still, this change will reduce bureaucratic silos, improve the speed of decision-making, and provide someone the president can hold accountable for results.

6. Take steps to reduce patron-client problems so that the United States is not held hostage by partners.

The United States needs to develop ways to address chronic patron-client issues that emerge from corrupt and predatory host nation governments. As its legitimacy in the eyes of the local population becomes more tenuous, the regime's dependency on the intervening power increases, creating entrapment. The vicious cycle of eroding legitimacy and increasing dependence undermines the prospects of success and heightens the danger that leaving will damage American security interests. I do not recall any interagency deliberations on these issues, except the quickly disproven hope that a withdrawal timeline would force the Afghan government to get serious about reform. I could find little discussion of such topics regarding Iraq or Vietnam either. Patron-client problems are not easy to disentangle, but the United States can take steps to limit the downsides.

First, develop a strategy with the host nation. The United States did not create a coordinated approach with the Afghan, Iraqi, or South Vietnamese governments. The United States should coordinate with the host nation in advance of intervention or immediately after a new government takes power so that the United States can gauge the prospects of the relationship and reduce misalignment. The strategy should address the critical factors framework and war termination.

Second, improve strategic empathy among US officials so that they are able to gauge host nation players' motivations and goals, recognize that their goals and motivations may differ from those of the United States, and may indeed conflict with those of the United States in ways that will produce friction and noncooperation.

Third, boost conditionality on agreed benchmarks and associated incentives and penalties for political and economic reforms. These incentives should target institutional and individual levels. Recent studies show that improvements are more likely when the patron enforces compelling penalties than when only inducements and encouragement are used.[12] Making smart choices on conditionality requires much more intensive monitoring of how the host nation uses US resources and capabilities, and greater political will to apply sanctions.

Fourth, address the pitfalls of developing host nation security forces. Host nation militaries do not exist in a political vacuum. The cases discussed in this book show some of the ways host nation politics affect the incentives of security officials. The United States and most western militaries take pride in the *objective control* relationship with their political leadership. In this relationship, military leaders stay out of politics and get a free hand in training and developing the armed forces. The military is professional. In most of the world, political leaders co-opt the military to ensure political support and muscle in times of domestic crisis. *Subjective control* occurs when military forces are primarily instruments of internal political power. US and western officials mirror imaged their professional viewpoints onto their host nation counterparts and were blind to the corruption. The perverse incentives eroded readiness and capacity faster than western officials could build them. Developing host nation security forces is a complex problem and needs to be addressed in the broader political and diplomatic context.

7. Develop expertise in wartime negotiations so that the United States can stop making bad deals while getting manipulated by corrupt partners.

In both Iraq and Afghanistan, the United States tired of the war and signaled a desire to withdraw. This combination created bargaining asymmetries in which the United States, as the stronger party, saw its negotiating leverage diminish. The steps here can help reduce the risk of quagmires and encourage the United States to place a higher priority on political and diplomatic approaches to intractable national security problems. For these to be effective, American diplomats need to institutionalize a body of professional knowledge about wartime negotiations and conditionality. As of this writing, US State Department officials tell me that their department provides no professional curriculum or specialty on such matters.

The lack of expertise could be prolonging American wars. Afghanistan showed the consequences of ineffective coordination, misaligned interests, and information asymmetries as the United States first avoided possible negotiated outcomes and then rushed to failure in talks after 2010. Vietnam shows how the offer to suspend US military actions in exchange for discussions with the DRV may have prolonged the conflict. Similarly, the 2015 and 2016 efforts by the United States to arrange cease-fires in Syria in the absence of any confidence-building have serially failed, which may be adding to cynicism about US intentions and credibility.[13] The US State Department should commission studies that create a body of expert knowledge on wartime negotiations and ways to use conditionality effectively. The interagency framework described here should include how the various elements of national power can support talks.

These recommendations will not turn unwinnable wars into winnable ones. Still, they will help the United States recognize the difference and avoid the preventable errors that turn the latter into the former.

33

Implications for Scholarship

This book has examined post-9/11 interventions in Afghanistan and Iraq in detail and provided a brief overview of the Vietnam War. The most obvious implication for the scholarship is the extent to which it is possible to generalize beyond these cases. This book has highlighted the limitations of viewing war termination as an interval between armed conflict and a peace agreement. By bringing the issue into the realm of strategy where it belongs, a wide array of opportunities opens for further research. I will use the three identified problem areas to frame the implications.

Implications for Strategy and Counterinsurgency

Examine if the presumption of decisive military victory heightens the risk of quagmires. The decisive victory presumption was a problem for the United States in Afghanistan, Iraq, and Vietnam. Cursory review of other cases suggests the problem is representative of a wider number of cases. The Soviets fell into the decisive victory trap in Afghanistan from 1979–1989, too. In 2011, the Obama administration demanded an end to the Assad regime and set a so-called red line regarding the use of chemical munitions against Syrian rebels. When Assad did so anyway, the United States backed down but later put a limited contingent of troops on the ground to fight ISIS and curb Iranian influence. As of early 2021, Assad seems highly likely to remain in power while the influence of Iran and Russia in Syria appears to be as strong as ever. A NATO intervention in Libya in 2011 ousted dictator Muammar al-Qaddafi but set the stage for a bloody civil war that still rages in 2021.

Not all interventions turn out poorly, of course. On the positive side of the ledger, the British response to Sierra Leone was mostly successful. NATO used various coercive measures to bring warring parties in Bosnia to the 1996 Dayton Accords and then deployed peacekeeping forces afterward. The threat of NATO military escalation convinced Serbia in 1999 to accede to Kosovo's secession. Do interventions that aim for negotiated settlements have a better track record than those seeking decisive victory?

Applying the three major war termination problem areas to the Soviet experience in Afghanistan and the French wars in Indo-China and Algeria could reveal the extent of these patterns beyond the United States. Analyzing the successful intervention in the Philippines in the early 1900s could illustrate the extent to which the growth in post–World War II national security bureaucracies has undermined America's ability to wage irregular war. Likewise, a comparison with the British intervention in Sierra Leone could reveal whether the United Kingdom has a better handle on these challenges than the United States.

Code outcomes by war termination method. Longitudinal studies produced by RAND tend to code conflict in zero-sum results.[1] More nuanced coding that presented variable-sums classified by war termination method might reveal the extent to which an intervening power protected its interests despite the host nation's government losing or winning. A negotiated outcome in the current Afghanistan war, for instance, will likely require more substantial Afghan government concessions to the Taliban than the reverse—a government loss. A credible commitment by the Taliban that prevented al Qaeda and other international terrorist groups from using Afghan soil to plan and execute large-scale terror attacks would achieve America's principal war aim while preventing the violent overthrow of the Afghan government. This outcome would be successful for the United States, albeit far bloodier and more expensive than had the Bush administration accepted the Taliban's surrender offer in 2001.

Conversely, the United States succeeded in preventing ISIS from overthrowing the Iraqi government. Still, Iranian influence is far more significant in 2021 than in Saddam's Iraq. Meanwhile, Iran's influence in Syria appears to have strengthened. The rise of ISIS and the Syrian civil war, which resulted in destabilizing refugee flows to Europe, were byproducts of America's 2003 intervention in Iraq. The refugee crisis strengthened right-wing parties in many European countries and provided opportunities for Russia to undermine NATO's cohesion. There are limits to coding variable-sum outcomes.

Scholarship on counterinsurgency should distinguish more effectively between doctrine and strategy. Critics tend to conflate doctrine and strategy in claiming that counterinsurgency does not work.[2] As this book has outlined, current US doctrine presumes that the host nation's government acts in the best interests of its citizens and that its officials are selfless public servants. These unrealistic characterizations can create blind spots that undermine effective strategy and coordination. This doctrine needs an overhaul.

Similarly, a counterinsurgency strategy is no more one-size-fits-all than is a conventional war strategy. Counterinsurgency does not presume large-scale interventions. An approach that places the political or diplomatic instrument as the top priority is likely to require fewer troops than a decisive victory. A small military footprint, however, does not necessarily mean that other agencies assume a higher priority. Foreign internal defense missions or partner capacity-building missions that rely on military advisors to help host nations fight an insurgency can suffer from an absence of strategy. Such efforts can damage the external power's credibility if people believe it provides weapons and training to predatory governments. Most countries that receive the highest levels of US security force assistance fare poorly on the Transparency International Corruption Perceptions Index. Several have populations that view the United States unfavorably.[3] The causal direction of such unfavorability merits analysis.

Differentiate levels of war in counterinsurgency. Most scholarship tends toward the tactical level with a heavy emphasis on the military.[4] Such works discuss the importance of political legitimacy and economic support alongside the need to conduct military operations against insurgents, but they rarely examine the dynamic interaction of these factors and how they affect the prospects of success. The tactical level is relatively easy to define as local and the strategic level as national, but what constitutes the operational level of war? How do counterinsurgents organize campaigns? The US military made some limited efforts to do this by coordinating military efforts in various "belts" around Baghdad. They also tried to focus military forces first in southern Afghanistan and then shift them to eastern Afghanistan—a campaign that ran out of time due to the drawdown.[5]

However, if irregular war requires the proper integration of political, diplomatic, economic, and military efforts, then campaigns at subnational levels should do so. The geographic shifting of security forces or the transfer of security responsibilities from foreign counterinsurgent to host nation forces seems inadequate conceptually and practically. Similar questions apply to the strategic level of war. Insufficient understanding of these levels may be leading senior leaders to obsess about tactics instead of setting their sights higher. Tactical victories in irregular warfare do not necessarily create a successful campaign. A series of successful campaigns might not add up to strategic success.

The challenges are complex. Factors such as governance, institutional integrity, and insurgent sanctuary, for instance, might have nonlinear effects on strategic outcomes. Successful tactical and operational efforts can have a limited and temporary impact if critical factors at the strategic level are problematic.

Understanding and Addressing Political, Bureaucratic, and Patron-Client Frictions

Examine the impact of modern bureaucracies on waging irregular war. Graham Allison and Philip Zelikow show how bureaucratic frictions affected decision-making during the Cuban Missile Crisis.[6] Robert Komer discussed this problem in Vietnam.[7] Bureaucratic silos in Afghanistan and Iraq created seams and fault lines that damaged the prospects of success and undermined the US government's ability to learn and adapt. Management scholars have discussed the chronic problem of organizational silos in business, and their efforts could be useful for reducing silos' impact on the conduct of war.

Military and management literature discuss micromanagement and over-centralization. The advance of information technology could be making micromanagement from national capitals more likely, especially as the leadership requirements at the strategic and operational levels remain opaque. A systemic study of these issues could identify ways to adapt national security structures to twenty-first-century realities.[8]

Manage patron-client challenges. The patron-client issues discussed in this book can help political scientists build upon the principal-agent theory in national security issues. The latter, introduced in chapter 8, addresses how dissimilarities in incentives between principal and agent can affect behaviors. Differences in strategy and sociocultural context between supporting country (patron) and supported country (client) can add complexity. Subtle differences can create friction. The United States wanted to win quickly and leave Afghanistan and Iraq. Host nation elites in both conflicts, however, manipulated American forces and resources to consolidate their grip on power, which pushed aggrieved citizens to the insurgencies and prolonged the war. The Iraqi government promoted the view that Sunni Arab insurgents loyal to tribal sheiks were the same as AQI and Ba'athists. The Afghan government played on western presumptions that the Taliban and al Qaeda were the same to convince the United States to destroy the Taliban, thus eliminating a rival political group.

Similarly, the Afghan government's insistence that the Taliban were Pakistan-controlled terrorists who needed to be killed or induced to defect conflicted with the post-2009 US assessment that the Taliban were an insurgent group who could support negotiations. The disconnect between Washington and Kabul undermined the 2010–2012 talks. Ironically, the negotiating position of the United States and the Afghan government was much more robust in 2011 and 2012, when US and Afghan officials in Kabul resisted talks, than in 2019, when the Trump administration committed to them. Strategic empathy and the effects of misaligned strategies between patron and client need more study.[9]

Assess the factors that make transition a realistic outcome. Systemic patron-client problems raise questions about the efficacy of the crossover point concept in counterinsurgency. This idea underpinned how the United States tried to limit its commitments to Afghanistan and Iraq, and Vietnam.[10] Security forces tend to reflect the political nature of the host nation. Kleptocratic governments, for instance, are likely to ensure their security forces participate in self-enrichment, which creates perverse incentives that erode readiness. Rather than linear growth in performance, as the US military tends to assume, local troops tend to hit a glass ceiling or even degrade as their size increases. Increased resources do not create symmetrical outcomes. Perhaps no amount of external support can enable a toxic host nation to reach the crossover point.

The impacts of sociocultural and historical context on the development of host nation security forces have been underappreciated. For instance, in Afghanistan, the US military was mirror imaging when designing the Afghan National Army. The use of western-style tactics, personnel, logistics, and command and control systems often grated against the capabilities and norms of their Afghan counterparts. Western systems, based on well-educated soldiers at the junior enlisted levels, were being fitted onto a nascent Afghan force that was mostly illiterate and led by people who expected highly centralized control and discouraged initiative and risk-taking. The result has been a high degree of dependency on western forces even after 15 years of capacity-building.[11] Studies of what kinds of tactical, logistical, and command and control systems work better for developing world militaries could advance our understanding of patron-client problems and address them.

Wartime Negotiations and Strategic Bargaining

Develop a theory of wartime negotiations for irregular war. Every actor wants to negotiate at the highest possible leverage, but how well can one predict when that point might occur? As the cases discussed here have shown, unilateral efforts to start a peace process will likely fail (unless one side is capitulating). Diplomatic efforts that are episodic or aimed at early high-profile concessions, both American tendencies, could be increasing the failure rate. The diplomacy needed to end civil wars and insurgencies could be very different from that used in conventional wars. The former might need subtle, deliberately paced, and continuous efforts to be successful.

Examine patron-client challenges during wartime negotiations. In each case, the host nation government discouraged the United States from outreach and talks with the insurgency or preinsurgency opposition. This systemic pattern could be prolonging conflicts. Examining the outcomes of wars with early settlements could illuminate whether an external power tends to be better off negotiating up front—particularly before significant military intervention. Such a study could also determine if a country's status as a superpower makes it more vulnerable to quagmires because its less powerful allies have limited influence.

Civil-Military Relations in Contemporary War

More broadly, this book has implications for the future study of civil-military relations.

Clarify the political purpose of irregular war. Samuel Huntington, in *The Soldier and the State,* famously outlined the difference between subjective control of the military, in which civilian rulers co-opt the military, thereby reducing its professionalism, and objective control, in which military professionalism can thrive far removed from politics.[12] In the latter model, while still under civilian control, the military enjoys a significant degree of professional autonomy in its unique field of expertise—the art and science of war. This principle of civilian control is unquestioned among senior officials in the United States and NATO countries. That autonomy, however, is not absolute. Statesmen must demand that military operations support political objectives. Eliot Cohen discusses how successful wartime leaders such as Abraham Lincoln, Georges Clemenceau, Winston Churchill, and David Ben-Gurion challenged and sometimes overrode the generals.[13] Intervening powers, however, may have a more difficult time articulating a clear political purpose because the stakes for the intervening power are rarely, if ever, existential and may be indirect. Because political and diplomatic factors are so decisive for irregular wars, there might not exist any military objectives that could plausibly deliver policy outcomes.

Refine agency theory in civil-military relations. Peter Feaver uses the principal-agent method to challenge Huntington.[14] He argues that the military operates under incentives like any other "agent" based on the levels of monitoring and expectations of punishment for shirking (not fully obeying civilian orders and guidance). In his view, uniformed officials have the right to provide advice in their field of specialized expertise, but not the right to question or circumvent even foolish orders from civilian leaders.[15] This formulation reduces the role of the military to technicians who follow orders, even if obeying them has catastrophic consequences. This method tightens civilian control but removes the relationship's moral and ethical dimensions.

These models are incomplete. Afghanistan and Iraq show examples of military leaders interpreting civilian guidance in ways that reinforce their existing views and practices, but that may have been at odds with the intentions of the president or secretary of defense. The Huntington model does not address this kind of problem. The Cohen model falls short as well: while he might expect civilian leaders to question more rigorously the military's execution of guidance, an overloaded NSC in the highly centralized national security structure might not have the bandwidth to recognize subtle deviations. Agency theory might describe this as evidence of shirking. That characterization could be accurate in some instances, but such behavior could also be a product of cognitive bias by the military or false cognates used by political leaders. How can civilians properly control the military instrument when there is no official language and set of concepts that enable them to articulate objectives, develop sound strategies, and govern civil-military integration? In each case presented in this book, civil-military miscommunication damaged outcomes.

Differentiate the military's role. More profoundly, scholars should question the military's elevated position in wartime. Each case has illustrated how the military was necessary but not sufficient for success and showed that the political, diplomatic, and economic domains operated concurrently with the military, rather than sequentially (as is often the norm in conventional war). The conventional wisdom that the military possesses unique professional expertise in strategy is likely part of the reason that each president relied on them for options. The choices framed by Defense were inadequate because they emphasized only a single instrument of national power.

The scholarship on civil-military relations should thus differentiate the military's role in war-waging versus combat. For the former, the military might be a co-equal partner with other elements of national power. The military's professional expertise is more precisely in warfighting. Refining the military's role in war-waging may put civil-military relations on a sounder footing and improve America's ability to successfully develop and implement strategies that have a reasonable chance of success.

Abbreviations

ANA	Afghan National Army
ANSF	Afghan National Security Forces (aka Afghan National Defense and Security Forces)
AQI	al Qaeda in Iraq
ATA	Afghan Transitional Administration
BSA	Bilateral Security Agreement
CENTCOM	US Central Command
COIN	Counterinsurgency
CPA	Coalition Provisional Authority
GTMO	Guantanamo Bay Naval Base
IGC	Iraqi Governing Council
IIG	Interim Iraqi Government
ISA	Islamic State of Afghanistan
ISAF	International Security Assistance Force
ISF	Iraqi security forces
ISIS/IS/ISIL	Islamic State of Iraq and Syria/the Levant
ITG	Iraqi Transitional Government
JAM	Jaish al-Mahdi
MNF-I	Multi-National Force—Iraq
NSC	National Security Council
OEF	Operation Enduring Freedom
OIF	Operation Iraqi Freedom
OPLAN 1003V	see OIF
ORHA	Office of Reconstruction and Humanitarian Assistance (US)
PACC	Pakistan-Afghanistan Coordination Cell
SCIRI	Supreme Council for the Islamic Revolution in Iraq

SFA	Strategic Framework Agreement
SOFA	Status of Forces Agreement
SRAP	Special Representative for Afghanistan and Pakistan
UNAMA	United Nations Assistance Mission in Afghanistan
UNAMI	United Nations Assistance Mission for Iraq

Key Events in the Afghanistan Conflict

1979	Soviet invasion of Afghanistan (simultaneous covert US support for rebels until 1989).	
1992	Najibullah's Afghan communist government overthrown by the mujahideen parties; the Peshawar Accords create the Islamic State of Afghanistan (aka the warlord government); ISA-HiG civil war until 1996.	
1994	Taliban emerge.	
1996	Taliban seize control of Kabul and establish the Islamic Emirate of Afghanistan; the Northern Alliance continues fighting the new Taliban government.	
1998	Al Qaeda attacks American embassies in Kenya and Tanzania.	
2000	Al Qaeda attacks warship USS *Cole*.	
2001		
	September 9	Al Qaeda kills *Shura-e-Nazar* leader Ahmad Shah Massoud.
	September 11	Terrorist attacks on US World Trade Center; President Bush demands that the Taliban government extradite Osama bin Laden and other al Qaeda members.
	October 7	Operation Enduring Freedom begins.
	November 13	Northern Alliance takes control of Kabul.
	December 5	Bonn Agreement is signed, which establishes Hamid Karzai as head of the interim administration.
	December 9	Taliban surrender Kandahar; United States

		rejects peace overture; Taliban leader Mullah Omar flees.
	December 20	UN Security Council Resolution 1386 establishes ISAF.
2002		
	April	Adoption of lead nation concept; United States to oversee Afghan National Army development.
	June 7	Emergency Loya Jirga elects Hamid Karzai as head of the Afghan Transitional Administration.
2003		
	May 1	Rumsfeld declares end of "major combat" in Afghanistan.
	June	United States approves "Accelerating Success" concept.
	August	NATO assumes control of ISAF, replacing the rotating national command; it is NATO's first deployment outside of Europe and North America.
	December	Loya Jirga ratifies new constitution.
2004		
	October	Presidential elections; Karzai becomes first democratically elected head of Afghanistan.
2005		
	September 18	Parliamentary elections; major victories for warlords and local strongmen.
2006		
	Throughout	Regular Taliban offensives and anti-United States demonstrations occur in Afghanistan.
	September 11	US embassy and military command in Afghanistan publishes "Strategic Directive for Afghanistan."
2007		
	February	Bush updates Afghanistan policy.
2009		
	January 20	Obama is inaugurated as 44th US president.

	January 22	Richard Holbrooke is selected for the newly created position of special representative for Afghanistan.
	March 27	Obama announces new strategy ("to disrupt, dismantle, and defeat al Qaeda and its safe havens in Pakistan, and to prevent their return to Pakistan or Afghanistan") and sends an additional 4,000 soldiers.
	June 15	General Stanley A. McChrystal replaces General David D. McKiernan as commander in Afghanistan.
	August 20	Afghan presidential elections; Karzai is declared as winner in November.
	September 20	McChrystal assessment becomes public (leaked to press).
	December 1	Obama announces surge and sets July 2011 to start troop drawdown.
2010		
	June 23	General David Petraeus replaces McChrystal as commander.
	September 18	Afghan parliamentary elections.
	November	NATO summit in Lisbon agrees to hand control of security to ANSF by end of 2014.
	December 1	2010 Afghanistan-Pakistan Annual Review outlines five lines of efforts; one of them is reconciliation.
	December 13	Richard Holbrooke dies.
2011		
	Throughout	Exploratory talks between the United States and Taliban.
	February	Mark Grossman follows Richard Holbrooke as SRAP.
	May 1	Osama bin Laden is killed by US forces.
	June 22	Obama announces troop withdrawal timeline.
	September 13	Taliban attack on US embassy in Kabul.
	December 5	Second Bonn Conference.

2012

December	Grossman resigns as SRAP; his deputy David D. Pearce takes over as acting SRAP.

2013

May	Jim Dobbins takes over as SRAP.
June 18	Abortive opening of Taliban office in Doha.
June 19	Karzai suspends talks on bilateral security agreement with United States (which resumed later).
November	Karzai refuses to sign BSA.

2014

April 5	Afghan presidential election; second round held on June 14 between Ashraf Ghani and Abdullah Abdullah.
July	Jim Dobbins retires from his SRAP position; Jarrett Blanc becomes acting SRAP.
September 21	Ghani is declared the winner of Afghan presidential election, but Abdullah disputes outcome. Some on Abdullah's side threaten civil war. At US insistence, the two candidates, Ashraf Ghani and Abdullah Abdullah, form a unity government; Ghani becomes president and Abdullah chief executive.
October	BSA is signed.
December	ISAF mandate ends.

2015

January 1	Operation Resolute Support succeeds ISAF to train, advise, and assist the ANSF.
March	Obama delays troop withdrawal.
July 29	Taliban confirm death of Mullah Mohammad Omar, announce Mullah Akhtar Mansour as his successor.
September 28	Taliban temporarily capture Kunduz.
November 17	Richard G. Olson is appointed as SRAP.

2016

May 21	Taliban leader Mansour is killed in an air strike in Pakistan; Mawlawi Haibatullah Akhundzada becomes successor.
July 6	Obama announces that 8,400 US troops will remain in Afghanistan; NATO extends "Resolute Support Mission" beyond 2016.
September 22	The Afghan government and Hizb-e-Islami sign a peace deal.

2017

January 20	Trump is inaugurated as 45th US president.
April 21	Taliban attack Afghan army headquarters in Mazar-i-Sharif, killing 140.
April 29	300 Marines are deployed to Helmand, where the Taliban are gaining control over an increasing number of districts.
May 31	A bomb attack in Kabul close to the presidential palace and foreign embassies kills 150 people.
August 21	Trump outlines his Afghanistan policy.

2018

January 4	Trump suspends nearly all US security aid to Pakistan.
February 14	Taliban issues "Letter to the American People" asking for peace talks.
February 28	President Ashraf Ghani says the Afghan government is willing to recognize the Taliban as a legitimate political party as part of a potential cease-fire agreement.
June 8–10	After President Ghani announced a unilateral cease-fire for the end of Ramadan, the Taliban followed with their own three-day cease-fire over Eid al-Fitr.
July	NATO extends "Resolute Support Mission" "until conditions indicate a change is appropriate."

	September 5	Zalmay Khalilzad is appointed as US special representative for Afghanistan reconciliation; US-Taliban talks begin.
2019		
	September 7	After nine rounds of US-Taliban talks and an alleged agreement in principle, President Trump suspends the process due to objections from his administration and the Afghan government.
2020		
	February 29	US and Taliban sign agreement.
	September 12	Afghan government begins talks with Taliban.

Key Events in the Iraq Conflict

1990	Iraq invades Kuwait; international community places economic sanctions on Iraq.	
1991	United States launches Operation Desert Storm to free Kuwait; Iraq is subjected to a UN weapons inspection program afterward.	
1998		
	October 31	Iraq Liberation Act makes support of regime change in Iraq official US policy.
	December 16	Dissatisfied with Iraqi compliance with UN Security Council resolutions and weapons inspections, the United States and the United Kingdom launch the four-day bombing campaign Desert Fox; Iraq bans weapons inspectors.
2002		
	January 29	President Bush names Iraq a member of the "axis of evil" in his State of the Union address.
	October 16	US Congress authorizes the use of armed force to defend the United States "against the continuing threat posed by Iraq."
	November 8	UN weapons inspectors return to Iraq (UN Security Council Resolution 1441).
2003		
	March 17	United States issues ultimatum to Saddam Hussein.
	March 20	United States and United Kingdom launch Operation Iraqi Freedom.

	April 9	Hussein regime falls; United States establishes the Coalition Provisional Authority, to be headed by Paul Bremer, and the Iraqi Governing Council.
	May 23	CPA orders de-Ba'athification and the disbanding of the Iraqi Army.
	August 19	A suicide attack at the UN compound in Baghdad kills 17 people, among them Sergio Vieira de Mello, the UN Secretary General's special representative to Iraq.
	December 13	Saddam Hussein is captured in Tikrit.
2004		
	March 31	Blackwater employees are killed and hung on a bridge in Fallujah.
	April 28	Abu Ghraib torture and prisoner abuse become public.
	June 28	United States hands sovereignty to Interim Iraqi Government headed by Prime Minister Ayad Allawi; the CPA disbands.
	November	Major US-led campaign against Iraqi insurgents in Fallujah.
2005		
	January 30	Elections for Transitional National Assembly.
	April 7	Kurdish leader Jalal Talabani becomes president; Ibrahim al-Jaafari, a Shi'a, is named as prime minister.
	October 15	Referendum approves new constitution.
2006		
	February 22	Bombing of Shi'a mosque in Samarra unleashes a new wave of violence.
	May 20	Nouri al-Maliki becomes prime minister.
	June 7	AQI leader Abu Musab al-Zarqawi is killed in US air strike.
	October	Islamic State of Iraq emerges out of ongoing insurgency.
	November 8	Secretary of Defense Rumsfeld resigns; he is succeeded by Robert Gates.

	December 30	Saddam Hussein is executed for crimes against humanity.
2007		
	January 10	Bush announces surge, committing more than 20,000 additional troops to Iraq; David Petraeus replaces George Casey as commander in Iraq; Ryan Crocker replaces Zalmay Khalilzad as chief of mission in Baghdad.
	Fall	Anbar Awakening spreads throughout Sunni Arab communities; violence levels reduce substantially in the fall.
	October	United States starts troop drawdown.
2008		
	March	Start of negotiations on Status of Forces Agreement and Strategic Framework Agreement between Iraq and United States.
	November 27	SOFA and SFA are signed; US troops must leave the country by the end of 2011.
2009		
	January 20	Obama becomes president.
	April	The Obama administration crafts a new strategy.
	June	US troops withdraw from towns and cities in Iraq.
2010		
	March	Parliamentary elections in Iraq.
	August 31	Obama declares end to Operation Iraqi Freedom.
2011		
	December 18	The last US troops leave Iraq.
2013		
	January	Pro-government protests emerge as a reaction to anti-government protests; the government uses security forces to control protests.
	April	The Sunni insurgency intensifies; Maliki warns of a new sectarian war.

	May 10	Abu Bakr al-Baghdadi is appointed the new leader of ISIS.
2014		
	January	ISIS seizes control of large parts of Fallujah and Ramadi.
	April 30	Iraqi parliamentary elections; Prime Minister al-Maliki fails to win a majority of seats.
	June	ISIS takes control of Mosul and Tikrit. Iraq officially asks United States for support.
	June 29	ISIS declares a caliphate across parts of Iraq and Syria.
2015		
	April	Iraqi security forces regain control of Tikrit.
	December	Iraqi security forces recapture Ramadi.
2016		
	February 26	Muqtada al-Sadr leads a mass demonstration in Baghdad to protest against government corruption and the failure to deliver on reform plans.
	June	Iraqi security forces retake Fallujah.
2017		
	July 10	Iraqi and militia forces recapture Mosul.
	September 25	Iraqi Kurds hold a referendum on independence; more than 90 percent of voters are in favor; after Iraqi forces are deployed to Kirkuk, the Kurdish regional government agrees to "freeze" the result.
	October 18	Raqqa, Syria, self-declared capital of the ISIS caliphate, is taken by US-backed Syrian Democratic Forces (SDF).
	December 9	Iraq announces the defeat of ISIS in Iraq.
2019		
	March 23	Syrian Democratic Forces announce the defeat of ISIS.

Notes

Key to Interview Abbreviations

Interviewee A. Regional expert and former MNF-I senior adviser.
Interviewee B. Former Bush administration senior official.
Interviewee C. Former Obama administration senior official.
Interviewee D. Former MNF-I senior adviser.
Interviewee E. Former US military official in Kabul.
Interviewee F. Former US intelligence official.
Interviewee G. Former US senior defense official.
Interviewee H. Former State Department senior adviser.
Interviewee I. An Afghanistan expert in frequent contact with the Taliban.
Interviewee J. Former State Department senior adviser.
Interviewee K. Former Bush administration senior official.
Interviewee L. Former Obama administration senior official.
Interviewee M. Former Obama administration senior official.
Interviewee N. Former senior US military official in ISAF.
Interviewee O. Former senior US military official in MNF-I.
Interviewee P. Former senior US military official in ISAF.
Interviewee Q. Former senior US military official in ISAF.
Interviewee R. Former Afghan government senior official.
Interviewee S. Former commander in the Afghan special operations forces.
Interviewee T. Former senior EU official in Kabul.
Interviewee U. Former senior NATO official in Kabul.
Interviewee V. Former senior NATO official in Kabul.
Interviewee W. Former senior Pentagon official.
Interviewee X. Former senior Pentagon official, Obama administration.
Interviewee Y. Former Taliban senior official.
Interviewee Z. Former UNAMA senior official.
Interviewee AA. Former senior Afghan official.
Interviewee AB. Former senior Afghan official.

Introduction

1. General Stanley A. McChrystal with David Silverman, Chris Fussell, and Tantum Collins, *Team of Teams: New Rules of Engagement for a Complex World* (New York: Penguin, 2015), 6, 18.

2. Rupert Smith, *The Utility of Force: The Art of War in the Modern World* (New York: Vintage, 2005), x–xii.

3. Daniel Bolger, *Why We Lost: A General's Inside Account of the Iraq and Afghanistan Wars* (New York: Houghton Mifflin, 2014).

4. Brendan R. Gallagher, *The Day After: Why America Wins the War but Loses the Peace* (Ithaca: Cornell Univ. Press, 2019).

5. William C. Martel, *Victory in War: Foundations of Modern Strategy* (Cambridge: Cambridge Univ. Press, 2011), 3–4.

6. Ibid., 342.

7. Kyle Rempfer, "H.R. McMaster Says the Public Is Fed a 'War-Weariness' Narrative That Hurts US Strategy," *Military Times,* May 8, 2019.

8. Christopher D. Kolenda, "America's Generals Are Out of Ideas for Afghanistan," *Survival* 59, no. 5 (2017): 37–46.

9. Department of Defense, *The National Military Strategy of the United States of America 2015* (Washington, D.C.: US Government Printing Office, 2015), 5 (emphasis added).

10. Headquarters, Department of the Army, *Army Doctrine Publication 1,* ADP 1 (Washington, D.C., July 31, 2019), 3-2 (emphasis added). For earlier descriptions of the decisive victory construct, see Headquarters, Department of the Army, *Field Manual 100-5 (Operations),* FM 100-5 (Washington, D.C.: US Government Printing Office, 1993), iv: "The mission of the United States Army is to protect and defend the Constitution of the United States of America. The Army does this by deterring war and, when deterrence fails, *by achieving quick, decisive victory on and off the battlefield* anywhere in the world and under virtually any conditions as part of a joint team" (emphasis added).

11. Lawrence Freedman, *Strategy: A History* (Oxford: Oxford Univ. Press, 2013), ix–x.

12. Bradford A. Lee, "Winning the War but Losing the Peace? The United States and the Strategic Issues of War Termination," in *Strategic Logic and Political Rationality: Essays in Honor of Michael I. Handel,* ed. Bradford A. Lee and Karl F. Walling (London: Frank Cass, 2003), 249–273, 255.

13. See Headquarters, Department of the Army, *Army Doctrine Publication 1,* ADP 1 (September 17, 2012), para. 1-15; 2-3–2-6; 3-2, and Headquarters, Department of the Army, ADP 1 (2019); see also Department of Defense, *The National Military Strategy of the United States* 2015, and Department of Defense, *Summary of the 2018 National Defense Strategy of the United States of America: Sharpening the American Military's Competitive Edge* (Washington, D.C.: US Government Printing

Office, 2018), and the Joint Staff, *Description of the National Military Strategy 2018* (Washington, D.C.: US Government Printing Office, 2019).

14. Gideon Rose, *How Wars End: Why We Always Fight the Last Battle* (New York: Simon and Schuster, 2010); Fred Charles Iklé, *Every War Must End* (New York: Columbia Univ. Press, 1991); Dan Reiter, *How Wars End* (Princeton: Princeton Univ. Press, 2009); Richard Caplan, ed. *Exit Strategies and State Building* (Oxford: Oxford Univ. Press, 2012); Bradford A. Lee and Karl F. Walling, *Strategic Logic and Political Rationality: Essays in Honor of Michael I. Handel* (London: Frank Cass, 2003); Elizabeth A. Stanley, "Ending the Korean War: The Role of Domestic Coalition Shifts in Overcoming Obstacles to Peace," *International Security* 34, no. 1 (2009): 42–82; and Elizabeth A. Stanley, *Paths to Peace: Domestic Coalition Shifts, War Termination, and the Korean War* (Stanford: Stanford Univ. Press, 2009); Matthew Moten, ed., *Between War and Peace: How America Ends Its Wars* (New York: Free Press, 2011).

15. Moten, *Between War and Peace,* xi (emphasis added).

16. Lee and Walling, *Strategic Logic and Political Rationality,* 245; see also Caplan, *Exit Strategies and State Building,* 3–4.

17. Caplan, *Exit Strategies and State Building,* 21.

18. Iklé, *Every War Must End,* ix; Colin L. Powell, *My American Journey* (New York: Random House, 1995), 519, 521.

19. Michael R. Gordon and Bernard E. Trainor, *The Generals' War: The Inside Story of the Conflict in the Gulf* (New York: Little, Brown, 1995), xiv–xv, 463–478.

20. Rose, *How Wars End,* 4.

21. Ibid.

22. Ivan Arreguín-Toft, *How the Weak Win Wars: A Theory of Asymmetric Conflict* (Cambridge: Cambridge Univ. Press, 2005), 3–4.

23. Lee and Walling, *Strategic Logic and Political Rationality,* 13.

24. Arreguín-Toft, *How the Weak Win Wars,* 3–5.

25. Christopher Paul, Colin P. Clarke, Beth Grill, and Molly Dunigan, *Paths to Victory: Lessons from Modern Insurgencies* (Washington, D.C.: RAND, 2013); Martin C. Libicki, "Eighty-Nine Insurgencies: Outcomes and Endings," in *War by Other Means: Building Complete and Balanced Capabilities for Counterinsurgency,* ed. David C. Gompert and John Gordon IV (Santa Monica, Calif.: RAND Corporation, 2008); James D. Fearon and David D. Laitin, "Ethnicity, Insurgency, and Civil War," *American Political Science Review* 97, no. 1 (February 2003): 75–90.

26. See John J. Mearsheimer, *The Tragedy of Great Power Politics* (New York: Norton, 2003); Kenneth Waltz, *Theory of International Politics* (New York: McGraw-Hill, 1979); and Kenneth Waltz, *Man, the State, and War: A Theoretical Analysis* (New York: Columbia Univ. Press, 2001).

27. Elizabeth A. Stanley and John P. Sawyer, "The Equifinality of War Termination: Multiple Paths to Ending War," *Journal of Conflict Resolution* 53, no. 5 (2009): 651–676; Iklé, *Every War Must End,* xv, 2–6.

28. Reiter, *How Wars End,* 6–7.

29. Stanley, "Ending the Korean War," argues that changes in domestic coalitions are often required to end a war. See also Stanley, *Paths to Peace*; H. E. Goemans, "Fighting for Survival: The Fate of Leaders and the Duration of War," *Journal of Conflict Resolution* 44, no. 5 (2000): 555–579; and H. E. Goemans, *War and Punishment: The Causes of War Termination and the First World War* (Princeton: Princeton Univ. Press, 2000), 15–22, which suggests that regime type is the most salient determinant of war termination behavior.

30. Daniel Kahneman, "Maps of Bounded Rationality: Psychology for Behavioral Economics," *American Economic Review* 93, no. 5 (2003): 1449–1475; Daniel Kahneman, *Thinking, Fast and Slow* (New York: Farrar, Strauss, and Giroux, 2013); Amos Tversky and Daniel Kahneman, "Rational Choice and the Framing of Decisions," *Journal of Business* 59, no. 4, part 2 (October 1986): S251–S278; Johanna Etner, Meglena Jeleva, and Jean Marc Tallon, "Decision Theory under Ambiguity," *Journal of Economic Surveys* 26, no. 2 (2012): 234–270; Dan Arielly, *Predictably Irrational: The Hidden Forces that Shape Our Decisions* (New York: HarperCollins, 2008); Tim Harford, *The Logic of Life: The Rational Economics of an Irrational World* (New York: Random House, 2008); Steven D. Levitt and Stephen J. Dubner, *Freakonomics: A Rogue Economist Explores the Hidden Side of Everything* (New York: HarperCollins, 2009); Jean-Jacques Laffont and David Martimort, *The Theory of Incentives: The Principal-Agent Model* (Princeton: Princeton Univ. Press, 2002); Walter C. Ladwig III, "Influencing Clients in Counterinsurgency U.S. Involvement in El Salvador's Civil War, 1979–92," *International Security* 41, no. 1 (summer 2016): 99–146.

31. Thomas Schelling, *The Strategy of Conflict* (Cambridge: Harvard Univ. Press, 1960), 5.

32. Roger Spiller, "Six Propositions," in *Between War and Peace: How America Ends Its Wars,* ed. Matthew Moten (New York: Free Press, 2011), 4

33. Reiter, *How Wars End,* 2–5, 16.

34. I. William Zartman, "The Timing of Peace Initiatives: Hurting Stalemates and Ripe Moments," *Global Review of Ethnopolitics* 1, no. 1 (2001): 8–18, 8.

35. Martel, *Victory in War,* 46.

36. I am indebted to Lieutenant General (Ret.) James Dubik, USA, for the war-waging versus war-fighting formulation, which was discussed on several occasions and which he captures in his 2016 book *Just War Reconsidered: Strategy, Ethics, and Theory,* Battles and Campaigns Series (Lexington: Univ. Press of Kentucky, 2018).

37. Colin Gray, "Concept Failure? COIN, Counterinsurgency, and Strategic Theory," *Prism* 3, no. 3 (2012): 17–32, 22.

38. Lee and Walling, *Strategic Logic and Political Rationality,* 13. See also Martel, *Victory in War,* 342.

39. Christopher D. Kolenda, "Slow Failure: Understanding America's Quagmire in Afghanistan," *Journal of Strategic Studies* 42, no. 7 (September 2019): 992–1014.

40. John Stuart Mill, *A System of Logic* (1843; reprint, Proquest E-book, 2016), 280.

41. Martha Finnemore, *The Purpose of Intervention: Changing Beliefs about the Use of Force* (Ithaca, N.Y.: Cornell Univ. Press, 2003), 13; John G. Ruggie, *Con-*

structing the World Polity (New York: Routledge, 1998), 94; Charles S. Peirce, *Philosophical Writings of Peirce,* ed. Justus Buchler (New York: Dover, 1955).

42. Anna Dubois and Lars-Erik Gadde, "Systematic Combining: An Abductive Approach to Case Research," *Journal of Business Research* 55, no. 7 (2002): 553–560, 555; Brian D. Haig, "An Abductive Theory of Scientific Method," *Psychological Methods* 10, no. 4 (2005): 371–388, 376–379.

43. "On background" means the interviewee granted me permission to use the information but not to name them directly.

The Past as Prologue

1. George C. Herring, "American Strategy in Vietnam: The Postwar Debate," *Military Review* 46, no. 2 (April 1982): 57–63; Andrew F. Krepinevich Jr., *The Army and Vietnam* (Baltimore, Md.: Johns Hopkins Univ. Press, 1986); Lewis Sorley, *A Better War: The Unexamined Victories and Final Tragedy of America's Last Years in Vietnam* (New York: Harcourt Brace, 1999); Guenter Lewy, *America in Vietnam* (New York: Oxford Univ. Press, 1978); John A. Nagl, *Learning to Eat Soup with a Knife: Counterinsurgency Lessons from Malaya and Vietnam* (Chicago: Univ. of Chicago Press, 2005).

2. See General William Westmoreland, "A Military War of Attrition," in *The Lessons of Vietnam,* ed. W. Scott Thompson and Donaldson D. Frizzell (New York: Crane, Russak, 1976); and William C. Westmoreland, *A Soldier Reports* (Garden City, N.Y.: Doubleday, 1976); John M. Carland, "Winning the Vietnam War: Westmoreland's Approach in Two Documents," *Journal of Military History* 68, no. 2 (2004): 533–574; Dale Andrade, "Westmoreland Was Right: Learning the Wrong Lessons from the Vietnam War," *Small Wars and Insurgencies* 19, no. 2 (2008): 145–181; Andrew J. Birtle, "PROVN, Westmoreland, and the Historians: A Reappraisal," *Journal of Military History* 72, no. 4 (2008): 1213–1247; Mark Moyar, *Triumph Forsaken: The Vietnam War, 1954–1965* (Cambridge: Cambridge Univ. Press, 2006): 335–336.

3. Harry G. Summers Jr., *On Strategy: A Critical Analysis of the Vietnam War* (Novato, Calif.: Presidio, 1982); Jonathan D. Caverley, "The Myth of Military Myopia: Democracy, Small Wars, and Vietnam," *International Security* 34, no. 3 (winter 2009/2010): 119–157; H. R. McMaster, *Dereliction of Duty: Johnson, McNamara, the Joint Chiefs of Staff, and the Lies That Led to Vietnam,* Kindle ed. (New York: HarperCollins, 1997), 1791 of 10792; Stanley Karnow, *Vietnam: A History* (New York: Penguin Press, 1997), 511–525; Herbert Y. Schandler, *America in Vietnam: The War that Couldn't Be Won* (Lanham, Md.: Rowman and Littlefield, 2009), 3.

4. For the evolution of US strategy, see the following documents: Edward C. Keefer, ed., *Foreign Relations of the United States, 1961–1963,* vol. 4, *Vietnam, August–December 1963* (Washington, D.C.: US Government Printing Office, 1991, e-book): document 331, "National Security Action Memorandum [NSAM] 273," November 26, 1963, 637–640; document 374, "Memorandum from the Secretary

of Defense (McNamara) to President Johnson, 'Vietnam Situation,'" December 21, 1963, 732–733; document 380, "Telegram From the Department of State to the Embassy in Vietnam, "Letter from President Johnson to General Minh," December 31, 1963, 745–746; Jack Schulimson, ed., *The Joint Chiefs of Staff and the War in Vietnam, 1960–1968,* part 1 (Washington, D.C.: Office of the Chairman of the Joint Chiefs of Staff, 2011), 199–239; McMaster, *Dereliction of Duty,* 1440–1559 of 10792; Edward C. Keefer and Charles S. Sampson, eds., *Foreign Relations of the United States, 1964–1968,* vol. 1, *Vietnam, 1964* (Washington, D.C.: US Government Printing Office, 1992, e-book): document 84, "Memorandum From the Secretary of Defense (McNamara) to the President," March 16, 1964, 154, 160; document 87, "National Security Action Memorandum [NSAM] No. 288," March 17, 1964, 172–173; document 201, "Paper Prepared for the President by the Secretary of Defense (McNamara)," June 5, 1964, 462–465; document 420, "Memorandum From the Joint Chiefs of Staff to the Secretary of Defense (McNamara), SUBJECT: Courses of Action in Southeast Asia," November 23, 1964, 932–935; document 424, "Memorandum of the Meeting of the Executive Committee," November 24, 1964, 943–945; document 428, "Memorandum of the Meeting of the Executive Committee," November 27, 1964, 958–960; document 433, "Paper Prepared by the Executive Committee, Washington, December 2, 1964, POSITION PAPER ON SOUTHEAST ASIA," December 2, 1964, 969–974; document 435, "Instructions From the President to the Ambassador to Vietnam (Taylor)," December 3, 1964, 974–978; David C. Humphrey, Edward C. Keefer, and Louis J. Smith, eds., *Foreign Relations of the United States, 1964–1968,* vol. 3, *Vietnam, June–December 1965* (Washington, D.C.: US Government Printing Office, 1996, e-book), document 93, "Document Summary Notes of the 553d Meeting of the National Security Council, Washington, July 27, 1965, 5:40 p.m.–6:20 p.m. SUBJECT: Deployment of Additional U.S. Troops to Vietnam," July 27, 1965, 262–263. A negotiated outcome was discussed as early as 1962. See Allan E. Goodman, *The Search for a Negotiated Settlement of the Vietnam War,* Institute of East Asian Studies (Berkeley: Univ. of California Press, 1986), 1–2.

5. McNamara used graduated pressure to convince the Soviet Union to abandon plans to put nuclear ballistic missiles in Cuba while avoiding the risk of a wider war. He overrode highly aggressive military advice that might have escalated the conflict out of control. Cyrus Vance, "Oral History Transcript," section 3, *LBJ* Library, March 9, 1970, 11; McMaster, *Dereliction of Duty,* 643, 1549 of 10792.

6. Keefer and Sampson, eds., *Foreign Relations of the United States, 1964–1968,* vol. 1, *Vietnam,* document 201; document 304, "Telegram from the Department of State to the Embassy in Canada," August 8, 1964, 651–653.

7. *H. K. Johnson Papers,* Notes on Meetings of the Joint Chiefs of Staff, January–April 1964, "Notes on JCS Meeting for 8 January 1964," box 126, General Harold K. Johnson; *The Pentagon Papers,* vol. 3, Joint Chiefs Memorandum 46–64, "Vietnam and Southeast Asia," January 22, 1964, 496–499; Herbert Y. Schandler, "America and Vietnam: The Failure of Strategy, 1964–67," in *Vietnam as History,*

ed. Peter Braestrup (Washington, D.C.: Univ. Press of America, 1984), 23–24; Keefer and Sampson, eds., *Foreign Relations of the United States, 1964–1968,* vol. 1, *Vietnam,* document 66, "Memorandum From the Joint Chiefs of Staff to the Secretary of Defense (McNamara) JCSM-174-64 SUBJECT: Vietnam," March 2, 1964, 116–117; document 70 (March 4, 1964), 129; document 191, "Memorandum From the Chairman of the Joint Chiefs of Staff (Taylor) to the Secretary of Defense (McNamara)," CM-1450-64, SUBJECT: Transmittal of JCSM-471-64, "Objectives and Courses of Action-Southeast Asia," June 2, 1964, 437–441; Schulimson, ed., *The Joint Chiefs of Staff and the War in Vietnam,* chapter 9, 241–264; McMaster, *Dereliction of Duty,* location 1351, 1491, 1914 of 10792. The so-called "Rostow Thesis" hypothesized that graduated military actions reinforced by political and economic pressures could cause a nation to reduce or eliminate support for an insurgency. This notion for North Vietnam was tested in a September 1964 wargame called SIGMA II, which concluded that such a bombing campaign would more likely stiffen North Vietnam's resolve, while eroding public support in the United States. The conclusions, however, had little effect on US strategy; McMaster, *Dereliction of Duty,* 3221–3277 of 10792.

8. Marilyn B. Young, *The Vietnam Wars, 1945–1990* (New York: HarperPerennial, 1991), 60–88; Karnow, *Vietnam,* 240–251; *The Pentagon Papers,* "Memorandum for Secretary of Defense from Brigadier General Edward G. Lansdale, 'Vietnam,'" January 17, 1961, 69, 72–73.

9. William Duiker, *Sacred War: Nationalism and Revolution in a Divided Vietnam* (New York: McGraw-Hill, 1995), 164–165.

10. Keefer and Sampson, eds., *Foreign Relations of the United States, 1964–1968,* vol. 1, *Vietnam,* document 339, "Telegram From the Embassy in Vietnam to the Department of State," Saigon, September 6, 1964, 733–736; and Humphrey et al., eds., *Foreign Relations of the United States, 1964–1968,* vol. 3, *Vietnam,* document 40, "Paper by the Under Secretary of State (Ball), 'A Compromise Solution for South Viet-nam," undated, 108.

11. See *The Pentagon Papers,* Special National Intelligence Estimate, "Evolution of the War. Origins of the Insurgency," part IV. A. 5, August 1960; Maxwell B. Taylor, *Swords into Ploughshares* (New York: Norton, 1972), 301; Keefer and Sampson, eds., *Foreign Relations of the United States, 1964–1968,* document 84, 155–158, 156; document 156, "Summary Record of the 532d Meeting of the National Security Council, Washington, May 15, 1964," 328–329; document 341, "Special National Intelligence Estimate, Washington, September 8, 1964, SNIE 53-64: 'Chances for a Stable Government in South Vietnam," September 8, 1964, 742–746; Robert S. McNamara with James Blight, Robert Brigham, Thomas Biersteker, Herbert Schandler, *Argument without End: In Search of Answers to the Vietnam Tragedy* (New York: Public Affairs, 1999), 369; Neil Sheehan, *A Bright Shining Lie: John Paul Vann and America in Vietnam* (New York: Vintage, 1989), 201–266; James Gibson, *The Perfect War: Technowar in Vietnam* (Boston: Atlantic Monthly Press, 1986), 88; Vincent H. Demma, "The U.S. Army in Vietnam," in *American Military History*

(Washington, D.C.: US Army Center of Military History, 1989), 619–694; Karnow, *Vietnam,* 339.

12. Johnson Library, "Memorandum for the Record, Washington, February 3, 1965, 'SUBJECT Discussion with the President re South Vietnam,'" John McCone memoranda of meetings with the president, dictated by McCone and transcribed in his office, February 3, 1965; Secret; Eyes Only; David C. Humphrey, Ronald D. Landa, and Louis J. Smith, eds., *Foreign Relations of the United States, 1964–1968,* vol. 2, *Vietnam, January–June 1965* (Washington, D.C.: US Government Printing Office, 1996, e-book), document 42, "Memorandum from the President's Special Assistant for National Security Affairs (Bundy) to President Johnson, 'Basic Policy in Vietnam,'" Washington, January 27, 1965, 95–97; document 84, "Memorandum from the President's Special Assistant for National Security Affairs (Bundy) to President Johnson, En route from Saigon to Washington, February 7, 1965, 175–181; Humphrey et al., eds., *Foreign Relations of the United States, 1964–1968,* vol. 3, document 235, "Notes of Meeting," Washington, December 18, 1965, 662.

13. Humphrey et al., eds., *Foreign Relations of the United States, 1964–1968,* vol. 3, document 189, "Draft Memorandum from Secretary of Defense McNamara to President Johnson," Washington, November 3, 1965, 514–528; document 194, "Memorandum for President Johnson, Subject: Courses of Action in Viet-Nam," Washington, November 9, 1965, 535–554; document 238, "Notes of Meeting," Washington, December 21, 1965, 677.

14. Ibid., document 228, "Memorandum From the President's Special Assistant (Califano) to President Johnson," Washington, December 13, 1965, 638; document 237, "Telegram From the Chairman of the Joint Chiefs of Staff (Wheeler) to Secretary of Defense McNamara," Saigon, December 21, 1965, 673.

15. Ibid., document 208, "Memorandum from the President's Special Assistant for National Security Affairs (Bundy) to President Johnson, SUBJECT: 'Once more on the pause,'" Washington, November 27, 1965, 582; document 223, "McGeorge Bundy, Personal Notes of Meeting with President Johnson. LBJ Ranch," Texas, December 7, 1965, 620–621.

16. Ibid., document 231, "Notes of Meeting," Washington, December 17, 1965, 647.

17. Ibid.

18. Ibid., document 238, "Notes of Meeting," Washington, December 21, 1965, 677.

19. For examples, see ibid., document 262, "Memorandum of Telephone Conversation Between the Under Secretary of State (Ball) and President Johnson," December 28, 1965, 732–734; document 265, "Telegram From the Embassy in Burma to the Department of State," Rangoon, December 29, 1965, 736–737; documents 266 and 267, "Telegram from the Embassy in Poland to the Department of State," Warsaw, December 29, 1965, 738; documents 268 and 269, "Telegram from the Department of State to the Embassy in Vietnam," Washington, December 29, 1965, 739–740; document 271, "Telegram from the Embassy in Italy to the Depart-

ment of State," Rome, December 29, 1965, 744–747; document 272, "Telegram from the Embassy in Thailand to the Department of State," Bangkok, December 30, 1965, 748.

20. Ibid., document 247, "Paper by Secretary of State Rusk, 'The Heart of the Matter in Viet-Nam,'" Washington, December 27, 1965, 707.

21. Ibid., document 199, "Telegram from the Embassy in Vietnam to the Department of State," Saigon, November 11, 1965, 463–468; document 148, "Special National Intelligence Estimate, SNIE 10-11-65, 'Probable Communist Reactions to a US Course of Action,'" Washington, September 22, 1965, 403 (note: the Special NIE does suggest that recent US bombing and willingness to escalate shook DRV and VC confidence); document 184, "Intelligence Memorandum, No. 2391/65, 'An Appraisal of the Bombing of North Vietnam,'" Washington, October 27, 1965, 500–504; document 212, "Memorandum From Secretary of Defense McNamara to President Johnson," Washington, November 30, 1965, 592; document 239, "Special Intelligence Supplement," Washington, December 21, 1965, 680–685, "While the air strikes against logistics facilities and sensitive lines of communications are causing major distribution problems, these operations have not significantly reduced the DRV capability to continue to support the Communist forces in Laos and South Vietnam."

22. Goodman, *The Search for a Negotiated Settlement of the Vietnam War,* ix.

23. Ibid., 9, 19, 37–46, 93–112.

24. McMaster, *Dereliction of Duty,* 3063, 3210 3678, 3875, of 10792; Keefer and Sampson, eds., *Foreign Relations of the United States,* vol. 1, document 331, "Memorandum from the Joint Chiefs of Staff to the Secretary of Defense (McNamara), 'SUBJECT: Recommended Courses of Action—Southeast Asia,' TCSM-746-64," Washington, August 27, 1964, 413–417.

25. Paul et al., *Paths to Victory,* 33.

26. *The Pentagon Papers,* part IV. C. 5. "Evolution of the War. Phase I in the Build-up of U.S. Forces: March–July 1965," 116–123, and part IV. C. 6. a. "Evolution of the War. U.S. Ground Strategy and Force Deployments: 1965–1967," vol. 1, Phase II, Program 3, Program 4, a, 1–7; McNamara, *Argument without End,* 353–354; John A. Nagl, "Counterinsurgency in Vietnam," in *Counterinsurgency in Modern Warfare,* ed. Daniel Marston and Carter Malkasian (Oxford, UK: Osprey Publishing, 2008), 131–148.

27. Karnow, *Vietnam,* 399.

28. Ibid., 18; McMaster, *Dereliction of Duty*; General William C. Westmoreland, "Address: National Press Club," Washington, D.C., November 21, 1967; Lewis Sorely, *Thunderbolt: From the Battle of the Bulge to Vietnam and Beyond: General Creighton Abrams and the Army of His Times* (New York: Simon and Schuster, 1992), 192–200.

29. Karnow, *Vietnam,* 488–527; Sorely, *Thunderbolt,* 199–200; Iklé, *Every War Must End,* 59, 84–85.

30. Sheehan, *A Bright Shining Lie,* 684.

31. Karnow, *Vietnam,* 546; Don Oberdorfer, *Tet! The Turning Point in the Vietnam War* (Baltimore, Md.: Johns Hopkins Univ. Press, 2001), 251; McNamara, *Argument without End,* 366–367.

32. John A. Farrell, "Nixon's Vietnam Treachery," *New York Times,* December 31, 2016.

33. Edward C. Keefer and Carolyn Yee, eds., *Foreign Relations of the United States, 1969–1976,* vol. 6, *Vietnam,* January 1969–July 1970 (Washington, D.C.: US Government Printing Office, 2006, e-book), document 8, "Memorandum from the President's Assistant for National Security Affairs (Kissinger) to President Nixon," January 24, 1969, 18–22; document 46, "Memorandum from the President's Assistant for National Security Affairs (Kissinger) to President Nixon," undated, 154–161; document 49, "Minutes of National Security Council Meeting," March 28, 1969, 164–176.

34. Ibid., document 106, "Memorandum from the President's Assistant for National Security Affairs (Kissinger) to President Nixon, 'Subject: Meeting in Paris with North Vietnamese,'" August 6, 1969, 330–343.

35. Ibid., document 117, "Memorandum From the President's Assistant for National Security Affairs (Kissinger) to President Nixon," September 10, 1969, 370–374.

36. Ambassador Edward Brynn, "Preface," in *Foreign Relations of the United States, 1969–1976,* vol. 9, *Vietnam, October 1972–January 1973,* ed. John M. Carland (Washington, D.C.: US Government Printing Office, 2010), iv–vi.

37. "The Paris Agreement on Vietnam: Twenty-five Years Later," Conference Transcript, The Nixon Center, Washington, D.C., April 1998, https://www.mtholyoke.edu/acad/intrel/paris.htm.

38. Peter Church, ed., *A Short History of South-East Asia.* (Singapore. John Wiley and Sons, 2006), 193–194.

1. Further Defining War Termination

1. Martel, *Victory in War,* 342; 370.

2. Thomas C. Schelling, *The Strategy of Conflict* (Cambridge: Harvard Univ. Press, 1980), 4–5.

3. Ibid., 5.

4. Carl von Clausewitz, *On War,* ed. and trans. Michael Howard and Pater Paret (Princeton: Princeton Univ. Press, 1984), 92 (emphasis added), 95, 99.

5. The White House, "President Addresses Nation, Discusses Iraq, War on Terror," June 28, 2005.

6. Clausewitz, *On War,* 584.

7. Freedman, *Strategy,* ix–x.

8. Lee and Walling, *Strategic Logic and Political Rationality,* 2–3. See also Michael I. Handel, *Masters of War: Classical Strategic Thought* (London: Frank Cass,

2001), 19–32; Clausewitz, *On War,* 81; Sun Tzu, *The Art of War,* trans. Samuel B. Griffith II (Oxford: Oxford Univ. Press, 1980), 77–79.

9. Kahneman, *Thinking, Fast and Slow,* 411–412.

10. Clausewitz, *On War,* 89.

11. Ibid., 84, 117–118; Daniel Kahneman and Amos Tversky, "Subjective Probability: A Judgment of Representativeness," *Cognitive Psychology* 3, no. 3 (1972): 430–454; Daniel Kahneman, Paul Slovik, and Amos Tversky, *Judgment Under Uncertainty: Heuristics and Biases* (Cambridge: Cambridge Univ. Press, 1987); Graham T. Allison, "Conceptual Models and the Cuban Missile Crisis," *American Political Science Review* 63, no. 3 (1969): 689–718; Graham T. Allison and Philip Zelikow, *Essence of Decision: Explaining the Cuban Missile Crisis* (New York: Longman, 1999). For a critique of Allison and Zelikow, see Jonathan Bendor and Thomas H. Hammond, "Rethinking Allison's Models," *American Political Science Review* 86, no. 2 (1992): 301–322.

12. Lee and Walling, *Strategic Logic and Political Rationality,* 3.

2. The Decisive Victory Paradigm Undermines Strategy for Irregular War

1. The Department of Defense defines irregular warfare as "A violent struggle among state and nonstate actors for legitimacy and influence over the relevant populations. IW favors indirect and asymmetric approaches, though it may employ the full range of military and other capabilities, in order to erode an adversary's power, influence, and will." *Irregular Warfare, Joint Operating Concept* (Washington, D.C.: US Government Printing Office, 2007), 6.

2. Clausewitz, *On War,* 75, 80–81, 87, 134–139. "The political object is the goal, war is the means of reaching it, and means can never be considered in isolation from their purpose," 87.

3. Mao Tse-Tung, *On Guerilla Warfare,* trans. Samuel B. Griffith II (Champaign: Univ. of Illinois Press, 1961); Robert Taber, *War of the Flea: The Classic Study of Guerilla Warfare* (New York: Brassey's, 2002); Bard E. O'Neill, *Insurgency and Terrorism: From Revolution to Apocalypse,* 2nd ed., rev. (Dulles, Va.: Potomac Books, 2005); David Betz, "The Virtual Dimension of Contemporary Insurgency and Counterinsurgency," *Small Wars and Insurgencies* 19, no. 4 (2008): 510–540; and David Betz, *Carnage and Connectivity: Landmarks in the Decline of Conventional Military Power* (Oxford: Oxford Univ. Press, 2015).

4. David Kilcullen, *Counterinsurgency* (Oxford: Oxford Univ. Press, 2010); David Galula, *Counterinsurgency Warfare: Theory and Practice* (Westport, Conn.: Praeger Press, 2006); Christopher D. Kolenda, *The Counterinsurgency Challenge* (Harrisburg, Pa.: Stackpole Books, 2012); Marston and Malkasian, eds., *Counterinsurgency in Modern Warfare.*

5. It is important to note that this book does not seek to address the extent to which these problems affect democracies more than autocracies. In his analysis of

286 insurgencies from 1800–2005, Lyall argues that "democracy appears to exert almost no causal effect on either war outcomes or duration." Jason Lyall, "Do Democracies Make Inferior Counterinsurgents? Reassessing Democracy's Impact on War Outcomes and Duration," *International Organization* 64, no. 1 (winter 2010): 167–192, 168.

6. Paul et al., *Paths to Victory,* 18. Others using statistical analysis include Libicki, "Eighty-Nine Insurgencies," 373–396. Libicki's 89 insurgencies reach back to 1934 and include ongoing ones. He classifies 28 as government wins, 25 cases as government defeat, 20 as mixed outcomes, and 16 as ongoing. Paul et al. added four cases that appeared to meet Libicki's criteria for inclusion, eliminated 17 that were ongoing or unresolved, cut out insurgencies prior to World War II, as well as four others that Paul et al. considered were not insurgencies. Libicki's original list was drawn from Fearon and Laitin, "Ethnicity, Insurgency, and Civil War." A key difference in the assessment of wins and losses is that Paul et al. assign a winner to the mixed outcome, depending on which side appeared to get the better outcome. For Paul et al.'s assessment criteria, see Paul et al., *Paths to Victory,* 16–20. I will rely primarily on Paul et al.'s study because it focuses on concluded insurgencies where outcomes can be assessed and correlations drawn more precisely. The specificity of Paul et al.'s analytic categories is more useful for the purposes of this study.

7. Paul et al., *Paths to Victory,* xxi–xxvii. Paul et al.'s 71 cases include 12 that he argues are unfit for comparative purposes because the governments in question were "fighting against the tide of history" (end of colonialism, end of apartheid, etc.). The authors use these 59 in determining the 15 good practices and 11 bad practices.

8. Paul et al., *Paths to Victory,* 149.

9. Ibid., xxiii–xxiv.

10. Ibid.

11. Lyall, "Do Democracies Make Inferior Counterinsurgents?," 188–189. The other two are the status of the external power as an occupier and its degree of mechanization (i.e., is the counterinsurgent comfortable operating among the people or more tied to machines). Lyall observes that democracies "do struggle to defeat insurgencies—but not because they are democracies." Libicki's statistics also show significant correlation between outside support and insurgent success. Libicki, "Eighty-Nine Insurgencies," 387–388.

12. Paul et al., *Paths to Victory,* 130–132.

13. Ibid., 129.

14. Ibid., xxiv. In cases where an external force demonstrated resolve but the host nation government and forces failed to do so, the result was a loss. Libicki, "Eighty-Nine Insurgencies," also notes the strong outcome correlations with government popularity and competence (388–391).

15. Paul et al., *Paths to Victory,* 156.

16. Ibid., xxix; Libicki, "Eighty-Nine Insurgencies," 391–392.

17. Paul et al., *Paths to Victory,* xxix; Christopher Paul, Colin P. Clarke, and Beth Grill, *Victory Has a Thousand Fathers: Sources of Success in Counterinsurgency, Case*

Studies (Washington, D.C.: RAND, 2010), 11–24 (Afghanistan 1978–1992), 126–135 (Liberia), 136–146 (Rwanda); Paul et al., *Paths to Victory,* 177–197 (Vietnam). For Iraq, see Michael R. Gordon and Bernard E. Trainor, *The Endgame: The Inside Story of the Struggle for Iraq, from George W. Bush to Barack Obama* (New York: Vintage, 2012).

18. Paul et al., *Victory Has a Thousand Fathers,* 158–167 (Sierra Leone), 108–116 (Uganda), 87–97 (Turkey against PKK); Paul et al., *Paths to Victory,* 54–55, 58. For further reading on the Turkey case, see Aliza Marcus, *Blood and Belief: The PKK and the Kurdish Fight for Independence* (New York: New York Univ. Press, 2007); Andrew Mango, *Turkey and the War on Terror: For Forty Years We Fought Alone* (London: Routledge, 2006). For Uganda, see Paul Nantulya, "Exclusion, Identity and Armed Conflict: A Historical Survey of the Politics of Confrontation in Uganda with Specific Reference to the Independence Era," in *Politics of Identity and Exclusion in Africa: From Violent Confrontation to Peaceful Cooperation,* Conference proceedings, Senate Hall, University of Pretoria, ed. Konrad Adenauer Stiftung (July 25–26, 2001); Thomas Ofcansky, "Musevenis War and the Ugandan Conflict," *Journal of Conflict Studies* 19, no. 1 (spring 1999). For Sierra Leone, see Dena Montague, "The Business of War and the Prospects for Peace in Sierra Leone," *Brown Journal of World Affairs* 9, no. 1 (spring 2002): 229–237; Lansana Gberie, *A Dirty War in West Africa: The RUF and the Destruction of Sierra Leone* (Bloomington: Indiana Univ. Press, 2005); Paul Richards, *Fighting for the Rainforest: War, Youth, and Resources in Sierra Leone* (New York: Oxford Univ. Press, 1996); Funmi Olonisakin, *Peacekeeping in Sierra Leone: The Story of UNAMSIL* (Boulder, Colo.: Lynne Rienner Publishers, 2008); David Richards, *Taking Command* (London: Headline Publishing, 2014).

19. Paul et al., *Victory Has a Thousand Fathers,* 77–86 (Senegal); 57–66 (Peru); Paul et al., *Paths to Victory,* 363–373 (Angola). For Senegal, see Andrew Manley, "Guinea Bissau/Senegal: War, Civil War and the Casamance Question," Writenet/United Nations High Commissioner for Refugees, November 1998; Linda Beck, Robert Charlick, Dominique Gomis, Geneviève Manga, Nana Grey Johnson, and Cheiban Coulibaly, *West Africa: Civil Society Strengthening for Conflict Prevention Study, Conflict Prevention and Peace Building Case Study: The Casamance Conflict and Peace Process (1982–2001)* (Burlington, Vt.: ARD, December 2001); Pierre Englebert, "Compliance and Defiance to National Integration in Barotseland and Casamance," *Afrika Spectrum* 39, no. 1 (2005): 29–59. For Peru, see Philip Mauceri, "State Development and Counter-Insurgency in Peru," in *The Counter-Insurgent State: Guerrilla Warfare and State Building in the Twentieth Century,* ed. Paul B. Rich and Richard Stubbs (London: Macmillan, 1997). For Angola, see Fernando Andresen Guimarães, *The Origins of the Angolan Civil War: Foreign Intervention and Domestic Political Conflict* (Basingstoke, UK: Macmillan, 1998); Assis Malaquias, "UNITA's Insurgency Lifecycle in Angola," in *Violent Non-State Actors in World Politics,* ed. Klejda Mulaj (New York: Columbia Univ. Press, 2010); Alex Vines and Bereni Oruitemeka, "Beyond Bullets and Ballots: The Reintegration of UNITA in

Angola," in *Reintegrating Armed Groups After Conflict: Politics, Violence and Transition,* ed. Mats Berdal and David H. Ucko (London: Routledge, 2009).

20. Paul et al., *Paths to Victory,* 177–178.

21. Ibid., 158–159.

Part II. The Pursuit of Decisive Victory in Afghanistan

1. Bette Dam, *A Man and a Motorcycle: How Hamid Karzai Came to Power* (IpsoFacto Publishers, 2014), 188–191.

2. Thom Shanker, "Rumsfeld Pays Call on Troops and Afghans," *New York Times,* December 17, 2001.

3. Light Footprints to a Long War

1. Peter Tomsen, *The Wars of Afghanistan: Messianic Terrorism, Tribal Conflicts, and the Failures of Great Powers* (New York: Perseus, 2011), 587–618; Thomas Barfield, *Afghanistan: A Cultural and Political History* (Princeton: Princeton Univ. Press, 2010), 283–285; Jon Lee Anderson, "The Surrender: Double Agents, Defectors, Disaffected Taliban, and a Motley Army Battle for Kunduz," *New Yorker,* December 10, 2001.

2. For US concerns about Northern Alliance forces in Kabul, see Bob Woodward, *State of Denial: Bush at War, Part III* (New York: Simon and Schuster, 2006), 230–234, 236–241, 306–311.

3. Vanda Felbab-Brown, "Slip-Sliding on a Yellow Brick Road: Stabilization Efforts in Afghanistan," *Stability: International Journal of Security and Development* 1, no. 1 (2012): 4–19. According to Woodward, *State of Denial,* 52, 114, the CIA briefed that the Taliban and al Qaeda were "joined at the hip." Nonetheless, the administration's initial approach was to allow the Taliban "time to do the right thing" and turn over bin Laden, 121–130; for discussion about inducing a Taliban coup against Mullah Omar, see 128–129.

4. James Dobbins and Carter Malkasian, "Time to Negotiate in Afghanistan: How to Talk to the Taliban," *Foreign Affairs* 94, no. 4 (July–August 2015): 53–64.

5. Seth Jones, *In the Graveyard of Empires: America's War in Afghanistan* (New York: Norton, 2009), 163–182.

6. Sarah Chayes, *Thieves of State: Why Corruption Threatens Global Security* (New York: Norton, 2015), 3–66, 135–155.

7. See, for instance, Ahmed Rashid, *Taliban: Militant Islam, Oil and Fundamentalism in Central Asia* (New Haven, Conn.: Yale Univ. Press, 2000); Khalid Hosseini, *The Kite Runner* (New York: Riverhead, 2003).

8. Davood Moradian, "Ethnic Polarisation: Afghanistan's Emerging Threat," *al Jazeera,* July 26, 2016.

9. Carlotta Gall, *The Wrong Enemy: America in Afghanistan, 2001–2014* (New York: First Mariner, 2014), xxi. See also Husain Haqqani, *Magnificent Delusions:*

Pakistan, the United States, and an Epic History of Misunderstanding (New York: Perseus, 2013).

10. See also Ahmed Rashid, *Descent into Chaos: The U.S. and the Disaster in Pakistan, Afghanistan, and Central Asia* (New York: Viking Press, 2008), and Ahmed Rashid, *Pakistan on the Brink: The Future of America, Pakistan, and Afghanistan* (New York: Penguin Books, 2012).

11. Barnett R. Rubin, "Saving Afghanistan," *Foreign Affairs* 86, no. 1 (January–February 2007): 57–78; and Barnett R. Rubin, *Afghanistan from the Cold War through the War on Terror* (New York: Oxford Univ. Press, 2013); see also Christina Lamb, *Farewell Kabul: From Afghanistan to a More Dangerous World* (London: HarperCollins, 2015); Jack Fairweather, *The Good War: Why We Couldn't Win the War or the Peace in Afghanistan,* Kindle ed. (New York: Basic Books, 2014); Tomsen, *The Wars of Afghanistan*; Rajiv Chandrasekaran, *Little America: The War within the War for Afghanistan* (New York: Vintage, 2012.); Vali Nasr, *The Dispensable Nation: American Foreign Policy in Retreat* (New York: Anchor Books, 2013), 1–94; Zalmay Khalilzad, "Here's What I Think Went Wrong in Afghanistan after I Left There," *Foreign Policy,* March 24, 2016.

12. Sarah Chayes, *The Punishment of Virtue: Inside Afghanistan after the Taliban* (New York: Penguin Press, 2006), and Chayes, *Thieves of State.*

13. Stephen M. Walt, "The REAL Reason the U.S. Failed in Afghanistan," *Foreign Policy,* March 15, 2013; Bolger, *Why We Lost.*

14. Mark Bowden, *Black Hawk Down: A Story of Modern War* (New York: Grove Press, 1999); Afyare Abdi Elmi, *Understanding the Somalia Conflagration: Identity, Political Islam and Peacebuilding* (New York: Pluto Press, 2010).

15. Nina M. Sefarino, "Peacekeeping: Issues of U.S. Military Involvement," Congressional Research Service, August 2000; Jack Spencer, "The Facts About Military Readiness," Heritage.org, September 15, 2000; Frederick H. Fleitz Jr., *Peacekeeping Fiascos of the 1990s: Causes, Solutions, and U.S. Interests* (Westview, Conn.: Praeger, 2002).

16. Michael R. Gordon, "The 2000 Campaign: The Military; Bush Would Stop U.S. Peacekeeping in Balkan Fights," *New York Times,* October 21, 2000.

17. Stephen Robinson, "Nation-Building Ambition Baffles Bush," *The Telegraph,* December 10, 2001.

18. Jones, *In the Graveyard of Empires.*

19. See Richard B. Andres, Craig Wills, and Thomas Griffith Jr., "Winning with Allies: The Strategic Value of the Afghan Model," *International Security* 30, no. 3. (2006): 124–160; Donald H. Rumsfeld, *Known and Unknown: A Memoir* (New York: Sentinel, 2011), 360.

20. Donald H. Rumsfeld, "Transforming the Military," *Foreign Affairs* 81, no. 3 (May–June 2002): 20–32; Paul C. Light, *Rumsfeld's Revolution at Defense,* Policy Brief #142, Brookings Institution, July 2005; Arthur K. Cebrowski and John J. Garstka, "Network-Centric Warfare: Its Origin and Future," *U.S. Naval Institute Proceedings* 124, no. 1 (January 1998): 161–188; Stuart E. Johnson and Martin C.

Libicki, eds., *Dominant Battlespace Knowledge* (Washington, D.C.: National Defense Univ. Press, 1995). For critiques, see Christopher D. Kolenda, "Transforming How We Fight: A Conceptual Approach," *U.S. Naval War College Review* 56, no. 2 (2003): 100–122; Frederick W. Kagan, *Finding the Target: The Transformation of American Military Policy* (New York: Encounter Books, 2006); Donald Kagan and Frederick W. Kagan, *While America Sleeps: Self-Delusion, Military Weakness, and the Threat to Peace Today* (New York: St. Martin's Press, 2000).

21. Stephen A. Biddle, "Afghanistan and the Future of Warfare," *Foreign Affairs* 82, no. 2 (2003): 31–46; Woodward, *State of Denial,* 53.

22. Interviewees M. and J; Rumsfeld, *Known and Unknown,* 372–373; see, for instance, Richard Holbrooke, "Four Things Afghanistan Needs," Reuters, November 15, 2001: "As for the United States, it would not be in anyone's interests for it to supply more than a limited number of logistics and communications support troops. Its presence in fixed positions on the ground in Afghanistan would be just the target the next generation of suicide bombers would most welcome."

23. Lakhdar Brahimi, "State Building in Crisis and Post-Conflict Countries," 7th Global Forum on Reinventing Government: Building Trust in Government, June 26–29, 2007, Vienna, Austria, 4, 16–17. Acclaimed UN envoy for Afghanistan Lakhdar Brahimi discussed the concept of "light footprint" during a UN conference in 2007. He argued that the idea should not fixate on a small international presence. Instead, he proposed that international experts should avoid creating parallel structures and engaging in capacity-substitution efforts that undermine legitimacy and create dependency. "A golden principle for international assistance," he said, "should be that everyone shall do everything possible to work himself or herself out of a job as early as possible." See also the description by Rubin, "Saving Afghanistan," 65; Barbara J. Stapleton and Michael Keating, "Military and Civilian Assistance to Afghanistan 2001–14: An Incoherent Approach," Briefing, *Afghanistan: Opportunity in Crisis Series No.10,* Asia Programme, Chatham House, July 2015.

24. George W. Bush, "Presidential Address to the Nation," October 7, 2001.

25. Sanjiv Miglani, "12 Die in Indian Parliament Attack," *The Guardian,* December 14, 2001; "Timeline of Attacks in India," *Wall Street Journal,* July 13, 2011. Notably, both attacks were actually part of a longstanding conflict between India and Pakistan.

26. Bob Woodward, *Plan of Attack* (New York: Simon and Schuster 2004), 24–27; Joby Warrick, *Black Flags: The Rise of ISIS* (New York: Penguin, 2015), 101–125.

27. Bush, "Presidential Address to the Nation."

28. The White House, "President Delivers State of the Union Address," January 29, 2002.

29. As noted in a November 3 US State Department cable outlining discussion points to be conveyed to other capitals, "The president has made clear that the campaign against terrorism is a sustained campaign that will outlast the immediate efforts in Afghanistan, but our immediate focus is the *al-Qaida* network and its base

in Afghanistan." US Department of State cable, "Afghanistan's Future—Next Steps," November 3, 2001.

4. Plans Hit Reality

1. Cognitive bias refers to systemic errors in thinking people use when interpreting information. Availability heuristics are mental shortcuts that rely on recent examples that people use to understand a subject. For instance, a recent airline crash might cause someone to prefer to drive rather than fly, even though flying is safer. Intuitive decision-making emphasizes the use of emotion or "gut-feel" or other subjective factors over objective analysis. Kahneman, *Thinking, Fast and Slow,* 252; Amos Tversky and Daniel Kahneman, "Judgment Under Uncertainty: Heuristics and Biases," *Science* 185, 4157 (1974): 1124–1131; Daniel Kahneman and Amos Tversky, "Variants of Uncertainty," *Cognition* 11, no. 2 (1982): 143–157; and Daniel Kahneman and Amos Tversky, "Advances in Prospect Theory: Cumulative Representation of Uncertainty," *Journal of Risk and Uncertainty* 5, no. 4 (1992): 297–323.

2. For instance, "Imagine that you face the following pair of concurrent decisions. First examine both decisions, then make your choices. Decision (i) Choose between A: sure gain of $240; B: 25 percent chance to gain $1,000 and 75 percent chance to gain nothing; Decision (ii): Choose between C: sure loss of $750; D: 75 percent chance to lose $1,000 and 25 percent chance to lose nothing." (The best combination is B and C, but most [73 percent] choose A and D). Kahneman, *Thinking, Fast and Slow,* 334–335.

3. In fact, the entire discipline of behavioral economics grew out of the persistence of such human decision-making. See, for instance, Levitt and Dubner, *Freakonomics,* and Steven D. Levitt and Stephen J. Dubner, *Super Freakonomics: Global Cooling, Patriotic Prostitutes and Why Suicide Bombers Should Buy Life Insurance* (New York: HarperCollins, 2011); Arielly, *Predictably Irrational,* and Dan Arielly, *The Upside of Irrationality: The Unexpected Benefits of Defying Logic at Work and Home* (New York: HarperCollins, 2010); Tim Harford, *The Undercover Economist* (Oxford: Oxford Univ. Press, 2006) and Harford, *The Logic of Life.*

4. Kahneman, *Thinking, Fast and Slow,* 253; Branislav L. Slantchev and Ahmer Tarar, "Mutual Optimism as a Rationalist Explanation of War." *American Journal of Political Science* 55, no.1 (2011): 135–148.

5. Kahneman, *Thinking, Fast and Slow,* 250–252. Risk deals with decisions in which the probabilities associated with possible outcomes are known. In uncertainty, such probabilities are not known. Amos Tversky and Craig R. Fox, "Weighing Risk and Uncertainty," *Psychological Review* 102, no. 2 (1995): 269–283, 269.

6. Clausewitz, *On War,* 75–123; Handel, *Masters of War,* xxiii, 26–32; Alan D. Beyerchen, "Clausewitz, Nonlinearity, and the Unpredictability of War," *International Security* 17, no. 3 (1992): 59–90.

7. Gareth Price, "India's Policy towards Afghanistan," Chatham House, August 2013; Tomsen, *The Wars of Afghanistan,* 91–96, 455; Satinder K. Lambah, "The US Needs to Change Its Attitude towards Indo-Afghan Relations," *The Wire,* November 12, 2015; "New Priorities in South Asia U.S. Policy Toward India, Pakistan, and Afghanistan," Council on Foreign Relations, October 2003; Jayshree Bajoria, "India-Afghanistan Relations," Council on Foreign Relations, July 22, 2009.

8. Nasr, *The Dispensable Nation,* 47–48; Barnett Rubin and Ahmed Rashid, "From Great Game to Grand Bargain: Ending Chaos in Afghanistan and Pakistan," *Foreign Affairs* 87, no. 6 (November–December 2008): 30–44, 35–38.

9. Brad L. Brasseur, "Recognizing the Durand Line: A Way Forward for Afghanistan and Pakistan?," EastWest Institute, November 7, 2011.

10. See Abubakar Siddique, "The Durand Line: Afghanistan's Controversial, Colonial-Era Border," *The Atlantic* October 25, 2012; for the Pashtun nationalist viewpoint, see Afghanland.com.

11. Treaty between the British and Afghan Governments, February 6, 1922 (London: His Majesty's Stationary Office, 1922).

12. Tomsen, *The Wars of Afghanistan,* 41–42, 237–240; Joseph V. Micallef, "Afghanistan and Pakistan: The Poisoned Legacy of the Durand Line," *Huffington Post,* November 21, 2015.

13. Price, "India's Policy towards Afghanistan"; Tomsen, *The Wars of Afghanistan,* 91–96, 455.

14. Stephen Philip Cohen, *The Idea of Pakistan* (Washington, D.C.: Brookings Institution, 2004), 8, 55, 62–63, 302.

15. Tomsen, *The Wars of Afghanistan,* 105–106.

16. Steve Coll, *Ghost Wars: The Secret History of the CIA, Afghanistan, and Bin Laden, from the Soviet Invasion to September 10, 2001* (New York: Penguin Books, 2005), 113–114; Tomsen, *The Wars of Afghanistan,* 237–241.

17. George Crile, *Charlie Wilson's War: The Extraordinary Story of How the Wildest Man in Congress and a Rogue CIA Agent Changed the History of Our Times* (New York: Grover Press, 2003); Coll, *Ghost Wars,* 19–186.

18. Rosanne Klass, "Afghanistan: The Accords," *Foreign Affairs* 66, no. 5 (summer 1998): 922–945.

19. Coll, *Ghost Wars,* 187–238.

20. Barnett R. Rubin, *The Fragmentation of Afghanistan* (New Haven, Conn.: Yale Univ. Press, 1995), 247–264; Coll, *Ghost Wars,* 235–239; 262–265; Thomas Barfield, *Afghanistan: A Cultural and Political History* (Princeton: Princeton Univ. Press, 2010), 164–271.

21. Coll, *Ghost Wars,* 289–290; Marvin G. Weinbaum, "Afghanistan and Its Neighbors," US Institute of Peace, June 2006; Nicholas Howenstein and Sumit Ganguly, "India-Pakistan Rivalry in Afghanistan," *Journal of International Affairs* 63, no. 1 (2009): 127–140.

22. Qais Akbar Omar, *A Fort of Nine Towers: An Afghan Family Story* (New York: Farrar, Strauss and Giroux, 2013).

23. Roy Gutman, *How We Missed the Story: Osama bin Laden, the Taliban, and the Hijacking of Afghanistan* (Washington, D.C.: US Institute of Peace, 2007), 89–90. Rob Crilly, "Former Islamist Warlord Who Brought bin Laden to Afghanistan to Run for President," *The Telegraph,* October 3, 2013; "Ustad Abdul Rasul Sayyaf," GlobalSecurity.org, April 5, 2014.

24. Abdul Salaam Zaeef, *My Life with the Taliban* (New York: Columbia Univ. Press, 2010).

25. US Department of State cable, "The Political Future of Afghanistan (Corrected copy)," October 11, 2001.

5. The Fall of the Taliban and the Bonn Conference

1. James Gleick, *Chaos: Making a New Science* (New York: Penguin, 2008), 20–21.

2. Ibid., 23.

3. M. Mitchell Waldrop, *Complexity: The Emerging Science at the Edge of Order and Chaos* (New York: Simon and Schuster, 1992), 11.

4. McChrystal et al., *Team of Teams,* 69

5. Ibid., 56.

6. Ibid., 132.

7. A *loya jirga* is a traditional governance council in Afghanistan, in which elders from across the country representing the various ethnicities, tribes, and communities come together to discuss and decide upon issues of national importance.

8. James Dobbins, *After the Taliban: Nation-Building in Afghanistan* (Washington, D.C.: Potomac Books, 2008), 51, 54, 85, 87–88; "Kabul Falls to Northern Alliance," BBC News, November 13, 2001.

9. John Kifner with Eric Schmitt, "U.S. Officials Say Al Qaeda Is Routed from Afghanistan," *New York Times,* December 17, 2001.

10. Woodward, *Plan of Attack,* notes that the administration was concerned about a security vacuum after the Taliban were overthrown, citing the parallels in the early 1990s that led to the rise of the Taliban, 192–193, 214–219, 321. For a critical view of overreliance on the Northern Alliance allies (but an overly optimistic view of a particular Pashtun alternative), see Lucy Morgan Edwards, *The Afghan Solution: The Inside Story of Abdul Haq, the CIA and How Western Hubris Lost Afghanistan* (London: Bactria Press, 2011).

11. Terrence K. Kelly, Nora Bensahel, and Olga Oliker, *Security Force Assistance in Afghanistan* (Washington, D.C.: RAND, 2011); Paul O'Brien and Paul Baker, "Old Questions Needing New Answers: A Fresh Look at Security Needs in Afghanistan," paper presented at the Bonn International Center for Conversion (BICC) e-conference Afghanistan: Assessing the Progress of Security Sector Reform, One Year after the Geneva Conference, June 4–11, 2003; "Yunus Qanooni / Muhammad Yunus Qanuni," Global Security.org.

12. The so-called "Rome Group" that represented the deposed Afghan king Zahir Shah was also in attendance but had limited influence.

13. US Embassy Rome cable, "Afghanistan: Zahir Shah's Future Plans, Closing out ESF Grant to Zahir Shah Foundation," April 3, 2002.

14. Thomas Ruttig, "Flash to the Past: Power Play before the 2002 Emergency Loya Jirga," Afghan Analysts Network, April 27, 2012, 3.

15. Dam, *A Man and a Motorcycle*; Robert L. Grenier, *88 Days to Kandahar: A CIA Diary* (New York: Simon and Schuster, 2015). For background on former Popalzai king Shah Shuja, see William Dalrymple, *Return of a King: The Battle for Afghanistan, 1839–42* (New York: Vintage Books, 2013).

16. Coll, *Ghost Wars,* 286–287.

17. "Hizb-i-Islami Gulbuddin (HiG)," Institute for the Study of War, https://understandingwar.org/hizb-i-Islami-gulbuddin-hig.

18. "US Designates Hekmatyar as a Terrorist," *Dawn,* February 20, 2003; "Profile: Gulbuddin Hekmatyar," BBC News, March 23, 2010; "Hizb-i-Islami," GlobalSecurity.org.

19. "Hekmatyar's Party Joins Peace Process," Reuters, May 4, 2004; Thomas Ruttig, "The Battle for Afghanistan: Negotiations with the Taliban: History and Prospects for the Future," New America Foundation, May 2011.

20. Interviewees J and Z.

21. Ahmed Rashid, "The Mess in Afghanistan," *New York Review of Books,* February 12, 2004.

22. Ruttig, "Flash to the Past."

23. See *The Rumsfeld Papers,* Donald Rumsfeld Email to Doug Feith, "Strategy," October 30, 2001, https://papers.rumsfeld.com; Colin Powell letter to Donald Rumsfeld, April 16, 2002; Donald Rumsfeld letter to Colin Powell, April 8, 2002; James Dao, "Bush Sets Role for U.S. in Afghan Rebuilding," *New York Times,* April 17, 2002.

24. Anders Fänge, "The Emergency Loya Jirga," in *Snapshots of an Intervention: The Unlearned Lessons of Afghanistan's Decade of Assistance (2001–2011),* ed. Martine van Bijlert and Sari Kouvo (Kabul, Afghanistan: Afghan Analysts Network, 2012), 2–4.

25. Ibid.; "Loya Jirga Dispute Prompts Mass Walk-Out," *The Guardian,* June 17, 2002; Philip Smucker, "Afghans Put Off Key Decisions," *Christian Science Monitor,* June 18, 2002.

26. "The September 2005 Parliamentary and Provincial Council Elections in Afghanistan," National Democratic Institute, 2006, 2–7.

27. Thomas Ruttig, "The Failure of Airborne Democracy: The Bonn Agreement and Afghanistan's Stagnating Democratisation," in *Snapshots of an Intervention: The Unlearned Lessons of Afghanistan's Decade of Assistance (2001–2011),* ed. Martine van Bijlert and Sari Kouvo (Kabul, Afghanistan: Afghan Analysts Network, 2012).

28. Ibid. The SNTV system can be used to fill multiple seats in a single electoral district and can help to ensure better minority representation. It also rewards better

organized parties, who can earn multiple seats in a single district, while splitting votes of non-party-affiliated candidates.

29. Rubin, "Saving Afghanistan," 66.

30. US Embassy Rome cable, "GOI Thinks Some Kind of Resolution on Afghanistan Is Inevitable," March 28, 2003.

31. Interviewee R; these views were corroborated by interviewees J and Z.

32. Theo Farrell, *Unwinnable: Britain's War in Afghanistan, 2001–2014* (London: The Bodley Head, 2017), chapter 3; Alan Sipress and Peter Finn, "US Says 'Not Yet' to Patrol by Allies," *Washington Post,* November 30, 2001.

33. Sipress and Finn, "US Says 'Not Yet' to Patrol by Allies."

34. Farrell, *Unwinnable,* chapter 3; Sean Naylor, *Not a Good Day to Die: The Untold Story of Operation Anaconda* (New York: Penguin, 2005).

35. US Embassy New Delhi cable, "Afghan FM Abdullah Discusses Taliban Violence, US-Afghan Partnership with GOI," July 5, 2005; US Embassy New Delhi cable, "GOI Wants to Help US in Afghanistan," July 14, 2005.

36. Rubin and Rashid, "From Great Game to Grand Bargain," 37.

37. Tomsen, *The Wars of Afghanistan,* 588–595; Barfield, *Afghanistan,* 326–330.

38. For Afghanistan claims on Pakistani territory, see Nasr, *The Dispensable Nation,* 44.

39. C. Christine Fair, *Fighting to the End: The Pakistan Army's Way of War* (Oxford: Oxford Univ. Press, 2014), 103–135; Stephen Tankel, "Beyond the Double Game: Lessons from Pakistan's Approach to Islamist Militancy," *Journal of Strategic Studies* 41, no. 4 (2018): 545–575.

40. Nasr, *The Dispensable Nation,* 50–52.

41. US Embassy Kabul cable, "Northern Views on Afghanistan's Future," May 15, 2004; US Embassy Rome cable, "Ex-King Determined to Go Home, But Concerned About Arrangements," February 27, 2002; Michael Kugelman, "The Iran Factor in Afghanistan," *Foreign Policy,* July 10, 2014; Mohsen Milani, "Iran and Afghanistan," US Institute of Peace, October 5, 2010.

6. America's Bureaucratic Way of War

1. Russell F. Weigley, *The American Way of War: A History of United States Military Strategy and Policy* (Bloomington: Indiana Univ. Press, 1973).

2. Max Boot, "The New American Way of War," *Foreign Affairs* 82, no. 4 (July/August 2003): 41–58.

3. Paul et al., *Paths to Victory*; Libicki, "Eighty-Nine Insurgencies"; Fearon and Laitin, "Ethnicity, Insurgency, and Civil War."

4. Lee and Walling, *Strategic Logic and Political Rationality,* 13.

5. Such integration seems particularly difficult for modern states with professional national security bureaucracies. Ivan Arreguín-Toft, a scholar who studies the challenges of irregular warfare, argues that stronger powers have been losing to

weaker ones increasingly often since the 19th century. Arreguín-Toft, *How the Weak Win Wars,* 3–5.

6. US Department of State cable, "Afghanistan's Future."

7. The White House, "Joint Statement on New Partnership Between U.S. and Afghanistan," January 28, 2002, "We [President Bush and Chairman Karzai] agree that the United States will work with Afghanistan's friends in the international community to help Afghanistan stand up and train a national military and police, as well as address Afghanistan's short-term security needs, including through demining assistance."

8. George W. Bush, *Decision Points* (New York: Random House, 2010), 205. According to Woodward, *State of Denial,* 160, 192–193, 220, Bush was adamant from the beginning about humanitarian assistance to Afghanistan, but cabinet officials were less enthusiastic.

9. Condoleezza Rice, *No Higher Honor: A Memoir of My Years in Washington* (New York: Random House, 2011), 96, 148; Rumsfeld, *Known and Unknown,* 398; Dobbins, *After the Taliban,* 137; Woodward, *State of Denial,* 192–195, 220, 237; Sten Rynning, "ISAF and NATO: Campaign Innovation and Organizational Adaptation," in *Military Adaptation in Afghanistan,* ed. Theo Farrell, Frans Osinga, and James A. Russell (Stanford: Stanford Univ. Press, 2013), 83–107; David Rohde and David E. Sanger, "How a 'Good War' in Afghanistan Went Bad," *New York Times,* August 12, 2007; Dao, "Bush Sets Role for U.S. in Afghan Rebuilding"; Jones, *In the Graveyard of Empires,* 116–117.

10. *The Rumsfeld Papers,* "Rumsfeld Memo to Powell, Rice, 'U.S. Financial Commitment,'" April 8, 2002.

11. *The Rumsfeld Papers,* "Powell Letter to Rumsfeld," April 16, 2002.

12. The White House, "President Meets with Afghan Interim Authority Chairman," January 28, 2002; "President Bush Speaks at VMI, Addresses Middle East Conflict," CNN, April 17, 2002; David Barno, "Fighting 'The Other War': Counterinsurgency Strategy in Afghanistan, 2003–2005," *Military Review,* September–October 2007; Farrell, *Unwinnable,* chapter 3.

13. Barnett R. Rubin, *Afghanistan from the Cold War through the War on Terror* (New York: Oxford Univ. Press, 2013), 311–312; Alison Laporte-Shapiro, "From Militants to Policemen: Three Lessons from U.S. Experience with DDR and SSR," US Institute of Peace, November 17, 2011, 3; Robert M. Perito, "Afghanistan's Police: The Weak Link in Security Sector Reform," Special Report 227 (Washington, D.C.: US Institute of Peace, August 2009), 2.

14. United Nations Security Council Resolution 1444, November 27, 2002; Lieutenant General James M. Dubik, "Afghan National Police (ANP)," Institute for the Study of War, 2009.

15. *The Rumsfeld Papers,* "Donald Rumsfeld Note to General Franks, 'Afghan National Army,'" January 28, 2002.

16. Antonio Giustozzi, "Military Reform in Afghanistan," paper presented at the Bonn International Center for Conversion (BICC) e-conference Afghanistan:

Assessing the Progress of Security Sector Reform, One Year after the Geneva Conference, June 4–11, 2003.

17. Kelly et al., *Security Force Assistance in Afghanistan,* 5; Jason Howk, "A Case Study in Security Sector Reform: Learning from Security Sector Reform/Building in Afghanistan (October 2002–September 2003)," Strategic Studies Institute, November 2009; see also US Embassy Abu Dhabi cable, "CENTCOM Commander Discusses Iraq, Afghanistan, Regional Threats," July 28, 2004.

18. Antonio Giustozzi, *The Army of Afghanistan: A Political History of a Fragile Institution* (London: Hurst, 2016.), 125–132; O'Brien and Baker, "Old Questions Needing New Answers."

19. US Embassy Ottawa cable, "ISAF: Canadian Defense Minister's Visit to Afghanistan," July 7, 2003; James Risen and Mark Landler, "Accused of Drug Ties, Afghan Official Worries U.S.," *New York Times,* August 26, 2009.

20. The Afghan National Police Working Group, "The Police Challenge: Advancing Afghan National Police Training," Project 2049 Institute, June 2011, 6–7.

21. *The Rumsfeld Papers,* "Donald Rumsfeld Letter to Doug Feith and General Myers, 'Afghanistan,'" May 2, 2003; Rumsfeld, *Known and Unknown,* 685.

22. Perito, "Afghanistan's Police."

23. The Afghan National Police Working Group, "The Police Challenge," 6–8; Perito, "Afghanistan's Police"; "Reforming the Afghan National Police," The Royal United Service Institute and the Foreign Policy Research Institute, 2009, 7–13; Chayes, *Thieves of State,* 20–38.

24. Vanda Felbab-Brown, "No Easy Exit: Drugs and Counternarcotics Policies in Afghanistan," Brookings Institution, 2016, 1, 6–10; William Byrd and David Mansfield, "Afghanistan's Opium Economy: An Agricultural, Livelihoods, and Governance Perspective," World Bank, June 23, 2014; David Mansfield and Adam Pain, "Counter-Narcotics in Afghanistan: The Failure of Success?" Afghan Research and Evaluation Unit, December 2008.

25. Danny Singh, "Explaining Varieties of Corruption in the Afghan Justice Sector," *Journal of Intervention and Statebuilding* 9, no. 2 (2015): 231–255; "Reforming Afghanistan's Broken Judiciary," International Crisis Group, November 17, 2010.

26. See, for instance, US Embassy Kabul cable, "Northern Views on Afghanistan's Future"; DeeDee Derksen, "The Politics of Disarmament and Rearmament in Afghanistan," US Institute of Peace, May 2015; Ruttig, "The Failure of Airborne Democracy."

27. Mark Sedra, "Police Reform in Afghanistan: An Overview," paper presented at the Bonn International Center for Conversion (BICC) e-conference Afghanistan: Assessing the Progress of Security Sector Reform, One Year After the Geneva Conference 2003, June 4–11, 2003, 34; Antonio Giustozzi, "'Good' State Vs. 'Bad' Warlords? A Critique of State-Building Strategies in Afghanistan," Crisis States Research Centre, London School of Economics, October 2004, 12–15.

28. Anand Gopal, *No Good Men Among the Living: America, the Taliban, and the War through Afghan Eyes* (New York: Metropolitan Books, 2014).

29. Christopher D. Kolenda, Rachel Reid, and Christopher Rogers, "The Strategic Costs of Civilian Harm: Applying Lessons from Afghanistan to Current and Future Conflicts," Open Society Foundations, June 2016, 17–28.

30. Zalmay Khalilzad, *The Envoy: From Kabul to the White House, My Journey Through a Turbulent World,* Kindle ed. (New York: St. Martin's Press, 2016), 2618 of 7203. See also *The Rumsfeld Papers,* "Donald Rumsfeld Letter to Douglas Feith and Condoleezza Rice," April 1, 2002, and "Principles for Afghanistan: Policy Guidelines," July 7, 2003, 2.

31. *The Rumsfeld Papers,* "Rumsfeld to Abizaid, 'Karzai's Strategy on Warlordism,'" September 15, 2003.

32. US Embassy Kabul cable, "Congressman Rohrabacher's April 16 Meeting with President Karzai," April 23, 2003.

33. "Killing You Is a Very Easy Thing for Us: Human Rights Abuses in Southeast Afghanistan," Human Rights Watch, July 2003; DeeDee Derksen, "Non-State Security Providers and Political Formation in Afghanistan," Centre for Security Governance, March 2016.

34. Dobbins and Malkasian, "Time to Negotiate in Afghanistan"; Kolenda et al., "The Strategic Costs of Civilian Harm."

35. Antonio Giustozzi, *The Taliban at War: 2001–2018* (London: Hurst, 2019), 238.

36. United Nations News Centre, Press Briefing by Manoel de Almeida e Silva, Spokesman for the Special Representative of the Secretary-General on Afghanistan, July 31, 2003. See also Elizabeth Olson, "UN Official Calls for Larger International Force in Afghanistan," *New York Times,* March 28, 2002; International Crisis Group, "Securing Afghanistan: The Need for More International Action" (Kabul and Brussels: International Crisis Group, 2002); Alan Sipress, "Peacekeepers Won't Go Beyond Kabul, Cheney Says," *Washington Post,* March 20, 2002.

37. Rubin and Rashid, "From Great Game to Grand Bargain," 37–38.

38. Dobbins and Malkasian, "Time to Negotiate in Afghanistan"; Brian Knowlton, "Rumsfeld Rejects Plan to Allow Mullah Omar 'To Live in Dignity': Taliban Fighters Agree to Surrender Kandahar," *New York Times,* December 7, 2001.

39. Rubin, "Saving Afghanistan," 58.

40. Pamela Constable, "U.S. Hopes to Attract Moderates in Taliban," *Washington Post,* October 17, 2001; Dexter Filkins, "Rebel Leader Rejects Role for Taliban in New Regime," *New York Times,* October 17, 2001.

41. Knowlton, "Rumsfeld Rejects Plan to Allow Mullah Omar 'To Live in Dignity'"; Barnett R. Rubin, "An Assassination That Could Bring War or Peace," *New Yorker,* June 4, 2016.

42. Thom Shanker, "Rumsfeld Pays Call on Troops and Afghans," *New York Times,* December 17, 2001.

43. Brahimi, "State Building in Crisis and Post-Conflict Countries," 13.

44. The Constitution of the Islamic Republic of Afghanistan, January 26, 2004; interviewee J. The "jihad" is associated with the Soviet war, the "resistance" is code for the fight against the Taliban.

Conclusion to Part II

1. Interview with Lieutenant General Terry A. Wolff, Washington, D.C., May 2, 2016.

2. See Susan David, *Emotional Agility: Get Unstuck, Embrace Change, and Thrive in Work and Life* (New York: Penguin, 2016); Zachary Shore, *A Sense of the Enemy: The High Stakes History of Reading Your Rival's Mind* (Oxford: Oxford Univ. Press, 2014).

3. Sun Tzu, *Art of War,* 84, 145–146.

4. H. R. McMaster, "How China Sees the World: And How We Should See China," *The Atlantic* (May 2020), https://www.theatlantic.com/magazine/archive/2020/05/mcmaster-china-strategy/609088/.

5. Interviewee Y.

6. Liz Sly, "Rumsfeld, Karzai Declare Taliban No Longer a Threat," *Chicago Tribune,* February 27, 2004.

Part III. Persisting in a Failing Approach

1. Jones, *In the Graveyard of Empires,* 163–182.

2. Kolenda et al., "The Strategic Costs of Civilian Harm," 17–28.

3. Chayes, *Thieves of State,* 20–38.

4. For example, US Department of State, "President Bush Discusses Progress in Afghanistan, Global War on Terror," February 15, 2007; NATO, "Progress in Afghanistan," Bucharest Summit, April 2–4, 2008. Secretary of Defense Gates told the House Armed Services Committee, "Our progress in Afghanistan is real but fragile," Robert M. Gates, "Statement to the House Armed Services Committee," December 11, 2007.

7. Accelerating Success, 2003–2007

1. Stanley and Sawyer, "The Equifinality of War Termination," 657.

2. Nate Silver, *The Signal and the Noise: Why So Many Predictions Fail—But Some Don't* (New York: Penguin, 2012), 12–13; McChrystal et al., *Team of Teams,* 233.

3. Kahneman, *Thinking, Fast and Slow,* 81.

4. Christopher Graves, "Why Debunking Myths About Vaccines Hasn't Convinced Dubious Parents," *Harvard Business Review,* February 20, 2015.

5. Iklé, *Every War Must End,* 17–37; Stanley, "Ending the Korean War," 53–55.

6. For more on organizational silos, see Mikkel Vedby Rasmussen, *The Military's Business: Designing Military Power for the Future* (Cambridge: Cambridge Univ. Press, 2015); "Improving Performance by Breaking Down Organizational Silos: Understanding Organizational Barriers," Select Strategy, 2002; Gillian Tett, *The Silo Effect: The Peril of Expertise and the Promise of Breaking Down Barriers* (New York: Simon and Schuster, 2015), 25–138; McChrystal et al., *Team of Teams,* 20, 118.

7. Farrell, *Unwinnable,* chapter 4; Khalilzad, *The Envoy,* 78–180; 183–186; Jack Fairweather, *The Good War: Why We Couldn't Win the War or the Peace in Afghanistan,* Kindle ed. (New York: Basic Books, 2014), 122 of 9507; Jones, *In the Graveyard of Empires,* 139–142.

8. Rohde and Sanger, "How a 'Good War' in Afghanistan Went Bad."

9. US National Security Council memorandum, "Accelerating Success in Afghanistan in 2004: An Assessment," January 18, 2005 (emphasis added).

10. For the UN Security Risk Maps, see US Government Accountability Office, "Afghanistan Reconstruction: Despite Some Progress, Deteriorating Security and Other Obstacles Continue to Threaten Achievement of U.S. Goals," *GAO-05-742,* July 2005, 56.

11. Rohde and Sanger, "How a 'Good War' in Afghanistan Went Bad."

12. Eric Schmitt and David S. Cloud, "U.S. May Start Pulling Out of Afghanistan Next Spring," *New York Times,* September 14, 2005.

13. See also Khalilzad, *The Envoy,* 2618 of 7203.

14. Vanda Felbab-Brown, "Afghan National Security Forces: Afghan Corruption and the Development of an Effective Fighting Force," Testimony to House Armed Services Committee, August 2, 2012; Danny Singh, "Corruption and Clientelism in the Lower Levels of the Afghan Police," *Conflict, Security and Development* 14, no. 5 (2014): 621–650.

15. Interview with former SRAP senior advisor Barnett R. Rubin, September 12, 2016.

16. Carlotta Gall, "Dispute Prompts Afghan Leader to Delay Trip," *New York Times,* July 26, 2004; and Carlotta Gall, "Afghan Leader, in a Surprise, Picks a New Running Mate," *New York Times,* July 27, 2004; Bruce Pannier, "Afghanistan: Unexpected Candidate Emerges in Presidential Race," Radio Free Europe/Radio Liberty, July 26, 2004; Pamela Constable, "Karzai's Talks Raise Fears about Elections," *Washington Post,* May 29, 2004.

17. US Embassy Brussels cable, "Coordinator for Afghanistan Quinn Meetings with European Commission, Council," January 21, 2005; Kenneth Katzman, "Afghanistan: Post-War Governance, Security, and U.S. Policy," Congressional Research Service, December 28, 2004, 14–16.

18. Carlotta Gall, "Protests Against U.S. Spread Across Afghanistan," *New York Times,* May 13, 2005; N. C. Aizenman and Robin Wright, "Afghan Protests Spread," *Washington Post,* May 14, 2005; "Afghanistan: Violence Surges, Karzai Needs More Support from U.S.," Human Rights Watch, May 24, 2005; Sonali Kolhatkar and

James Ingalls, *Bleeding Afghanistan: Washington, Warlords, and the Propaganda of Silence* (New York: Seven Stories, 2006), 117–168; Rohde and Sanger, "How a 'Good War' in Afghanistan Went Bad";"Lessons in Terror Attacks on Education in Afghanistan," Human Rights Watch, July 2006, 13–17; Duncan Campbell, "Afghan Warlords 'Bigger Threat than Taliban,'" *The Guardian,* July 13, 2004; "Conflict Analysis Afghanistan Update," Cooperation for Peace and Unity, May 2005.

19. Andrew Wilder, "A House Divided? Analysing the 2005 Afghan Elections," Afghan Research and Evaluation Unit, December 2005; Erich Marquardt, "Insurgents, Warlords and Opium Roil Afghanistan," EurasiaNet.org, November 17, 2005; Tom Lansford, *9/11 and the Wars in Afghanistan and Iraq: A Chronology and Reference Guide* (Santa Barbara, Calif.: ABC-CLIO, 2012), 146; "The September 2005 Parliamentary and Provincial Council Elections in Afghanistan," National Democratic Institute, 2006.

20. Interview with Barnett R. Rubin.

21. Chayes, *Thieves of State,* 3–66, 135–155.

22. *Afghanistan in 2018: A Survey of the Afghan People,* The Asia Foundation, 2018, 117–118.

23. See Matt Waldman, "Falling Short: Aid Effectiveness in Afghanistan," ACBAR Advocacy Series, Oxfam, March 2008, 11.

24. Discussion with Afghan elders, August 2009, Kabul.

25. Carlotta Gall, "Anti-U.S. Rioting Erupts in Kabul; at Least 14 Dead," *New York Times,* May 30, 2006.

26. "100 Dead in Afghanistan Air Strike," *The Guardian,* May 22, 2006.

27. US Embassy Dushanbe cable, "President Karzai Worries about Russian Intentions in Central Asia and Afghanistan," July 28, 2006.

28. Griff Witte and Ellen Nakashima, "Cartoon Protests Stoke Anti-American Mood Three Killed Outside U.S. Base in Bagram," *Washington Post,* February 7, 2006; Mirwair Harooni, "Muslim Protesters Rage at United States in Asia, Middle East," Reuters, September 17, 2006.

29. US Embassy Kabul cables, "PRT/TARIN KOWT—Taliban Remain Potent Threat in Southern Afghanistan," March 14, 2006; "Karzai Nervous on Provincial Security: Resurrects Idea of Auxiliary Police, Swipes at Pakistan," April 17, 2006; "Karzai: A Lame Duck President?" July 17, 2006; "Karzai Comments on Counter Narcotics Policy," September 6, 2006.

30. US Embassy Kabul cable, "PAG Makes First Recommendations to President Karzai," August 21, 2006.

31. US Embassy Kabul cable, "Karzai Dissatisfied; Worries About Newsweek; Plans More War Against Narcotics," January 10, 2006. See also Fairweather, *The Good War,* 140–162 of 9507.

32. Carlotta Gall, "Taliban Surges as U.S. Shifts Some Tasks to NATO," *New York Times,* June 11, 2006.

33. *The Rumsfeld Papers,* "Donald Rumsfeld to General Peter Pace, 'General McCaffrey's Report on Afghanistan,'" June 15, 2006.

34. US Embassy Kabul cable, "PAG Makes First Recommendations to President Karzai."

35. US Embassy New Delhi cable, "Afghan FM Abdullah Discusses Taliban Violence"; US Embassy New Delhi cable,"GOI Wants to Help US in Afghanistan," July 14, 2005; Rohde and Sanger, "How a 'Good War' in Afghanistan Went Bad."

36. Rubin and Rashid, "From Great Game to Grand Bargain," 35–38; "Afghanistan's Forgotten War," editorial, *New York Times,* August 5, 2005.

37. US Embassy Kabul cables, "Karzai Nervous on Provincial Security"; "President Karzai's Visit to India," May 2, 2006; "Codel Hayes Meets Karzai, Wardak," June 15, 2006; "Scenesetter for President Karzai's Upcoming Visit to Washington," September 19, 2006.

38. US Embassy New Delhi cable, "India Strengthens Resolve in Afghanistan Despite Taliban Beheading of Engineer," May 1, 2006.

39. Lieutenant General Michael D. Maples, US Army, Director, Defense Intelligence Agency, "The Current Situation in Iraq and Afghanistan," Defense Intelligence Agency Statement for the Record, Senate Armed Services Committee, November 15, 2006, 7.

40. The United States continued wanting to limit the Afghan Army to a "more sustainable" 50,000.

41. The London Conference on Afghanistan, "Building on Success: The London Compact," January 31–February 1, 2006, 6–8. In the event, barely any of the milestones would be met by the end of 2010 or even the end of 2014.

42. US Embassy Kabul cable, "PAG Makes First Recommendations to President Karzai."

43. Ronald Neumann and Karl Eikenberry, "Strategic Directive for Afghanistan," US Embassy—Kabul, September 11, 2006, 5–7.

44. Ibid., 52, 55. See also "5-Year Plan for the Afghan National Security Forces (ANSF)," Office of Security Cooperation—Afghanistan, September 3, 2005, 3–15, 18–21.

45. Barry R. McCaffrey, "Academic Report: Trip to Afghanistan and Pakistan," US Military Academy Department of Social Sciences, June 3, 2006.

46. James Dobbins, John G. McGinn, Keith Crane, Seth G. Jones, Rollie Lal, Andrew Rathmell, Rachel M. Swanger, and Anga R. Timilsina, *America's Role in Nation-Building: From Germany to Iraq* (Santa Monica, Calif.: RAND Corporation, 2003), 25–26, Figure S-2; US Department of Defense, *Joint Publication 3-24: Counterinsurgency,* 2013, 1–13, para. 1-67.

47. Combined Security Transition Command—Afghanistan (CSTC-A), "Afghan National Security Forces (ANSF) End Strength Analysis," briefing slides, February 21, 2007, slides 4–8.

48. *The Rumsfeld Papers,* "General McCaffrey's Report on Afghanistan."

49. Vance Serchuck, "Don't Undercut the Afghan Army," *Washington Post,* June 2, 2006.

8. Failing to Keep Pace with the Insurgency, 2007–2009

1. The US Department of Defense defines insurgency as "the organized use of subversion and violence to seize, nullify, or challenge political control of a region." US Department of Defense, *Joint Publication 3-24: Counterinsurgency,* 2013, GL-5.

2. Conversation with political scientist Stephen Biddle, July 2009, Kabul, Afghanistan; Daniel Byman, *Going to War with the Allies You Have: Allies, Counterinsurgency, and the War on Terrorism* (Carlisle, Pa.: Strategic Studies Institute, 2006), 3.

3. Thomas A. Grant, "Government, Politics, and Low-Intensity Conflict," in *Low-Intensity Conflict: Old Threats in a New World,* ed. Edwin G. Corr and Stephen Sloan (Boulder, Colo.: Westview, 1992), 261.

4. Ladwig, "Influencing Clients in Counterinsurgency," 103.

5. Stephen Biddle, Julia Macdonald, and Ryan Baker, "Small Footprint, Small Payoff: The Military Effectiveness of Security Force Cooperation," typescript, George Washington University, 2016, 9.

6. Daniel Byman, "Friends Like These: Counterinsurgency and the War on Terrorism," *International Security* 31, no. 2 (fall 2006): 79–115, 82.

7. George W. Downs and David M. Rocke, "Conflict, Agency, and Gambling for Resurrection: The Principal-Agent Problem Goes to War," *American Journal of Political Science* 38, no. 2 (1994): 362–380; Peter D. Feaver, *Armed Servants: Agency, Oversight, and Civil-Military Relations,* Kindle ed. (Cambridge: Harvard Univ. Press, 2003); Idean Salehyan, "The Delegation of War to Rebel Organizations," *Journal of Conflict Resolution* 54, no. 3 (2010): 493–515.

8. Ladwig, "Influencing Clients in Counterinsurgency," 105.

9. Bruce Bueno de Mesquita and Alastair Smith, "Foreign Aid and Policy Concessions," *Journal of Conflict Resolution* 51, no. 2 (2007): 251–284, 254.

10. Ladwig, "Influencing Clients in Counterinsurgency," 99.

11. Biddle et al., "Small Footprint, Small Payoff," 9.

12. Stanley and Sawyer, "The Equifinality of War Termination," 657. See also Bruce Bueno de Mesquita and Randolph M. Siverson, "War and the Survival of Political Leaders: A Comparative Study of Regime Types and Political Accountability," *American Political Science Review* 89, no. 4 (1995): 841–855; Giacomo Chiozza and H. E. Goemans, "International Conflict and the Tenure of Leaders: Is War Still Ex Post Efficient?" *American Journal of Political Science* 48, no. 2 (2004): 604–619; Goemans, "Fighting for Survival."

13. Samuel Huntington, *The Soldier and the State: The Theory and Politics of Civil-Military Relations* (Harvard: Balknap Press, 1957), 80–83.

14. Biddle et al., "Small Footprint, Small Payoff," 10. See also Michael Desch, *Civilian Control of the Military: The Changing Security Environment* (Baltimore: Johns Hopkins Univ. Press, 2001).

15. Ladwig, "Influencing Clients in Counterinsurgency," 103.

16. William E. Odom, *On Internal War: American and Soviet Approaches to Third World Clients and Insurgents* (Durham, N.C.: Duke Univ. Press, 1992), 9, 103–104, 213–215; Ladwig, "Influencing Clients in Counterinsurgency," 104; Biddle et al., "Small Footprint, Small Payoff," 6–7.

17. Biddle et al., "Small Footprint, Small Payoff," 12; Ladwig, "Influencing Clients in Counterinsurgency," 104.

18. Biddle et al., "Small Footprint, Small Payoff," 12.

19. Ladwig, "Influencing Clients in Counterinsurgency," 105–108.

20. Biddle et al., "Small Footprint, Small Payoff." 11–13.

21. US Embassy Kabul cable, "Allegations of Secret Prisons in Kurdistan," September 19, 2006.

22. US Department of State, "President Bush Discusses Progress in Afghanistan."

23. Nasr, *The Dispensable Nation,* 17.

24. Christine Fair, "America's Pakistan Policy Is Sheer Madness," *The National Interest,* May 15, 2015; Tim Craig and Karen DeYoung, "Pakistan Fears That U.S. Will Slash Military Aid over Counterterror Efforts," *Washington Post,* August 20, 2015; "Pentagon Withholds $300 Million In Aid To Pakistan Over Haqqani Network," Gandhara, August 5, 2016.

25. Kelly et al., *Security Force Assistance in Afghanistan,* 35–45; US Government Accountability Office, "Afghanistan Security: Further Congressional Action May Be Needed to Ensure Completion of a Detailed Plan to Develop and Sustain Capable Afghan National Security Forces," *GAO-08-661,* June 18, 2008, 11, 27–30.

26. US Government Accountability Office, *GAO-08-661,* 3.

27. David H. Bayley and Robert Perito, *The Police in War: Fighting Insurgency, Terrorism, and Violent Crime* (New York: Lynne Reiner, 2010), 22; Kelly et al., *Security Force Assistance in Afghanistan,* 50; US Government Accountability Office, *GAO-08-661,* 4.

28. US Government Accountability Office, *GAO-08-661,* 17.

29. Interview with Lieutenant General James Dubik, Washington, D.C., February 3, 2017.

30. "Corruption Perceptions Index," Transparency International, 2005, 2009.

31. Astri Suhrke, *When More Is Less: The International Project in Afghanistan* (New York: Columbia Univ. Press, 2012), 133.

32. Chayes, *Thieves of State,* 59.

33. See Paul Collier, *The Bottom Billion: Why the Poorest Countries Are Failing and What Can Be Done About It* (Oxford: Oxford Univ. Press, 2007); Paul Collier, *Wars, Guns, and Votes: Democracy in Dangerous Places* (New York: HarperCollins, 2009); and Larry Diamond, *The Spirit of Democracy: The Struggle to Build Free Societies Throughout the World* (New York: Henry Holt, 2008).

34. US Department of Defense, *Report on Progress Toward Security and Stability in Afghanistan,* January 2009, 61.

35. United Nations Office on Drugs and Crime (UNODC), *Corruption in Afghanistan: Bribery as Reported by the Victims* (New York: UNODC, January 2010).

36. *Afghanistan in 2010: A Survey of the Afghan People,* The Asia Foundation, 2010, 3, 6, 85–90.

37. Nasr, *The Dispensable Nation,* 19; Kolenda et al., "The Strategic Costs of Civilian Harm," 17–28; Gopal, *No Good Men Among the Living.*

38. United Nations Assistance Mission in Afghanistan (UNAMA), *Protection of Civilians in Armed Conflict, Annual Report 2008,* January 2009, ii. Kolenda et al., "The Strategic Costs of Civilian Harm," 17.

39. For example, US Department of State, "President Bush Discusses Progress in Afghanistan"; NATO, "Progress in Afghanistan"; "The Situation in Afghanistan," hearing transcript, Senate Armed Services Committee (SASC), March 22, 2012.

40. This cycle is part of the reason highly fragile states rarely become self-reliant. See Lant Pritchett, Michael Woolcock, and Matt Andrews, "Capability Traps? The Mechanisms of Persistent Implementation Failure," Center for Global Development, December 2010.

9. The Good War Going Badly

1. Clausewitz, *On War,* 88; Sun Tzu, *The Art of War,* 63–71, 96–101; US Department of Defense, *Joint Publication 5-0: Joint Operations Planning,* 2011; US Department of the Army, *Army Doctrine Publication 5-0: The Operations Process,* May 2012.

2. Clausewitz, *On War,* 80–81.

3. US Department of Defense, *Joint Publication 5-0: Joint Operations Planning*; US Department of the Army, *Army Doctrine Publication 5-0: The Operations Process.*

4. Richard K. Betts, "Analysis, War, and Decision: Why Intelligence Failures Are Inevitable," *World Politics* 31, no. 1 (1978): 61–89; Richard K. Betts, "The New Politics of Intelligence: Will Reforms Work This Time?" *Foreign Affairs* 83, no. 3 (May/June 2004): 2–8; Michael T. Flynn, Matt Pottinger, and Paul D. Batchelor, "Fixing Intel: A Blueprint for Making Intelligence Relevant in Afghanistan," *Center for New American Security,* January 2010.

5. Betts, "The New Politics of Intelligence."

6. Nasr, *The Dispensable Nation,* 13–14.

7. The White House, "Remarks by the President on a New Strategy for Afghanistan and Pakistan," March 27, 2009.

8. McChrystal directed me to lead a team of civilian and military experts and draft the strategic assessment.

9. Stanley A. McChrystal, "COMISAF's Initial Assessment," August 30, 2009, 1-1; Stanley A. McChrystal, *My Share of the Task: A Memoir* (New York: Penguin, 2013), 292–315; Robert M. Gates, *Duty: Memoirs of a Secretary at War* (New York: Knopf, 2014), 367–368.

10. Karl Forsberg, "The Taliban's Campaign for Kandahar," Institute for the Study of War, December 2009.

11. McChrystal, "COMISAF's Initial Assessment," 1-1 to 1-4, 2-1 to 2-15; McChrystal, *My Share of the Task,* 316–338; Jones, *In the Graveyard of Empires,* ix.

12. McChrystal, "COMISAF's Initial Assessment," 2-22.

13. Matthew C. Brand, *General McChrystal's Strategic Assessment: Evaluating the Operating Environment in Afghanistan in the Summer of 2009* (Maxwell AFB: Air Univ. Press, 2011).

14. As the leader of the strategic assessment team and drafter of the document's core ideas and recommendations, I accept my responsibility for that shortcoming.

15. McChrystal, *My Share of the Task,* 331.

16. Eric Schmitt and Thom Shanker, "General Calls for More U.S. Troops to Avoid Afghan Failure," *New York Times,* September 21, 2009.

17. Rajiv Chandrasekaran, "McChrystal Preparing New Afghan War Strategy," *Washington Post,* July 31, 2009; Peter Baker and Dexter Filkins, "Obama to Weigh Buildup Option in Afghan War," *New York Times,* August 31, 2009; McChrystal, *My Share of the Task,* 344–345.

18. Personal knowledge as lead of the assessment team. We put the early drafts of the assessment on a NATO computer system that did not interface with American secret networks and carefully restricted access to prevent leaks. McChrystal sent the assessment to Defense Department senior leaders using a classified US email system. The assessment was leaked to the *Washington Post,* which published it on September 21. For an article speculating on the leak, see Ben Smith, "A D.C. Whodunit: Who Leaked and Why?" *Politico,* September 22, 2009; Gates, *Duty,* 367, said he was told that someone on McChrystal's staff leaked the assessment.

19. Michael Hastings, "The Runaway General," *Rolling Stone,* June 22, 2010.

20. McChrystal, *My Share of the Task,* 345–346.

21. Nasr, *The Dispensable Nation,* 22–25.

22. Holbrooke would publicly deny these allegations from Karzai and others, but Holbrooke's efforts were confirmed by Defense Secretary Robert Gates in his memoirs, Gates, *Duty,* 340–341, 358–359; McChrystal, *My Share of the Task,* 342–343; US Embassy Kabul cable, "Karzai on Elections and the Future: September 1 Meeting at the Palace," September 3, 2009.

23. US Embassy Kabul cable, "Karzai on the State of US-Afghan Relations," July 7, 2009.

24. See Shan Carter, Matthew Ericson, and Archie Tse, "Setting the Stage for the Recount," *New York Times,* October 16, 2009.

25. US Embassy Kabul cable, "Tensions Rise between Karzai and Abdullah Camps," September 10, 2009.

26. Sabrina Tavernise and Abdul Waheed Wafa, "U.N. Official Acknowledges 'Widespread Fraud' in Afghan Election," *New York Times,* October 11, 2009.

27. US Embassy Kabul cables, "Abdullah's Concerns: Fraud, Security, and His Future," October 6, 2009; "Balkh Governor Atta on Possible Karzai-Abdulla Power-Sharing Arrangements," October 16, 2009; "Karzai and IEC Announce Second Round," October 20, 2009; "The Day After: Meetings with the Candidates," Octo-

ber 21, 2009; "Abdullah's Second-Round Brinkmanship," October 26, 2009; "Abdullah Deflated," October 27, 2009; "Abdullah and Karzai to Meet Alone Oct 28," October 28, 2009; "Elections Endgame: Abdullah Close to Conceding," October 31, 2009; "Abdullah's Withdrawal: The Door Stays Open," November 1, 2009.

28. For an analysis of the election, see "The 2009 Presidential and Provincial Council Elections in Afghanistan," National Democratic Institute, 2010.

29. Helene Cooper and Jeff Zeleny, "Obama Warns Karzai to Focus on Tackling Corruption," *New York Times,* November 2, 2009.

30. Alissa J. Rubin and Mark Landler, "Karzai Sworn In for Second Term as President," *New York Times,* November 19, 2009. Interviewee L suggests that the United States recommended this language to Karzai.

31. Rod Nordland, "Afghan Presidential Rivals Finally Agree on Power-Sharing Deal," *New York Times,* September 20, 2014.

32. For descriptions of the process, see McChrystal, *My Share of the Task,* 354–357; Gates, *Duty,* 370–385; Nasr, *The Dispensable Nation,* 28.

33. Elisabeth Bumiller and Mark Landler, "U.S. Envoy's Cables Show Worries on Afghan Plans," *New York Times,* November 11, 2009.

34. Peter Baker, "Biden No Longer a Lone Voice on Afghanistan," *New York Times,* October 13, 2009.

35. Interviewee L; Mark Landler, "The Afghan War and the Evolution of Obama," *New York Times,* January 1, 2017.

36. Interviewees L, M, N, P, and X.

37. Interviewees L, M, N, P, and X; McChrystal, *My Share of the Task,* 357; Gates, *Duty,* 375, 379–380; Nasr, *The Dispensable Nation,* 21–22.

38. Gates, *Duty,* 378–379.

39. Interviewees L, M, N, P, and X; Nasr, *The Dispensable Nation,* 25.

40. Interviewees L, M, N, P, and X; Gates, *Duty,* 375, 382.

41. Interviewees L, M, N, P, and X.

42. Interviewee M.

43. Interview with Michèle A. Flournoy, Washington, D.C., March 9, 2016.

10. Surging into the Good War

1. James D. Fearon, "Domestic Political Audiences and the Escalation of International Disputes," *American Political Science Review* 88, no. 3 (1994): 577–592; Michael Tomz, "Domestic Audience Costs in International Relations: An Experimental Approach," *International Organization* 61, no. 4 (2007): 821–840; Han Dorussen and Mo Jongryn, "Ending Economic Sanctions: Audience Costs and Rent Seeking as Commitment Strategies," *Journal of Conflict Resolution* 45, no. 4 (2001): 395–426; Joe Eyerman and Robert A. Hart Jr., "An Empirical Test of the Audience Cost Proposition: Democracy Speaks Louder than Words," *Journal of Conflict Resolution* 40, no. 4 (1996): 597–616; Branislav L. Slantchev, "Politicians,

the Media, and Domestic Audience Costs," *International Studies Quarterly* 50, no. 2 (2006): 445–477; Alastair Smith, "To Intervene or Not to Intervene: A Biased Decision," *Journal of Conflict Resolution* 40, no. 1 (1996): 16–40.

2. Tomz, "Domestic Audience Costs in International Relations," 822.

3. Stephen M. Walt, "Rigor or Rigor Mortis? Rational Choice and Security Studies," *International Security* 23, no. 4 (1999): 5–48.

4. Tomz, "Domestic Audience Costs in International Relations," 836.

5. Ibid., 829.

6. Tomz, "Domestic Audience Costs in International Relations," 831–833; Brandice Canes-Wrone and Scott de Marchi, "Presidential Approval and Legislative Success," *Journal of Politics* 64, no. 2 (2002): 491–509; J. A. Krosnick and D. R. Kinder, "Altering the Foundations of Support for the President Through Priming," *American Political Science Review* 84, no. 2 (1990): 497–512.

7. This study is agnostic on the question of whether and to what extent democratic leaders are more affected than leaders of other regime types.

8. The White House, "Remarks by the President in Address to the Nation on the Way Forward in Afghanistan and Pakistan," December 1, 2009.

9. Karzai called for the same timeline in his 2009 inaugural speech, Rubin and Landler, "Karzai Sworn In for Second Term as President." This timeline was affirmed publicly in NATO's Lisbon Summit Declaration of November 20, 2010, paragraph 4.

10. Interviewees L and M.

11. Frances Z. Brown, *The U.S. Surge and Afghan Local Governance Lessons for Transition* (Washington, D.C.: US Institute of Peace, September 2012), 3.

12. Interview with Douglas E. Lute, Washington, D.C., September 20, 1016; interviewees M, N, and X.

13. Interviewees R, S, V, and U.

14. Sarah Chayes, "What Vali Nasr Gets Wrong," *Foreign Policy,* March 12, 2013.

15. Peter Erickson, Muhsin Hassan, Leigh Ann Killian, Gordon LaForge, Sarah Levit-Shore, Lauren Rhode, and Kenneth Sholes, "Lessons from the U.S. Civilian Surge in Afghanistan, 2009–2014," Princeton University, Woodrow Wilson School, January 2016; "Waiting on a Civilian Surge in Afghanistan," Council on Foreign Relations, March 31, 2010.

16. Interviewees H, J, L, M, and X; Chayes, "What Vali Nasr Gets Wrong."

17. Nasr, *The Dispensable Nation,* 45–46.

18. Eikenberry, for instance, cosigned 2009 and 2011 versions of the "Integrated Civilian–Military Campaign Plan for Support to Afghanistan," which led to improved coordination in some areas. Major issues such as strategic priorities and managing trade-offs were beyond its remit.

19. Nasr, *The Dispensable Nation,* 35–41.

20. Ibid., 25, 28.

21. Interviewees L, M, N, P, and X; Gates, *Duty,* 375, 382.

22. Gates, *Duty,* 299.

23. Interviewee L.

24. According to that poll, the percentage of those who thought the war was going well dropped from 54 percent in mid-2009 to 31 percent in December 2009; accordingly, the percentage of those who thought the war was going badly jumped from 43 percent to 66 percent. See Frank Newport, "Americans Divided on How Things Are Going in Afghanistan: Views Are More Positive Than in Recent Years," Gallup.com, April 8, 2011.

25. For Gates's recollection of the process, see Gates, *Duty,* 370–385.

11. More Shovels in the Quicksand

1. Gates, *Duty,* 375, 382.

2. McChrystal, *My Share of the Task,* 380–383.

3. Brett van Ess, "The Fight for Marjah: Recent Counterinsurgency Operations in Southern Afghanistan," *Small Wars Journal* (September 30, 2010): https://smallwarsjournal.com.

4. McChrystal, *My Share of the Task,* 373–375.

5. Dion Nissenbaum, "McChrystal Calls Marjah a 'Bleeding Ulcer' in Afghan Campaign," McClatchy, May 24, 2010.

6. Interview with General Stanley A. McChrystal, Alexandria, Virginia, January 23, 2017; interviewee E; personal recollections.

7. McChrystal, *My Share of the Task,* 353–354. The small team included me, one other US officer, and a couple of British officers.

8. Interviewee E.

9. Seth Jones, *Reintegrating Afghan Insurgents* (Washington, D.C.: RAND, 2011), ix; Jake Tapper, *The Outpost: An Untold Story of American Valor* (New York: Little, Brown, 2012), 330–340.

10. See also Emily Winterbotham, *Healing the Legacies of Conflict in Afghanistan: Community Voices on Justice, Peace and Reconciliation,* Afghan Research and Evaluation Unit, January 2012.

11. Interviewee E; interviewees H and J held similar views.

12. Interviewee E.

13. Interviewee E; personal recollection of discussions at ISAF at the time. See also Chayes, *Thieves of State,* 39–57.

14. Holbrooke's successor, Marc Grossman, described reintegration as "retail" politics and reconciliation as "wholesale" politics. Mark Grossman, "Talking to the Taliban 2010–2011: A Reflection," *Prism* 4, no. 4 (December 2013): 21–37, 31.

15. Interviewees E, H, and J; personal recollection of events at ISAF at the time. For a discussion of Holbrooke's views, see Nasr, *The Dispensable Nation,* 56.

16. Interview with General Stanley A. McChrystal.

17. Holbrooke had poor relations with Karzai and Kayani, recalled interviewees E, H, and J. See also Gates, *Duty,* 296; Ronald E. Neumann, "Failed Relations

between Hamid Karzai and the United States: What Can We Learn?," US Institute of Peace, May 2015, 6–7.

18. Hastings, "The Runaway General"; McChrystal, *My Share of the Task*, 387–388.

12. Misapplying the Iraq Formula

1. Kahneman, *Thinking, Fast and Slow*, 98.

2. Ibid., 138–140.

3. Bolger, *Why We Lost*, 366–367; Michael Hirsch and Jamie Tarabay, "Washington Losing Patience with Counterinsurgency in Afghanistan," *The Atlantic*, June 28, 2011.

4. Bolger, *Why We Lost*, 366–367.

5. I returned from Afghanistan in October 2010 to the Pentagon to continue work on reconciliation.

6. For skeptical views in the ISAF staff, see Chandrasekaran, *Little America*, 245–246.

7. Fred W. Baker III, "Petraeus Parallels Iraq, Afghanistan Strategies," American Forces Press Service, April 28, 2009. See also the Iraq Case Study below, especially chapter 29.

8. See the discussion in chapter 29.

9. Interviewees H, J, L, M, N, P, and X.

10. Stephen Biddle, Jeffrey A Friedman, and Jacob N. Shapiro, "Testing the Surge: Why Did Violence Decline in Iraq in 2007?," *International Security* 37, no. 1 (2012): 7–40.

11. Barfield, *Afghanistan*, 277, 285, 337, 339–342; Nathaniel C. Fick and Vikram Singh, "Winning the Battle, Losing the Faith," *New York Times*, October 4, 2008; Christopher D. Kolenda, "Winning Afghanistan at the Community Level," *Joint Force Quarterly* 56, 1st Quarter 2010: 25–31.

12. McChrystal, "COMISAF's Initial Assessment," 2-6 to 2-8; Antonio Giustozzi, *Decoding the New Taliban: Insights from the Afghan Field* (New York: Columbia Univ. Press, 2009), 293–300; Taliban Code of Conduct (Layha), "Taliban 2009 Rules and Regulations Booklet Seized by Coalition Forces on 15 Jul 2009 IVO Sangin Valley"; Stephanie Nijssen, "The Taliban's Shadow Government in Afghanistan," Civil-Military Fusion Centre, September 2011.

13. "The Future of the Afghan Local Police," Asia Report No. 268, International Crisis Group, June 4, 2015; "Just Don't Call It a Militia: Impunity, Militias, and the Afghan Local Police," Human Rights Watch, September 12, 2011.

14. DeeDee Derksen, "Impact or Illusion? Reintegration under the Afghanistan Peace and Reintegration Program," US Institute of Peace, September 22, 2011.

15. Ibid. Not all Afghan local police were raised in this manner. Most were developed in a coordinated effort between local coalition special operations forces and Afghan officials. Performance was uneven. Some Afghan local police improved local security. Others undermined it.

16. Briefings to me by ISAF officials in 2011 and 2012.

17. See, for instance, US Department of Defense, *Report on Progress Toward Security and Stability in Afghanistan,* July 2013, 37–43, and November 2013, 25–30.

13. Assessments and Risks

1. Tett, *The Silo Effect,* 43 of 5577.

2. Ibid., 3482 of 5577.

3. Ibid., 1257–1683 of 5577.

4. Robert W. Komer, *Bureaucracy Does Its Thing* (Washington, D.C.: RAND, 1972).

5. Joshua Partlow, "Elaborate Ruse Behind Vast Kabul Bank Fraud," *Washington Post,* June 30, 2011; Emma Graham-Harrison, "Afghan Elite Ransacked $900m from Kabul Bank, Inquiry Finds," *The Guardian,* November 28, 2010.

6. "Corruption Perceptions Index," Transparency International, 2010.

7. Interview with Michèle A. Flournoy.

8. Interview with General David H. Petraeus, Washington, D.C., December 8, 2016.

9. Elisabeth Bumiller, "Intelligence Reports Offer Dim View of Afghan War," *New York Times,* December 14, 2010.

10. Nasr, *The Dispensable Nation,* 26, 56–57.

11. Karen Parrish, "Gates: Afghanistan Progress Exceeds Expectations," Armed Forces Press Service, December 16, 2010.

12. Nasr, *The Dispensable Nation,* 26–27.

13. Gates, *Duty,* 385; interviewees H, M, and P, who were involved in the review; personal recollections as the Department of Defense lead strategist for the review.

14. Bumiller, "Intelligence Reports Offer Dim View of Afghan War."

15. The White House, "Statement by the President on the Afghanistan-Pakistan Annual Review," December 16, 2010; US Department of Defense, "Report on Progress Toward Security and Stability in Afghanistan," December 2012, 11; The White House, "Overview of the Afghanistan and Pakistan Annual Review," December 16, 2010; US Department of State, "Remarks by Secretary of State Hillary Clinton Launch of the Asia Society's Series of Richard C. Holbrooke Memorial Addresses," February 18, 2011.

16. Interviewees H, J, L, M, P, W, and X.

17. See the testimonies by Secretary of Defense Leon Panetta and Chairman of the Joint Chiefs of Staff Admiral Michael Mullen, "Hearing to Receive Testimony on the U.S. Strategy in Afghanistan and Iraq," Senate Armed Services Committee, September 22, 2011.

18. US Department of Defense, *Report on Progress Toward Security and Stability in Afghanistan,* October 2011, 1. See also subsequent reports through October 2014.

19. The military did make efforts to address corruption but could only do so effectively in military contracting. General David Petraeus tasked Brigadier General

H. R. McMaster to lead Task Force Shafafiyat (Transparency), but the effort had no overall impact on the kleptocratic nature of the Afghan government or its corrosive effects on the security forces.

20. Testimony by Chairman of the Joint Chiefs of Staff Admiral Michael Mullen, "Hearing to Receive Testimony on the U.S. Strategy in Afghanistan and Iraq," Senate Armed Services Committee.

21. Ibid. See also General David H. Petraeus, "Nomination of General David H. Petraeus to Be Director, Central Intelligence Agency," Hearing Before the Senate Select Committee on Intelligence, June 23, 2011, 18.

22. Mullen did note during his testimony: "We must work toward a reconciliation process internal to Afghanistan that provides for redress of grievances and a state-to-state interaction between Afghanistan and Pakistan to resolve matters of mutual concern."

23. Officials in the State Department were not given the same level of scrutiny by their congressional committees or by the White House, so the risks to success were never fully examined.

24. Interviewee M.

25. US Department of Defense, *Report on Progress Toward Security and Stability in Afghanistan,* October 2014, 8.

26. Interviewee S. Ghost soldiers are soldiers listed on the rosters, but who do not actually exist. Their being on the roster means pay is sent for them, but then the unit commanders keep those funds for themselves.

27. Jonathan Goodhan and Aziz Hakimi, "Counterinsurgency, Local Militias, and Statebuilding in Afghanistan," US Institute of Peace, January 2014; "The Future of the Afghan Local Police," Asia Report No. 268, International Crisis Group, June 4, 2015; "Just Don't Call It a Militia," Human Rights Watch; Mujib Mashal, Joseph Goldstein, and Jawad Sukhanyar, "Afghans Form Militias and Call on Warlords to Battle Taliban," *New York Times,* May 24, 2015.

28. Senate Foreign Relations Committee, "Evaluating U.S. Foreign Assistance To Afghanistan," June 8, 2011; "Report: U.S. Aid to Afghanistan Encouraging Dependency, Corruption," *PBS NewsHour,* June 8, 2011.

29. Nicolas Schmidle, "Getting Bin Laden: What Happened That Night in Abbottabad," *New Yorker,* August 8, 2011.

30. Mark Mazzetti, "How a Single Spy Helped Turn Pakistan Against the United States," *New York Times,* April 9, 2013.

31. Salman Masood and Eric Schmitt, "Tensions Flare Between U.S. and Pakistan After Strike," *New York Times,* November 26, 2011.

32. Elisabeth Bumiller and Jane Perlez, "Pakistan's Spy Agency Is Tied to Attack on U.S. Embassy," *New York Times,* September 22, 2011.

33. Opening statement by Senator Carl Levin, "Hearing to Receive Testimony on the U.S. Strategy in Afghanistan and Iraq," Senate Armed Service Committee, September 22, 2011. See also Tankel, "Beyond the Double Game."

34. "Corruption Perceptions Index," Transparency International, 2011.

35. Alissa J. Rubin, "Karzai Says Foreigners Are Responsible for Corruption," *New York Times,* December 11, 2011.

36. Testimony by Mullen, Mullen response to Senator Carl Levin, Senate Armed Service Committee, September 22, 2011.

37. Chayes, *Thieves of State,* 149–154.

38. Interviewees L, M, N, P, Q, W, and X.

39. NSC meetings are chaired by the president. Principals meetings are chaired by either the vice president or the national security advisor.

40. Interviewees L, M, and W.

41. Michael Pizzi, "As US Withdrawal Looms, Taliban Ponders Talking Peace," *al Jazeera,* March 21, 2014.

42. The White House, "Remarks by the President on the Way Forward in Afghanistan," June 22, 2011.

43. See US Department of Defense, *Report on Progress Toward Security and Stability in Afghanistan,* October 2014; United Nations, "The Situation in Afghanistan and Its Implications for International Peace and Security: Report of the Secretary-General," December 9, 2014.

44. Vanda Felbab-Brown, *Aspiration and Ambivalence: Strategies and Realities of Counterinsurgency and State-Building in Afghanistan* (Washington, D.C.: Brookings, 2013).

45. General John Allen, quoted in *The Economist,* "Outta here: After a Decade in Afghanistan, the United States Rushes for the Exit," February 4, 2012.

46. Interviewees F, S, U, and V; David Jolly, "U.S. to Send More Troops to Aid Afghan Forces Pressed by Taliban," *New York Times,* February 9, 2016.

47. Interviewee S.

48. Joseph Goldstein and Mujib Mashal, "Taliban Fighters Capture Kunduz City as Afghan Forces Retreat," *New York Times,* September 28, 2015.

49. Halimullah Kousary, "Taliban in Kunduz, ISIS in Nangarhar: Fiefdoms of Conflict in Afghanistan," *The Diplomat,* October 5, 2015.

50. Jessica Donati and Ehsanullah Amiri, "U.S. Military Moves to Clear 'Ghost Soldiers' from Afghan Payroll," *Wall Street Journal,* January 19, 2017.

51. Bill Roggio and Caleb Weiss, "Taliban Controls or Contests Scores of Districts in Afghanistan," *FDD's Long War Journal,* October 5, 2015; Sarah Almukhtar and Karen Yourish, "More Than 14 Years After U.S. Invasion, the Taliban Control Large Parts of Afghanistan," *New York Times,* April 19, 2016.

Conclusion to Part III

1. Interview with Douglas E. Lute.

2. Interviewee T.

3. Interviewee AB. Interviewees AA, R, S, and Z expressed similar sentiments.

Part IV. Ending the War in Afghanistan

1. Rod Nordland, Elisabeth Bumiller, and Matthew Rosenberg, "Karzai Calls on U.S. to Pull Back as Taliban Cancel Talks," *New York Times,* March 15, 2012.

2. Alissa J. Rubin and Rod Nordland, "U.S. Scrambles to Save Taliban Talks after Afghan Backlash," *New York Times,* June 19, 2013.

3. Rod Nordland, "Elders Back Security Pact That Karzai Won't Sign," *New York Times,* November 24, 2013.

4. Jon Boone, "Afghan Delegation Travels to Pakistan for First Known Talks with Taliban," *The Guardian,* July 7, 2015; Saeed Shah and Margherita Stancati, "Taliban Name Chief as Peace Talks Are Canceled," *Wall Street Journal,* July 30, 2015.

5. Halimullah Khousary, "The Afghan Peace Talks, QCG and China-Pakistan Role," *The Diplomat,* July 8, 2016.

6. Matthew Rosenberg and Michael D. Shear, "In Reversal, Obama Says U.S. Soldiers Will Stay in Afghanistan to 2017," *New York Times,* October 15, 2015; Missy Ryan and Karen DeYoung, "Obama Alters Afghanistan Exit Plan Once More, Will Leave 8,400 Troops," *Washington Post,* July 6, 2016.

7. Mark Landler, "Obama Says He Will Keep More Troops in Afghanistan Than Planned," *New York Times,* July 6, 2016.

14. Reconciliation versus Transition

1. Donald Wittman, "How a War Ends: A Rational Model Approach," *Journal of Conflict Resolution* 23, no. 4 (1979): 743–763, 743.

2. It is possible that to avoid the potential costs of war both sides would negotiate for lesser outcomes than they might achieve via conflict. See R. Harrison Wagner, "Bargaining and War," *American Journal of Political Science* 44, no. 3 (July 2000): 469–484, and Paul R. Pillar, *Negotiating Peace: War Termination as Bargaining Process* (Princeton: Princeton Univ. Press, 1983).

3. Reiter, *How Wars End,* 8–21; Branislav L. Slantchev, "The Principle of Convergence in Wartime Negotiations," *American Political Science Review* 97, no. 4. (2003): 621–632; Dan Reiter, "The Bargaining Model of War," *Perspectives on Politics* 1 (March 2003): 27–43; James D. Fearon, "Bargaining, Enforcement, and International Cooperation," *International Organization* 52, no. 2 (1998): 269–305; Darren Filson and Suzanne Werner, "A Bargaining Model of War and Peace: Anticipating the Onset, Duration, and Outcome of War," *American Journal of Political Science* 46 (October 2002): 819–838; Wagner, "Bargaining and War"; Pillar, *Negotiating Peace.*

4. Wagner, "Bargaining and War," 481. See also Thomas Schelling, *Arms and Influence* (New Haven, Conn.: Yale Univ. Press, 1966).

5. Clausewitz, *On War,* 91; Pillar, *Negotiating Peace,* 200; Wagner, "Bargaining and War."

6. Zartman, "The Timing of Peace Initiatives," 8; Wittman, "How a War Ends," 747–748, especially note 6; Reiter, *How Wars End,* 8–21.

7. Zartman, "The Timing of Peace Initiatives," 9.

8. Wagner, "Bargaining and War"; Slantchev, "The Principle of Convergence in Wartime Negotiations." A more mature insurgency could have a much higher expected utility, which would likely limit potential bargaining space even further.

9. See Tse-Tung, *On Guerilla Warfare*; Taber, *War of the Flea*; O'Neill, *Insurgency and Terrorism.*

10. For further discussion of incentives for negotiation, see Zartman, "The Timing of Peace Initiatives," 8; Wittman, "How a War Ends," 747–748, especially note 6; Reiter, *How Wars End,* 8–21.

11. For Vietnam: Henry Kissinger, *Ending the Vietnam War: A History of America's Involvement in and Extrication from the Vietnam War* (New York: Simon and Schuster, 2003), and Karnow, *Vietnam.* For the Soviets in Afghanistan: Rodrick Braithwaite, *Afghantsy: The Russians in Afghanistan, 1979–89* (Oxford: Oxford Univ. Press, 2011), and Tomsen, *The Wars of Afghanistan.* Note, these negotiations were with third-party states providing tangible support to the insurgency, not with the insurgent groups themselves.

12. Wittman, "How a War Ends," 750–754; Wagner, "Bargaining and War," 479; Zartman, "The Timing of Peace Initiatives."

13. Jack Holland, *Hope against History: The Course of Conflict in Northern Ireland* (London: Holt, 1999); Marianne Elliot, ed., *The Long Road to Peace in Northern Ireland: Peace Lectures from the Institute of Irish Studies at Liverpool University* (Liverpool, UK: University of Liverpool Institute of Irish Studies, Liverpool Univ. Press, 2007); David McKittrick and David McVea, *Making Sense of the Troubles: A History of the Northern Ireland Conflict* (Chicago: New Amsterdam Books, 2002); Richard English, *Armed Struggle: The History of the IRA* (Oxford: Oxford Univ. Press, 2003).

14. Dermot Keogh and Michael H. Haltzel, eds., *Northern Ireland and the Politics of Reconciliation* (Cambridge: Cambridge Univ. Press, 1994); Elliot, *The Long Road to Peace in Northern Ireland.*

15. Barbara Walter, "The Critical Barrier to Civil War Settlement," *International Organization* 51, no. 3 (1997): 335–364. See also Caplan, *Exit Strategies and State Building,* 316; James Dobbins and Laurel Miller, "Overcoming Obstacles to Peace," *Survival: Global Politics and Strategy* 55, no. 1 (February–March. 2013): 103–120.

16. Tomsen, *The Wars of Afghanistan,* 367–450; Coll, *Ghost Wars,* 189–224. Al Qaeda was invited to Afghanistan by the Rabbani government before the Taliban took power.

17. Nasr, *The Dispensable Nation,* 38; for a critique of Nasr's views, see Chayes, "What Vali Nasr Gets Wrong."

18. Thomas Waldman, "Reconciliation and Research in Afghanistan: An Analytical Narrative," *International Affairs* 90, no. 5 (2014): 1049–1068, 1062.

19. Chayes, "What Vali Nasr Gets Wrong."

20. Dobbins and Malkasian, "Time to Negotiate in Afghanistan."

21. James D. Fearon, "Iraq's Civil War," *Foreign Affairs* 86, no. 2 (March/April 2007): 2–15.

22. Paul et al., *Paths to Victory,* 16–21. The side which got the better end of the major concessions was declared the winner.

23. Fearon, "Iraq's Civil War."

24. Walter, "The Critical Barrier to Civil War Settlement."

25. For a detailed overview of scholarship on reconciliation, see Waldman, "Reconciliation and Research in Afghanistan."

26. Amin Saikal, "Don't Cave In to the Taliban," *New York Times,* October 18, 2007; William Maley, "Talking to the Taliban," *The World Today* 63, no. 11 (2007): 4–6. For an expert analysis of Northern Alliance views, see Bette Dam, "To Talk or Not to Talk? Abdullah Abdullah's Likely Stance on Negotiating with the Taliban," Norwegian Peacebuilding Resource Centre, August 2014.

27. Michael Semple, "Talking to the Taliban," *Foreign Policy,* January 10, 2013.

28. Dobbins and Malkasian, "Time to Negotiate in Afghanistan"; Knowlton, "Rumsfeld Rejects Plan to Allow Mullah Omar"; Rubin, "An Assassination That Could Bring War or Peace."

29. See, for instance, Rubin and Rashid, "From Great Game to Grand Bargain"; Mohammad Masoom Stanekzai, "Thwarting Afghanistan's Insurgency: A Pragmatic Approach Towards Peace and Reconciliation," Special Report 212, US Institute of Peace, September 2008; Adam Roberts, "Doctrine and Reality in Afghanistan," *Survival* 51, no. 1 (Feb.–March 2009): 29–60; Mariet D'Souza, "Talking to the Taliban: Will It Ensure 'Peace' in Afghanistan?," *Strategic Analysis* 33, no. 2 (2009): 254–272; Michael Semple, *Reconciliation in Afghanistan* (Washington, D.C.: USIP Press Books, 2009); Matt Waldman, "Dangerous Liaisons with the Afghan Taliban: The Feasibility and Risks of Negotiations," Special Report 256 (Washington, D.C.: US Institute for Peace, October 2010); Giles Dorronsoro, *Afghanistan at the Breaking Point* (Washington, D.C.: Carnegie Endowment for International Peace, 2010); Jonathan Steele, *Ghosts of Afghanistan: The Haunted Battleground* (London: Portobello, 2011), 340; "A New Way Forward: Rethinking US Strategy in Afghanistan," Afghanistan Study Group, 2010; Lakhdar Brahimi and Thomas Pickering, *Afghanistan: Negotiating Peace* (New York: Century Foundation International Task Force on Afghanistan and Its Regional and Multinational Dimensions, 2011); Matt Waldman and Thomas Ruttig, "Peace Offerings: Theories of Conflict Resolution and Their Applicability to Afghanistan," discussion paper, Afghan Analysts Network, 2011; Sherard Cowper-Coles, *Cables from Kabul: The Inside Story of the West's Afghanistan Campaign* (New York: HarperCollins, 2011).

30. Cowper-Coles, *Cables from Kabul*; personal recollections from the Obama review leading to the Reidel Report (February–March 2009) and discussions with UK officials during the McChrystal assessment (June–August 2009).

31. "A New Way Forward," Afghanistan Study Group.

32. Radio Free Europe/Radio Liberty, "Interview: McChrystal Says Solution in Afghanistan Is Developing Governance," August 30, 2009; Baker, "Petraeus Parallels Iraq, Afghanistan Strategies."

33. Frederick W. Kagan and Kimberly Kagan, "Why Negotiate with the Taliban?," *American Enterprise Institute,* March 17, 2010; Chayes, "What Vali Nasr Gets Wrong."

34. Discussion with a veteran of the Northern Ireland peace process, hosted by the British Foreign Commonwealth Office, October 15, 2014.

35. For other examples, see Stanley and Sawyer, "The Equifinality of War Termination," 656–657.

36. For Taliban views on the importance of sanctuary, see Theo Farrell and Michael Semple, "Making Peace with the Taliban," *Survival* 57, no. 6 (December 2015–January 2016): 79–110, 92.

37. Ibid.

38. Rubin and Rashid, "From Great Game to Grand Bargain," 35–38.

39. For ripeness theory, see Zartman, "The Timing of Peace Initiatives."

40. Interview with General David H. Petraeus.

41. Grossman, "Talking to the Taliban 2010–2011"; Dobbins and Malkasian, "Time to Negotiate in Afghanistan."

42. Interview with Douglas E. Lute. He argues that the best time to have advanced reconciliation was when 140,000 international troops were in country.

15. Reconciling Reconciliation

1. Iklé, *Every War Must End,* 59–83.

2. Tversky and Kahneman, "Rational Choice and the Framing of Decisions"; Benedetto De Martino, Dharshan Kumaran, Ben Seymour, and Raymond J. Dolan, "Frames, Biases, and Rational Decision-making in the Human Brain," *Science* 313, no. 5787 (2006): 684–687; William Samuelson and Richard Zeckhauser, "Status Quo Bias in Decision Making," *Journal of Risk and Uncertainty* 1, no. 1 (1988): 7–59; Alex Mintz, "Applied Decision Analysis: Utilizing Poliheuristic Theory to Explain and Predict Foreign Policy and National Security Decisions," *International Studies Perspectives* 6, no. 1 (2005): 94–98; Frank C. Zagare, "Analytic Narratives, Game Theory, and Peace Science," in *Frontiers of Peace Economics and Peace Science* (*Contributions to Conflict Management, Peace Economics and Development,* vol. 16, ed. M. Chatterji, C. Bo, and R. Misra (Bingley, UK: Emerald Group Publishing, 2011), 19–35; Graham T. Allison, "Conceptual Models and the Cuban Missile Crisis," *American Political Science Review* 63, no. 3 (1969): 689–718; Graham T. Allison and Philip Zelikow, *Essence of Decision: Explaining the Cuban Missile Crisis* (New York: Longman, 1999).

3. Iklé, *Every War Must End,* 84.

4. Allison, "Conceptual Models and the Cuban Missile Crisis"; Allison and Zelikow, *Essence of Decision,* 255–324; Bendor and Hammond, "Rethinking Allison's Models."

5. Kahneman, *Thinking, Fast and Slow,* 304–305; Iklé, *Every War Must End,* 84–105.

6. Stanley and Sawyer, "The Equifinality of War Termination," 657; Iklé, *Every War Must End*; Downs and Rocke, "Conflict, Agency, and Gambling for Resurrection"; Feaver, *Armed Servants*; Allison and Zelikow, *Essence of Decision*; Dominic D. P. Johnson and Dominic Tierney, *Failing to Win: Perceptions of Victory and Defeat in International Politics* (Cambridge: Harvard Univ. Press, 2006).

7. See also Goemans, "Fighting for Survival"; Edward D. Mansfield and Jack Snyder, "Democratization and the Danger of War," *International Security* 20, no. 1 (1995): 5–38.

8. Reiter, *How Wars End.* See also Stanley, "Ending the Korean War," 52–56.

9. Interviewees H and J; personal recollections as Department of Defense lead for the 2010 Afghanistan-Pakistan Annual Review.

10. Rubin, "An Assassination That Could Bring War or Peace"; Rashid, *Pakistan on the Brink,* 113–136.

11. Several US officials were concerned that the Taliban might use talks for deception (*taqiyya*) to undermine the Afghan government and/or relieve military pressure until the United States completed its drawdown. See Raymond Ibrahim, "How *Taqiyya* Alters Islam's Rules of War: Defeating Jihadist Terrorism," *Middle East Quarterly* (winter 2010): 3–13.

12. US Department of State, "Remarks by Secretary of State Hillary Clinton," February 18, 2011.

13. Ibid.

14. Nasr, *The Dispensable Nation,* 34–35.

15. Chandrasekaran, *Little America,* 247.

16. Interview with General David H. Petraeus.

17. General John Allen, quoted in "Outta Here," *The Economist.*

18. Rubin and Rashid, "From Great Game to Grand Bargain."

19. Nasr, *The Dispensable Nation,* 35–36.

20. Personal conversation with a senior SRAP official, March 2011. See also Nasr, *The Dispensable Nation,* 26–27.

21. Walter, "The Critical Barrier to Civil War Settlement"; Fearon, "Iraq's Civil War."

22. Nasr, *The Dispensable Nation,* 41.

23. Grossman, "Talking to the Taliban 2010–2011."

24. Such works included Rose, *How Wars End*; Iklé, *Every War Must End*; Reiter, *How Wars End*; Caplan, *Exit Strategies and State Building*; Lee and Walling, *Strategic Logic and Political Rationality*; Stanley, "Ending the Korean War"; Stanley, *Paths to Peace*; and Zartman, "The Timing of Peace Initiatives."

25. When Obama announced delays to the drawdown timeline after 2014, American public opinion did not register any political costs to Obama (in fact, those surveyed were less inclined to see the war as a mistake). See Andrew Dugan, "Fewer in U.S. View Iraq, Afghanistan Wars as Mistakes," Gallup, June 12, 2015.

26. Nasr, *The Dispensable Nation,* 38, notes that it was given a slow death in the interagency process.

27. Interviewees H, J, L, and M.

28. Grossman, "Talking to the Taliban 2010–2011," 30.

29. Confidential conversations with two ISAF commanders.

30. Interviewees H, J, M, and W. Grossman argues that he consulted closely with Defense officials about reconciliation, Grossman, "Talking to the Taliban 2010–2011," 30–31.

31. See, for instance, Henry Mintzberg and James A. Waters, "Of Strategies, Deliberate and Emergent," *Strategic Management Journal* 6 (1985): 257–272.

32. Interviewees H and M.

33. Nasr, *The Dispensable Nation,* 49.

34. Grossman, "Talking to the Taliban 2010–2011," 23.

35. Interview with Vikram Singh, Washington, D.C., December 10, 2016.

16. Competing Visions

1. Rahim Faiez and Kimberly Dozier, "Karzai Accuses U.S. of Collaborating with Taliban," Associated Press, March 10, 2013; "Hamid Karzai Says U.S.-Afghan Relationship 'Has Been at a Low Point for a Long Time,'" *Washington Post,* March 2, 2014.

2. Rubin, "An Assassination That Could Bring War or Peace."

3. Ernesto Londoño and Kevin Seiff, "Afghan Officials Accuse U.S. of Snatching Pakistani Taliban Leader from Their Custody," *Washington Post,* October 10, 2013; Matthew Rosenberg, "U.S. Disrupts Afghans' Tack on Militants," *New York Times,* October 23, 2013.

4. Amir Shah and Rahim Faiez, "Karzai: Afghan Government Should Lead Peace Talks," Associated Press, January 29, 2013; Alissa J. Rubin and Declan Walsh, "Renewed Push for Afghans to Make Peace with Taliban," *New York Times,* February 16, 2013.

5. Interviewees F, H, L, M, and W.

6. Bill Roggio, "Afghan Taliban Denounces Former Senior Official, Denies Involvement in Peace Talks," *FDD's Long War Journal,* February 22, 2014; Julian Borger, "Afghan Insurgents Want Peace Deal, Says Ex-Taliban Minister," *The Guardian,* September 20, 2014.

7. Dam, *A Man and a Motorcycle.*

8. "Too Many Missed Opportunities: Human Rights in Afghanistan under the Karzai Administration," Amnesty International, 2014, 5–6; Sari Kouvo, "After Two

Years in Legal Limbo: A First Glance at the Approved 'Amnesty Law,'" Afghan Analysts Network, February 22, 2010.

9. Rubin and Rashid, "From Great Game to Grand Bargain."

10. Carlotta Gall, "Afghan Peace Talks End with Plea to Combatants," *New York Times,* June 4, 2010; Islamic Republic of Afghanistan, "The Resolution Adopted at the Conclusion of the National Consultative Peace Jirga," June 2–4, 2010, Loya Jirga Tent, Kabul.

11. Interviewees F, G, H, I, J, and Y; personal conversations with former and then-current senior Taliban officials, 2011–2013. Clinton later explained them as end conditions.

12. For Taliban views on the intensity of the military pressure, see Farrell and Semple, "Making Peace with the Taliban," 82–83.

13. Interview with Barnett R. Rubin. The preamble notes, "Appreciating the sacrifices, historical struggles, jihad and just resistance of all the peoples of Afghanistan, admiring the supreme position of the martyrs of the country's freedom." According to Taliban experts and former Taliban officials, the jihad refers to the Soviet war and "just resistance" signifies opposition to the Taliban rule. For additional Taliban views on the Afghan constitution, see Michael Semple, Theo Farrell, Anatol Lieven, and Rudra Chaudhuri, "Taliban Perspectives on Reconciliation," Royal United Services Institute, September 2012, 4.

14. Personal conversations with former and then-current senior Taliban officials, 2011–2013.

15. Islamic Emirate of Afghanistan, "Statement of Islamic Emirate of Afghanistan Regarding Negotiations," January 3, 2012. See also Farrell and Semple, "Making Peace with the Taliban," 97–98.

16. "Letter from Mohammad Tayib," Bin Laden's Bookshelf database, Office of the Director of National Intelligence, released March 1, 2016.

17. Interviewees F, G, I, M, and S; Matthew Rosenberg and Alissa J. Rubin, "Taliban Step Toward Afghan Peace Talks Is Hailed by U.S.," *New York Times,* June 18, 2013.

18. Farrell and Semple, "Making Peace with the Taliban," 101; Semple et al., "Taliban Perspectives on Reconciliation."

19. Personal discussion with Tayyab Agha in Doha, September 2011.

20. Farrell and Semple, "Making Peace with the Taliban," 97–99.

21. Interviewees F, G, and I.

22. National Security Presidential Directive 12 (NSPD-12), "United States Citizens Taken Hostage Abroad," was signed by President George W. Bush on February 18, 2002, forbidding the United States from negotiating with terrorists. This was later modified by President Obama on June 24, 2015, with Presidential Policy Directive 30 (PPD-30), "Hostage Recovery Activities."

23. For a detailed discussion of the Taliban's relationship with al Qaeda, see Alex Strick van Linschoten and Felix Kuehn, *An Enemy We Created: The Myth of the Taliban/Al Qaeda Merger in Afghanistan, 1970–2010* (London: Hurst, 2012).

24. "Recommendations for the Mujahideen Entering Afghanistan," Bin Laden's Bookshelf database, Office of the Director of National Intelligence, released March 1, 2016. Wahabbism is an ultra-conservative religious movement prominent in the Gulf Arab countries, and supported by al Qaeda. Deobandis are Hanafi Muslims. Hanafi is one of the four Sunni schools of jurisprudence. They view the writings of scholar Abu Hanifa as authoritative. Wahabbis do not.

25. "Summary on Situation in Afghanistan and Pakistan," Bin Laden's Bookshelf database, Office of the Director of National Intelligence, released May 20, 2015.

26. "Letter from Mohammad Tayib," Bin Laden's Bookshelf database, Office of the Director of National Intelligence, released March 1, 2016.

27. "Updated letter RE: Afghanistan," Bin Laden's Bookshelf database, Office of the Director of National Intelligence, released May 20, 2015.

28. See also Semple et al., "Taliban Perspectives on Reconciliation," 3, 5–7. Verification, of course, would be a major challenge, as would prevention of cheating. For a good discussion of the challenges, see Waldman, "Dangerous Liaisons with the Afghan Taliban."

29. Islamic Emirate of Afghanistan, "Message of Felicitation of Amir-ul-Momineen on the Occasion of Eid-ul-Fitr," September 19, 2009.

30. Vahid Brown, "Al-Qaeda and the Afghan Taliban: 'Diametrically Opposed'?" *Foreign Policy,* October 22, 2009.

31. "Open Letter of the Islamic Emirate of Afghanistan to the Shanghai Summit," Voice of Jihad, October 14, 2009.

32. Islamic Emirate of Afghanistan, "Layha" (Code of Conduct) 2006, and 2009; Conference statements, June 29, 2012; December 24, 2012; May 5, 2015; Eid al-Adha and Eid al-Fitr statements since September 19, 2009. Alex Strick van Linschoten and Kuehn, Felix, *Separating the Taliban from Al-Qaeda: The Core of Success in Afghanistan* (New York: Center on International Cooperation, 2011); Dobbins and Malkasian, "Time to Negotiate in Afghanistan," 60–61. Bin Laden encouraged Mullah Omar to avoid shedding Muslim blood. See "Letter to Our Honored Commander of the Faithful," Bin Laden's Bookshelf database, Office of the Director of National Intelligence, released March 1, 2016

33. Semple et al., "Taliban Perspectives on Reconciliation," 12.

34. By contrast, the Taliban had to deal with internal turmoil when talks about a political office in Doha leaked to the press in January 2012.

35. Personal discussions with Tayyab Agha, 2011, Doha; interviewees F, H, J, and L; Rubin, "An Assassination That Could Bring War or Peace."

36. Mehran Kamrava, *Qatar: Small State, Big Politics* (Ithaca, N.Y.: Cornell Univ. Press, 2015).

37. Islamic Emirate of Afghanistan, "Text of Speech Enunciated by Islamic Emirate of Afghanistan at Research Conference in France," December 24, 2012; Graham Bowley and Matthew Rosenberg, "Afghan Officials Hail Talks with Insurgents," *New York Times,* June 28, 2012.

38. Tankel, "Beyond the Double Game," 1–2.

39. Ahley J. Tellis, "The Menace That Is Lashkar-e-Taiba," Carnegie Endowment for International Peace, March 2012.

40. US National Counter Terrorism Center, "Tehrik-e-Taliban Pakistan," *Counter Terrorism Guide,* January 26, 2017.

41. Interviewees F, H, I, and J; Hekmatullah Azamy, "It's Complicated: The Relationship Between Afghanistan, Pakistan, and the Taliban," *Foreign Policy,* March 2, 2015.

42. Farrell and Semple, "Making Peace with the Taliban," 92.

43. Fair, "America's Pakistan Policy Is Sheer Madness"; Peter Bergen, ed., *Talibanistan: Negotiating the Borders Between Terror, Politics, and Religion* (New York: Oxford Univ. Press, 2013); Vahid Brown and Don Rassler, *Fountainhead of Jihad: The Haqqani Nexus, 1973–2012* (New York: Columbia Univ. Press, 2013). For an alternative view, see Anatol Lieven, *Pakistan: A Hard Country* (New York: Perseus, 2011).

44. Barnett R. Rubin, "Afghanistan and the Taliban Need Pakistan for Peace," *al Jazeera,* January 10, 2016; Bruce Reidel, "Pakistan, Taliban and the Afghan Quagmire," Brookings Institution, August 24, 2013; Matt Waldman, *The Sun in the Sky: The Relationship Between Pakistan's ISI and Afghan Insurgents,* Crisis States Research Center, January 2010. Both Reidel and Waldman note significant limitations on Pakistan's influence. Azmat Khan, "Leaked NATO Report Alleges Pakistani Support for Taliban," *PBS Frontline,* February 1, 2012; "Afghan Army Chief: 'Pakistan Controls Taliban,'" Afghan Institute for Strategic Studies, July 3, 2013; Sanjay Kumar, "Afghanistan's Ex-Intelligence Chief Reflects on Mullah Omar's Death," *The Diplomat,* July 30, 2015; Lisa Curtis, "How Pakistan Is Tightening Its Grip on the Taliban," *National Interest,* August 15, 2015.

45. Abubakar Siddique, "Aziz Admits Pakistan Housing Afghan Taliban Leaders," *Dawn,* March 3, 2016.

46. Mujib Mashal, "How Peace Between Afghanistan and the Taliban Foundered," *New York Times,* December 26, 2016.

47. Dexter Filkins, "Pakistanis Tell of Motive in Taliban Leader's Arrest," *New York Times,* August 22, 2010.

48. For instance, the Taliban reportedly arrested members of Tayyab Agha's family in connection to his relocation to Doha. Interviewees F and J; Rubin, "An Assassination That Could Bring War or Peace."

49. Rubin and Rashid, "From Great Game to Grand Bargain," 36.

17. Exploratory Talks

1. The National Security Act of 1947, 80th Congress, 1st sess., Public Law 253, chapter 343, S. 758, July 26, 1947.

2. Allison and Zelikow, *Essence of Decision.*

3. Ibid., 143–196.

4. Ibid., 255–324.

5. See Rashid, *Pakistan on the Brink,* 113–136; Grossman, "Talking to the Taliban 2010–2011," 28.

6. Missy Ryan, Warren Strobel, and Mark Hosenball, "Exclusive: Secret U.S., Taliban Talks Reach Turning Point," Reuters, December 19, 2011; personal recollections from meetings in Doha, August and September 2011.

7. Grossman, "Talking to the Taliban 2010–2011," 28.

8. The White House, "Remarks by the President on the Way Forward in Afghanistan."

9. Grossman, "Talking to the Taliban 2010–2011," 28.

10. Islamic Emirate of Afghanistan, "Statement of Islamic Emirate of Afghanistan Regarding Negotiations," January 3, 2012; Islamic Emirate of Afghanistan, "Declaration About the Suspension of Dialogue with the Americans, the Office in Qatar, and Its Political Activity," March 15, 2012; Islamic Emirate of Afghanistan, "Regarding the Participation of Its Representative in an Academic Conference in Japan," June 29, 2012.

11. Interviewees F, J, I, Y, and Z.

12. Ali A. Jalali, "Afghanistan: Regaining Momentum," *Parameters* 37, no. 4 (winter 2007–2008): 5–19, 12; Theo Farrell and Antonio Giustozzi, "The Taliban at War: Inside the Helmand Insurgency, 2004–2012," *International Affairs* 89, no. 4 (2013): 845–871; interviewees F, G, J, I, Y, and Z.

13. For a detailed discussion of the evolution of Taliban command and control systems, see Claudio Franco and Antonio Giustozzi, "Revolution in the Counter-Revolution: Efforts to Centralize the Taliban's Military Leadership," *Central Asian Affairs* 3 (2016): 249–286. See also Rashid, *Taliban,* 41–42; Farrell and Semple, "Making Peace with the Taliban," 93–96.

14. Ryan, Strobel, and Hosenball, "Exclusive"; personal recollection as participant in discussions.

15. Interviewees F, H, J, L, M, P, and Q.

16. The Inteqal framework was established in 2010 to manage the Afghan-led transition process. The Joint Afghan-NATO Inteqal Board (JANIB) is responsible for approving transition implementation plans and recommending areas to enter or complete the transition process. See US Department of Defense, *Report on Progress Toward Security and Stability in Afghanistan,* December 2012, 27–31; NATO, "Inteqal: Transition to Afghan Lead," January 7, 2015. No similar effort was made to coordinate reconciliation.

17. Interview with Lieutenant General Terry A. Wolff.

18. Discussion with senior SRAP official, November 2011.

19. Neumann, "Failed Relations between Hamid Karzai and the United States," 11; Dobbins and Malkasian, "Time to Negotiate in Afghanistan," 58.

20. Robert D. Blackwill, "Plan B in Afghanistan: Why a De Facto Partition Is the Least Bad Option," *Foreign Affairs* 90, no. 1 (January/February 2011): 42–50.

21. Interviewees F, G, I, M, P, Q, and W; Dobbins and Malkasian, "Time to Negotiate in Afghanistan," 58.

22. Alissa J. Rubin, "Assassination Deals Blow to Peace Process in Afghanistan," *New York Times,* September 20, 2011.

23. Discussion with former senior Afghan official, January 2014; interviewees F, J, L, and M.

24. Rod Nordland and Sharifullah Sahak, "Afghan Rebuke of Qatar Sets Back Peace Talks," *New York Times,* December 15, 2011.

25. Nasr, *The Dispensable Nation,* 57–58.

26. Personal recollections from the meetings. For more on the discussions with Congress, see Ed O'Keefe, "White House First Discussed Bergdahl Prisoner Exchange with Lawmakers in 2011," *Washington Post,* June 3, 2014; Tim Mack, "Obama Shut Out Congress for 2 Years About Bergdahl Deal, Key Senator Says," *Daily Beast,* June 3, 2014; for the attack on the US embassy, see Alissa J. Rubin, Ray Rivera, and Jack Healy, "U.S. Embassy and NATO Headquarters Attacked in Kabul," *New York Times,* September 13, 2011.

27. See previous chapter.

28. Grossman, "Talking to the Taliban 2010–2011," 34; Deirdre Walsh and Ted Barrett, "Congressional Leaders Initially Pushed Back on Bergdahl Swap," CNN, June 4, 2014; Charlie Savage, "Negotiations with Taliban Could Hinge on Detainees," *New York Times,* June 2, 2013; Jason Leopold, "What Congress Really Told the White House About the Bowe Bergdahl Swap," *Vice News,* March 4, 2015; David E. Sanger and Matthew Rosenberg, "Critics of P.O.W. Swap Question the Absence of a Wider Agreement," *New York Times,* June 8, 2014.

29. See Dobbins and Malkasian, "Time to Negotiate in Afghanistan," 57.

30. National Defense Authorization Act for Fiscal Year 2013 (NDAA 2103), Public Law 112-239, Section 1033, "Requirements for Certifications Relating to the Transfer of Detainees at United States Naval Station, Guantanamo Bay, Cuba, to Foreign Countries and Other Foreign Entities," January 2, 2013.

31. Rafaella Wakeman, "The Senate Armed Services Committee's GTMO Transfer Provisions in the 2014 NDAA," *Lawfare,* June 25, 2013.

32. Interviewees J, L, M, and X.

18. Coming Off the Rails

1. Oren Dorell, "Taliban May Be Ready to Try Talking," December 27, 2011; "Taliban Will Open Office in Qatar for Peace Talks," *USA Today,* January 4, 2012; Matthew Rosenberg, "Taliban Opening Qatar Office, and Maybe Door to Talks," *New York Times,* January 4, 2012.

2. Interviewees F, G, I, and J; Rubin, "An Assassination That Could Bring War or Peace."

3. "Taliban Will Open Office in Qatar for Peace Talks." *USA Today.*

4. Islamic Emirate of Afghanistan, "Statement of Islamic Emirate of Afghanistan Regarding Negotiations"; Islamic Emirate of Afghanistan, "Declaration About the Suspension of Dialogue with the Americans, the Office in Qatar, and Its Political Activity," March 15, 2012.

5. See the Taliban's Voice of Jihad website, https://alemarahenglish.net.

6. Interviewees F and J. Farrell and Semple, "Making Peace with the Taliban," 98–99.

7. Grossman, "Talking to the Taliban 2010–2011," 29.

8. Personal recollection from the January 2012 meetings in Doha with Qatari officials and Taliban representative Tayyab Agha.

9. Interviewees F, J, L, and M.

10. Islamic Emirate of Afghanistan, "Declaration About the Suspension of Dialogue with the Americans"; Kate Clark, "The End of the Affair? Taleban Suspend Talks," Afghan Analysts Network, March 16, 2012.

11. The White House, "Remarks by President Obama in Address to the Nation from Afghanistan," May 1, 2012.

12. The White House, "Joint Statement by President Obama and President Karzai," January 11, 2013.

13. Interviewees J and L; personal recollections. Karzai spokesman Aimal Faizi incorrectly blamed Ambassador James Dobbins for rejecting an agreement between Qatar and the Afghan government. See Borhan Osman and Kate Clark, "Who Played Havoc with the Qatar Talks? Five Possible Scenarios to Explain the Mess," Afghan Analysts Network, July 9, 2013.

14. The letter's assurances were confirmed by a White House official to the *New York Times*. Rubin and Nordland, "U.S. Scrambles to Save Taliban Talks after Afghan Backlash"; "Faizi Reveals Details of Karzai's Letter to Obama," Pajhwok, June 27, 2013.

15. NATO, "NATO Secretary General in Kabul as Afghan Security Forces Take Lead Countrywide," June 18, 2013.

16. Personal recollections of meetings and discussions in the lead-up to the office opening; interviewees F, H, and J.

17. "Afghan Taliban Opens Qatar Office, Says Seeks Political Solution," Reuters, June 18, 2013. The Taliban have always referred to themselves as the Islamic Emirate of Afghanistan, and to have not used that term could have created unrest in the Taliban ranks.

18. Kate Clark, "The Opening of the Taleban Office in Qatar: A Propaganda Coup and an Angry Government," Afghan Analysts Network, June 19, 2013.

19. Dobbins and Malkasian, "Time to Negotiate in Afghanistan," 57.

20. Nasr, *The Dispensable Nation,* 58–59.

21. Clark, "The Opening of the Taleban Office in Qatar."

22. "Protests Staged in Kabul City Against Taliban Office in Qatar," *Khaama Press,* June 29, 2013.

23. Karen DeYoung, Tim Craig, and Ernest Londoño, "Despite Karzai's Ire, U.S. Confident That Talks with Taliban Will Be Held," *Washington Post,* June 19, 2013.

24. Osman and Clark, "Who Played Havoc with the Qatar Talks?"

25. Interview with Vikram Singh.

19. Fallout

1. Dalrymple, *Return of a King.*

2. Nordland, "Elders Back Security Pact That Karzai Won't Sign."

3. Embassy of Afghanistan, "President Obama's Letter to President Karzai on BSA," November 20, 2013.

4. William Booth, "Israel's Prisoner Swaps Have Been Far More Lopsided than Obama's Bergdahl Deal," *Washington Post,* June 5, 2014.

5. The United States included family support provisions in the detainee transfer agreement to make relocation in Doha even more attractive.

6. Kamrava, *Qatar.*

7. Charlie Savage and David E. Sanger, "Deal to Free Bowe Bergdahl Puts Obama on Defensive," *New York Times,* June 3, 2014.

8. Sanger and Rosenberg, "Critics of P.O.W. Swap Question the Absence of a Wider Agreement"; Dobbins and Malkasian, "Time to Negotiate in Afghanistan," 57. Some of the exchangees participated in 2018–2019 in talks with the United States.

9. "Transcript of Siraj Haqqani's Interview," BBC News, October 3, 2011.

10. Interviewees F and J.

11. William A. Byrd, Casey Garret Johnson, and Sanaullah Tasal, "Compounding Uncertainty in Afghanistan: Economic Consequences of Delay in Signing the Bilateral Security Agreement," US Institute of Peace, February 4, 2014; Jason Campbell, "The Pernicious Effects of Uncertainty in Afghanistan," *War on the Rocks,* March 12, 2014.

12. Vanda Felbab-Brown, "The Stakes, Politics, and Implications of the U.S.-Afghanistan Bilateral Security Agreement," Brookings Institution, November 17, 2013; General Joseph F. Dunford, "Statement of General Joseph F. Dunford Commander U.S. Forces-Afghanistan Before the Senate Armed Services Committee on the Situation in Afghanistan, 12 March 2014," Senate Armed Services Committee, March, 12, 2014, 8–9; Lieutenant General Michael Flynn, "Statement Before the Senate Armed Services Committee, United States Senate, 11 February 2014," Defense Intelligence Agency, February 11, 2014.

13. Margherita Stancati, Nathan Hodge, and Dion Nissenbaum, "Afghan Crisis Risks Splitting Country: Presidential Candidate Claims Victory, Defying Early Vote Count, and Considers Forming Own Government," *Wall Street Journal,* July 8, 2014; Matthew Rosenberg and Azam Ahmad, "Tentative Results in Afghan Presidential Runoff Spark Protests," *New York Times,* July 7, 2014.

14. Interview with Carter Malkasian, Washington, D.C., August 25, 2016.

15. Barnett Rubin and Georgette Gagnon, "The U.S. Presence and Afghanistan's National Unity Government: Preserving and Broadening the Political Settlement," Center on International Cooperation, August 2016.

16. Mujib Mashal, "Afghan Government Faces New Set of Rivals," *New York Times,* December 2, 2015; and Mujib Mashal, "Afghanistan Is in Chaos. Is That What Hamid Karzai Wants?" *New York Times,* August 5, 2016; Martine van Bijlert and Ali Yawar Adili, "When the Political Agreement Runs Out: On the Future of Afghanistan's National Unity Government," Afghan Analysts Network, May 29, 2016.

17. Farrell and Semple, "Making Peace with the Taliban," 88–89; Aimal Faizi, "Karzai's Stand for Afghan National Interests," Gandhara, June 4, 2015; note: Faizi is a Karzai spokesman. Frud Behzan, "Afghan Unity Government Split on Intelligence-Sharing Deal," Radio Free Europe/Radio Liberty, May 21, 2015.

18. Kay Johnson and Mehreen Zahra-Malik, "Taliban, Afghan Officials Hold Peace Talks, Agree to Meet Again," Reuters, July 8, 2015. Borhan Osman, "The Murree Process: Divisive Peace Talks Further Complicated by Mullah Omar's Death," Afghan Analysts Network, August 5, 2015.

19. Osman, "The Murree Process"; Mujib Mashal, "Taliban Were Authorized to Talk, Afghan Envoys Say," *New York Times,* July 9, 2015.

20. Rubin, "An Assassination That Could Bring War or Peace."

21. Statista, "Number of Fatalities among Western Coalition Soldiers Involved in the Execution of Operation Enduring Freedom from 2001 to 2020," https://www.statista.com/statistics/262894/western-coalition-soldiers-killed-in-afghanistan/ (accessed August 2, 2020).

Conclusion to Part IV

1. Interviewees H, J, L, M, W, and X. The White House did hold many meetings about the more tactical aspects of reconciliation, particularly in advance of meetings with the Taliban. These, however, mostly focused on coordinating talking points and sequencing of confidence-building measures.

2. See also Matt Waldman, "System Failure: The Underlying Causes of US Policy-Making Errors in Afghanistan," *International Affairs* 89, no. 4 (July 2013): 825–843, 829–832.

Part V. Pursuit of Decisive Victory in Iraq

1. "In Their Own Words: Iraq's 'Imminent' Threat," Center for American Progress, January 29, 2004; George W. Bush, "Third State of the Union Address," January 29, 2003.

2. For maps of the evolving situation, see "Operation Iraqi Freedom Maps," GlobalSecurity.org.

3. James P. Pfiffner, "US Blunders in Iraq: De-Baathification and Disbanding the Army," *Intelligence and National Security* 25, no. 1 (2010): 76–85.

4. Gordon and Trainor, *The Endgame,* 36–39; Thomas E. Ricks, *Fiasco: The American Military Adventure in Iraq* (New York: Penguin Press, 2006), 196–200.

5. Gordon and Trainor, *The Endgame,* 195.

6. Ibid., 267–295.

7. Ibid., 35–37.

8. Ibid., 15–16; Ricks, *Fiasco,* 158–166; James Dobbins, Seth G. Jones, Benjamin Runkle, and Siddharth Mohandas, *Occupying Iraq: A History of the Coalition Provisional Authority* (Santa Monica, Calif.: RAND Corporation, 2009); Nora Bensahel, Olga Oliker, Keith Crane, Richard R. Brennan Jr., Heather S Gregg, Thomas Sullivan, and Andrew Rathmell, *After Saddam: Prewar Planning and the Occupation of Iraq* (Santa Monica, Calif.: RAND, 2008); "The Lost Year in Iraq: Interview with Robert Blackwill," *PBS Frontline,* October 17, 2006; Larry Diamond, "What Went Wrong in Iraq," *Foreign Affairs* 83, no. 5 (September/October 2004): 34–56.

9. Dobbins et al., *Occupying Iraq*; Bensahel et al., *After Saddam,* xix; Michael R. Gordon and Bernard E. Trainor, *Cobra II: The Inside Story of the Invasion and Occupation of Iraq* (New York: Pantheon Books, 2006), 503–504; "The Lost Year in Iraq," *PBS Frontline*; Diamond, "What Went Wrong in Iraq."

10. Ricks, *Fiasco*; John A. Nagl, *Knife Fights: A Memoir of Modern War in Theory and Practice* (New York: Penguin Books, 2014); Kilcullen, *Counterinsurgency*; Bolger, *Why We Lost.*

11. Pfiffner, "US Blunders in Iraq"; Rajiv Chandrasekaran, *Imperial Life in the Emerald City: Inside Iraq's Green Zone* (New York: Vintage Books, 2006), 78–88.

12. Dobbins et al., *Occupying Iraq,* xv, xxii–xxiii, xxvi, 52–60, 107–119.

13. Ibid., xxxviii–xli. See also Dobbins et al., *America's Role in Nation-Building.* For a less favorable view of the CPA, see Chandrasekaran, *Imperial Life in the Emerald City.*

14. Dobbins et al., *Occupying Iraq,* xiii, 326–333; see also "The Lost Year in Iraq," *PBS Frontline*; Diamond, "What Went Wrong in Iraq."

15. Dobbins et al., *Occupying Iraq,* xli.

16. Ibid., xl.

20. Operation Iraqi Freedom

1. The White House, "President Discusses the Future of Iraq," February 26, 2003.

2. Thom Shanker and Eric Schmitt, "A Nation at War: The Pentagon; Rumsfeld Says Iraq Is Collapsing, Lists 8 Objectives of War," *New York Times,* March 22, 2003.

3. Strategic Studies Institute, "A War Examined: Operation Iraqi Freedom, 2003, A Discussion with Kevin Benson, COL (USA Retired)," *Parameters* 43, no. 4 (winter 2013–14): 119–123, 120.

4. General Tommy Franks, *American Soldier* (New York: Regan Books, 2004), 315.

5. Gordon and Trainor, *Cobra II,* 27.

6. Franks, *American Soldier,* 329; Bensahel et al., *After Saddam,* 6–7.

7. Bensahel et al., *After Saddam,* xviii.

8. Ibid., 7.

9. Gordon and Trainor, *Cobra II,* 38–54.

10. Franks, *American Soldier,* 366; Bensahel et al., *After Saddam,* 7. In the event, Turkey did not allow a ground assault from its soil.

11. "Generated Start" envisioned a large-scale 90-day build-up in Kuwait prior to invasion. "Running Start" envisioned a force-flow into Kuwait during an ongoing air campaign, with the ground invasion to commence about 25 days after the air strikes began. In the event, ground forces built up in advance of the war. Air and ground operations commenced simultaneously on March 19, 2003. Gordon and Trainor, *Cobra II,* 48–51, 551.

12. Strategic Studies Institute, "A War Examined," 120.

13. Franks, *American Soldier,* 351.

14. Ibid., 366; Bensahel et al., *After Saddam,* 8. These would include forces in theater but not on the ground in Iraq.

15. Franks, *American Soldier,* 422, 424.

16. Ibid., 366, 419; Bensahel et al., *After Saddam,* 9.

17. Franks, *American Soldier,* 419; Dobbins et al., *Occupying Iraq,* xli.

18. UNSCR 1441 demanded that Hussein disarm alleged stockpiles of weapons of mass destruction and long-range missiles.

19. Sir John Chilcot, Sir Lawrence Freedman, Sir Roderic Lyne, and Baroness Usha Prashar, *The Report of the Iraq Inquiry* (London: Williams Lea Group, 2016), vol. 3, section 3.6, 135–138.

20. Ibid., vol. 3, section 3.6, 135–138, 165; see also vol. 5, section 6.1, 175; section 6.5.

21. Ibid., vol. 3, section 3.6, 140–141.

22. Ibid., vol. 3, section 3.6, 147–148.

23. Ibid., vol. 5, section 6.4 and section 6.5, 333–338.

24. Ibid., vol. 5, section 6.5, 360–407.

25. Franks, *American Soldier,* 441 (emphasis in original).

26. Bensahel et al., *After Saddam,* xix.

27. Gordon and Trainor, *Cobra II,* 82–84.

28. A "Rock Drill" is simply a rehearsal, often using a model of the area of operations and icons representing friendly and enemy forces, in which commanders brief their actions in support of the campaign.

29. Gordon and Trainor, *Cobra II,* 87–94.

30. Ibid., 93–94.

31. Ibid., 92.

32. Ibid., 138–163. The Joint Staff conducted its own war game in the fall of 2002, Prominent Hammer II, which highlighted the need for a military headquarters for Phase IV.

33. Bensahel et al., *After Saddam,* 15.

34. Ibid.

35. Paul D. Wolfowitz, "Deputy Secretary of Defense Testimony to House Budget Committee," House Budget Committee, February 27, 2003.

36. The US government did outline a series of unclassified bullet points capturing the essence, but without using the ends-ways-means approach to strategy. *The Rumsfeld Papers,* "Principals Committee Review of Iraq Policy Paper," October 29, 2002.

37. Franks, *American Soldier,* 392–393.

38. Franks, *American Soldier,* 366; Bensahel et al., *After Saddam,* xx; Wolfowitz, "Deputy Secretary of Defense Testimony to House Budget Committee." Lieutenant General William Scott Wallace, the ground force commander, recalled, "But what in fact happened, which was unanticipated at least in [my mind], is that when [we] decapitated the regime, everything below it fell apart." "Interview with General William Scott Wallace for Frontline: The Invasion of Iraq," *PBS Frontline,* February 26, 2004.

39. Three days before the war, Vice President Richard Cheney clearly articulated this view, "My belief is we will, in fact, be greeted as liberators." Richard Cheney, "Remarks to Meet the Press," *NBC Meet the Press,* March 16, 2003. See also Joel Brinkley and Eric Schmitt, "Iraqi Leaders Say U.S. Was Warned of Disorder after Hussein, but Little Was Done," *New York Times,* November 30, 2003. Gordon and Trainor, *The Endgame,* 9–11.

40. Gordon and Trainor, *The Endgame,* 9–11.

41. Franks, *American Soldier,* 419. See also Woodward, *State of Denial,* 111–131.

42. Commission on Presidential Debates, "The Second Gore-Bush Presidential Debate Transcript," October 11, 2000; "Rumsfeld Opposed to Any U.S. Role in Nation Building," *Washington Times,* December 2, 2001; "Interview with John Hamre," *PBS Frontline,* July 23, 2004; Jonah Goldberg, "Bush, Gore, and Nation-Building," *The National Review,* October 23, 2000.

43. Gordon and Trainor, *Cobra II,* 503–504; Donald H. Rumsfeld, "Beyond 'Nation-Building,'" *Washington Post,* September 25, 2003.

44. Dobbins et al., *America's Role in Nation-Building.*

45. James T. Quinlivan, "Force Requirements in Stability Operations," *Parameters* 15, no. 4 (winter 1995–96): 59–69.

46. US Department of State, *Somalia 1992–3,* Office of the Historian, https://history.state.gov/milestones/1993-2000/somalia.

47. Laurent Dubois, *Haiti: The Aftershocks of History* (New York: Metropolitan Books, 2012); Jonathan M. Katz, *The Big Truck That Went By: How the World Came to Save Haiti and Left Behind a Disaster* (New York: Palgrave, 2013).

48. Bensahel et al., *After Saddam,* 17.

49. "Interview with James Fallows for Frontline: The Invasion of Iraq," *PBS Frontline,* February 26, 2004.

50. "American Forces in Afghanistan and Iraq," *New York Times,* June 22, 2011; "Iraq War in Figures," BBC, December 14, 2011.

51. Bensahel et al., *After Saddam,* 9; Franks, *American Soldier,* 366.

52. See Bensahel et al., *After Saddam,* 8, note 11.

53. Gordon and Trainor, *The Endgame,* 6–8.

54. Ibid., 9.

55. Bensahel et al., *After Saddam,* 31–33.

56. Ibid., 33.

57. "New State Department Releases on the 'Future of Iraq' Project," National Security Archive; "Turf Wars and the Future of Iraq," *PBS Frontline,* October 9, 2003; Bensahel et al., *After Saddam,* 29–33.

58. Bensahel et al., *After Saddam,* 32.

59. Jeffrey Goldberg, "A Little Learning," *New Yorker,* May 9, 2005.

60. "Turf Wars and the Future of Iraq," *PBS Frontline*; Bensahel et al., *After Saddam,* xx, 31. The Department of Defense, in fact, reportedly blocked the effort's leader, Tom Warrick, from becoming ORHA chief Jay Garner's deputy.

61. *The Rumsfeld Papers,* "Donald Rumsfeld: A Parade of Horribles," October 15, 2002.

62. Reported in Gordon and Trainor, *Cobra II,* 468, 570–571. For an excerpt from the report, see "A Long, Difficult, and Probably Turbulent Process," *New York Times,* October 20, 2004.

63. Gordon and Trainor, *The Endgame,* 9–10.

64. Bensahel et al., *After Saddam,* 27; Gordon and Trainor, *The Endgame,* 9–11.

65. Brent Scowcroft, "Don't Attack Saddam," *Wall Street Journal,* August 15, 2002.

66. Brinkley and Schmitt, "Iraqi Leaders Say U.S. Was Warned of Disorder after Hussein."

67. Bensahel et al., *After Saddam,* xxi–xxiii; for Task Force IV, see also 41–51, for ORHA, see 53–72.

68. Ibid., xix–xx.

21. A Complicated Approach to a Complex Situation

1. Gordon and Trainor, *Cobra II,* 478–479.

2. *The Rumsfeld Papers,* "Donald Rumsfeld: President's Goal," October 14, 2003.

3. Rajiv Chandrasekaran, "Exile Finds Ties to U.S. a Boon and a Barrier," *Washington Post,* April 27, 2003; Gordon and Trainor, *The Endgame,* 27–32.

4. Gordon and Trainor, *Cobra II,* 478–479.

5. Gordon and Trainor, *The Endgame,* 30; Bensahel et al., *After Saddam,* xiii–xiv; 106–107.

6. Dobbins et al., *Occupying Iraq,* 114–118.

7. Scott Wilson, "U.S. Delays Timeline for Iraqi Government," *Washington Post,* May 22, 2003; Bensahel et al., *After Saddam,* xxii–xxiii, 53–72.

8. Rajiv Chandrasekaran, "Iraqis Assail U.S. Plans for Council," *Washington Post,* June 3, 2003; Dobbins et al., *Occupying Iraq,* 268.

9. Rajiv Chandrasekaran, "U.S. to Appoint Council in Iraq," *Washington Post,* June 2, 2003.

10. Bensahel et al., *After Saddam,* 167. Sunni Arabs make up roughly 20 percent of the Iraqi population.

11. Dan Murphy, "Baghdad's Tale of Two Councils," *Christian Science Monitor,* October 29, 2003.

12. Dobbins et al., *Occupying Iraq,* xxvii. Dobbins notes that Bremer restrained more aggressive de-Ba'athification measures proposed by the IGC. Ricks, *Fiasco,* 154–155.

13. Gordon and Trainor, *Cobra II,* 476; Chandrasekaran, "Exile Finds Ties to U.S. a Boon and a Barrier."

14. Ricks, *Fiasco,* 162.

15. Dexter Filkins and Ian Fisher, "U.S. Is Now in Battle for Peace after Winning the War in Iraq," *New York Times,* May 3, 2003; Edmund L. Andrews and Patrick E. Tyler, "As Iraqis' Disaffection Grows, U.S. Offers Them a Greater Political Role," *New York Times,* June 7, 2003.

16. Ricks, *Fiasco,* 164.

17. Ibid.

18. "Two Killed in Baghdad Protest," CNN, June 18, 2003.

19. Gordon and Trainor, *Cobra II,* 484.

20. Gordon and Trainor, *The Endgame,* 26.

21. See L. Paul Bremer, *My Year in Iraq: The Struggle to Build a Future of Hope* (New York: Simon and Schuster, 2006), 23–49.

22. Dobbins et al., *Occupying Iraq,* xxxvi–xxxvii.

23. Ibid., xxiii–xxv.

24. Ibid., xxvi.

25. Diamond, "What Went Wrong in Iraq."

26. David Ignatius, "How ISIS Spread in the Middle East," *The Atlantic,* November 1, 2015.

27. US Embassy Baghdad cable, "Leading Sunni Arab Slate Rep on Elections, U.S. Role in Iraq," December 5, 2005.

28. "Violent Response: The U.S. Army in al-Falluja," Human Rights Watch, June 16, 2003.

29. Gordon and Trainor, *Cobra II,* 491–492.

30. Ricks, *Fiasco,* 232–240, 258–261; "Iraq: ICRC Explains Position over Detention Report and Treatment of Prisoners," International Committee of the Red Cross Resource Centre, May 8, 2004; David S. Cloud, Carla Anne Robbins, and Greg Jaffe, "Red Cross Found Widespread Abuse of Iraqi Prisoners," *Wall Street Journal,* May 7, 2004.

31. Dobbins et al., *Occupying Iraq,* 93.

32. McChrystal et al., *Team of Teams,* 2, 24, 27; McChrystal, *My Share of the Task,* 107–108, 119–122.

33. Dobbins et al., *Occupying Iraq,* 298–301.

34. Ibid., 298; *The Rumsfeld Papers,* "Info Memo from Hume Horan to the Administrator, 'Subject: Muqtada al-Sadr's Published Threats,'" July 31, 2003; *The Rumsfeld Papers,* "Memo from Secretary Rumsfeld to L. Paul Bremer, III, 'Re: CPA Issues,'" August 4, 2003.

35. Dobbins et al., *Occupying Iraq,* 301–307.

36. Ibid., 301

37. Coalition Provisional Authority, "Vision for Iraq," July 11, 2003.

38. Coalition Provisional Authority, "Achieving the Vision: Taking Forward the CPA Strategic Plan for Iraq," July 18, 2003, 1.

39. Coalition Provisional Authority, "Vision for Iraq."

40. Dexter Filkins, "Iraqi Council Picks a Cabinet to Run Key State Affairs," *New York Times,* September 2, 2003. One position, minister of information, remained unfilled. By contrast 14 of 24 members were Shi'a and 4 were Kurds.

41. Rajiv Chandrasekaran, "How Cleric Trumped U.S. Plan for Iraq," *Washington Post,* November 26, 2003.

42. Ian Sipress, "Once-Dominant Minority Forms Council to Counter Shiites and Negotiate Future," *Washington Post,* January 6, 2004. For the IGC vote, see Joel Brinkley, "Iraqi Council Agrees on National Elections," *New York Times,* December 1, 2003.

43. Dobbins et al., *Occupying Iraq,* 269–270.

44. Ibid., 114–119.

45. Yochi J. Dreazen, "Insurgents Turn Guns on Iraqis Backing Democracy," *Wall Street Journal,* December 10, 2003; Douglas Jehl, "CIA Report Suggests Iraqis Are Losing Faith in U.S. Efforts," *New York Times,* November 13, 2003; Gregg Zoroya, "Danger Puts Distance Between Council, People," *USA Today,* October 21, 2003; Scott Wilson, "Iraqi Council's Leader Is Slain," *Washington Post,* May 18, 2004.

46. Dexter Filkins and Richard A. Oppel Jr., "Truck Bombing; Huge Suicide Blast Demolishes U.N. Headquarters in Baghdad," *New York Times,* August 20, 2003.

47. Patrick E. Tyler, "Iraqi Factions Seek to Take Over Security Duties," *New York Times,* September 19, 2003; Alex Berenson, "Security: Use of Private Militias in Iraq Is Not Likely Soon, U.S. Says," *New York Times,* November 6, 2003.

48. Steven R. Hurst, "Shi'ite Picked to Be Iraq's First President," *Washington Times,* July 31, 2003.

49. "Arab League Nations Agree to Grant Seat to Iraq's Council," *New York Times,* September 9, 2003; John Daniszewski and Jailan Zayan, "Iraqi Council's Foreign Minister Takes a Seat at the Arab League's Table," *Los Angeles Times,* September 10, 2003; Bruce Stanley, "Iraq to Attend Next Week's OPEC Meeting," *New York Times,* September 17, 2003.

50. Rajiv Chandrasekaran, "Iraqi Council Denies Access to Two Arab Satellite Networks," *Washington Post,* September 24, 2003.

51. L. Paul Bremer, "Iraq's Path to Sovereignty," *Washington Post,* September 8, 2003. See also "Bremer's Seven-Step Plan for Iraqi Sovereignty," *PBS Frontline,* October 17, 2006.

52. Gordon and Trainor, *The Endgame,* 14–15; Dobbins et al., *Occupying Iraq,* xxxv.

53. "The Lost Year in Iraq," *PBS Frontline.*

54. Bremer, *My Year in Iraq,* 210–243; Rajiv Chandrasekaran, "U.N. Envoy Backs Iraqi Vote," *Washington Post,* February 13, 2004; Colum Lynch, "U.N. Plan for Iraq Transition Released," *Washington Post,* February 24, 2004; Nick Wadhams, "U.N. Says No to Iraq Elections Until at Least 2005," Associated Press Worldstream, February 23, 2004; Dexter Filkins, "Iraqi Ayatollah Insists on Vote by End of Year," *New York Times,* February 27, 2004; Tom Lasseter, "Top Cleric Spurns U.S. Plans," *Miami Herald,* January 26, 2004; Maggie Farley and Sonni Efron, "U.N. Envoy May Provide the Key to a Transfer of Power in Iraq," *Los Angeles Times,* April 14, 2004; Gordon and Trainor, *The Endgame,* 40–48; Edward Cody, "Influential Cleric Backs New Iraqi Government," *Washington Post,* June 4, 2004; William Douglas and John Walcott, "U.S. Focuses on Faster Handover to the Iraqis," *Philadelphia Inquirer,* November 13, 2003; Robin Wright and Daniel Williams, "U.S. to Back Re-Formed Iraq Body," *Washington Post,* November 13, 2003; Dobbins et al., *Occupying Iraq,* 271–273.

55. Gordon and Trainor, *The Endgame,* 20–21.

56. Ibid., 22–23, 703.

57. Mark Danner, "Abu Ghraib: The Hidden Story," *New York Review of Books,* October 7, 2004; Rajiv Chandrasekaran and Scott Wilson, "Mistreatment of Detainees Went Beyond Guards' Abuse," *Washington Post,* May 11, 2004.

58. Gordon and Trainor, *The Endgame,* 36–37.

59. Ignatius, "How ISIS Spread in the Middle East."

60. Gordon and Trainor, *The Endgame,* 57–58.

61. McChrystal, *My Share of the Task,* 89–109; Ignatius, "How ISIS Spread in the Middle East"; William McCants, *The ISIS Apocalypse: The History, Strategy, and Doomsday Vision of the Islamic State* (New York: St. Martin's Press, 2015).

62. Gordon and Trainor, *The Endgame,* 58–59.

63. Dobbins et al., *Occupying Iraq,* 307.

64. Gordon and Trainor, *The Endgame,* 66.

65. Ricks, *Fiasco,* 330–335; Gordon and Trainor, *The Endgame,* 56–66.

66. Ricardo Sánchez, *Wiser in Battle: A Soldier's Story* (New York: HarperCollins, 2008), 350–355.

67. Ibid., 364–365.

68. Gordon and Trainor, *The Endgame,* 67–73.

69. Dobbins et al., *Occupying Iraq,* 312–314; Gordon and Trainor, *The Endgame,* 66.

22. From Decisive Victory to Transition

1. David Sanger and Eric Schmitt, "Hot Topic: How U.S. Might Disengage in Iraq," *New York Times,* January 10, 2005.

2. Gordon and Trainor, *The Endgame,* 95–97.

3. Quoted in ibid., 97.

4. The White House, "President Addresses Nation, Discusses Iraq, War on Terror," June 28, 2005.

5. To deflect growing criticism about the direction of the war, the Bush administration released an unclassified version of its strategy. See US National Security Council, *National Strategy for Victory in Iraq,* November 30, 2005.

6. US Embassy Baghdad cable, "Sunni Leaders on Constitution Drafting," June 16, 2005.

7. Meghan O'Sullivan and Razzaq al-Saiedi, "Choosing an Electoral System: Iraq's Three Electoral Experiments, Their Results, and Their Political Implications," Harvard Kennedy School Belfer Center, Working Paper, April 29, 2014, 10–11. See also United Nations, "Iraq Electoral Factsheet," January 2005.

8. O'Sullivan and al-Saiedi, "Choosing an Electoral System," 10–11; Gordon and Trainor, *The Endgame,* 135.

9. Gordon and Trainor, *The Endgame,* 135.

10. "Election Guide: Republic of Iraq, 30 January 2005," International Foundation for Electoral Systems; Gordon and Trainor, *The Endgame,* 135.

11. US Embassy Ankara cables, "GOT and Turkmen Raise Grievances over Kirkuk," January 19, 2005, and "Iraqi Turkmen Groups Square Off in Turkey," June 6, 2005; US Embassy Baghdad cables, "Sunni Negotiators Stick to Call for Leaders Discuss Current Political Obstacles, Delay on Federalism Until New National Assembly," August 22, 2005; "New Election Law Highlights," September 16, 2005; "Shi'a Independent Alleges IECI Corruption," December 8, 2005, and "Electoral War Heats Up—Jazeera and Furat Channel Feuds Spark Demonstrations in Baghdad," December 15, 2005.

12. Chandrasekaran, *Imperial Life in the Emerald City,* 336.

13. US Embassy Baghdad cable, "AMB Urges Sunni Political Leaders to Denounce Violence, Help End the Insurgency," January 9, 2006; Gordon and Trainor, *The Endgame,* 140–141.

14. "173 Sunnis Freed from Secret Iraqi Torture Bunker," *Washington Times,* November 16, 2005; Gordon and Trainor, *The Endgame,* 144–150; 185–187.

15. Gordon and Trainor, *The Endgame,* 142–143, 151–157, 188–189, 191–195; Dobbins et al., *Occupying Iraq,* 327–328.

16. US Embassy Baghdad cable, "TNA Member Qasim Daoud on Constitution, Sistani, and Possible Breakup of United Iraqi Alliance," July 3, 2005.

17. US Embassy Baghdad cable, "New Election Law Highlights."

18. US Embassy Baghdad cable, "Watch Salah Ad-Din and Ninewa Provinces to Determine Fate of the Draft Constitution," October 11, 2005.

19. US Embassy Baghdad cables, "First Reactions to Referendum: Shi'a Confidence, Sunni Arabs Reflective," October 15, 2005, and "Official Result Shows the Constitution Passes—Mixed Sunni Arab Reactions," October 25, 2005.

20. US Embassy Baghdad cable, "Review of Election Campaign Status with U.S. NGO's," December 8, 2005; "Shi'a Independent Alleges IECI Corruption"; and "Election Update: Special Voting, Kirkuk Decision, and Complaints," December 13, 2005.

21. US Embassy Riyadh cable, "Ambassador Khalilzad Seeks Post-Iraqi Elections Support from Saudi Leaders," January 2, 2006.

22. US Embassy Baghdad cables, "Electoral War Heats Up"; "AMB Urges Sunni Political Leaders to Denounce Violence"; and "Constitution Review Should Be Delayed Says Sunni Hard-Line Leader Mutlak," January 20, 2006; Anthony Cordesman, *The Impact of the Iraqi Election: A Working Analysis,* Center for Strategic and International Studies, January 10, 2006, 4.

23. US Embassy Baghdad cable, "Sunni Leaders Fear Street Reaction to Election Review Report," January 21, 2006.

24. US Embassy Baghdad cable, "Shi'a Alliance Leaders Blame MNF-I and Sunnis for Current Security Situation," January 7, 2006, and "PM Ja'afari, MOI Jabr, MOD Dulime Blame Coalition for Current Security Situation," January 8, 2006.

25. Ayad Allawi, "How Iraq's Elections Set Back Democracy," *New York Times,* November 2, 2007.

26. "What the Iraqi Public Wants," WorldPublicOpinion.org, conducted by the Program on International Policy Attitudes (PIPA), January 31, 2006, 8.

27. Ali Khedery, "Why We Stuck with Maliki—and Lost Iraq," *Washington Post,* July 3, 2014.

28. Gordon and Trainor, *The Endgame,* 211–219. JAM was part of Maliki's governing coalition.

29. US Embassy Baghdad cable, "Leading Sunni Arab Slate Rep on Elections."

30. Gordon and Trainor, *The Endgame,* 216.

31. US Embassy Riyadh cable, "Ambassador Khalilzad Seeks Post-Iraqi Elections Support from Saudi Leaders."

32. Gordon and Trainor, *The Endgame,* 210.

33. "Iraq Index," Brookings Institution, September 27, 2007, 7–8.

34. Ellen Knickmeyer and Bassam Sebti, "Toll in Iraq's Deadly Surge: 1,300," *Washington Post,* February 28, 2006.

35. Gordon and Trainor, *The Endgame,* 211–212.

Conclusion to Part V

1. Bensahel et al., *After Saddam,* xvii; "The Lost Year in Iraq," *PBS Frontline*; Dobbins et al., *Occupying Iraq,* 326–333; Diamond, "What Went Wrong in Iraq." Bremer argued that failure to provide security was the biggest obstacle to progress

and lamented in 2006 the persistent lack of a military plan to defeat the Sunni insurgency. See Bremer, *My Year in Iraq,* 397–399.

2. Paul R. Pillar, "Intelligence, Policy, and the War in Iraq," *Foreign Affairs* 85, no. 2 (March/April 2006): 15–27, 18–19; Conrad C. Crane, "Phase IV Operations: Where Wars Are Really Won," *Military Review* (May–June 2005): 27–36.

3. Wolfowitz, "Deputy Secretary of Defense Testimony to House Budget Committee."

4. Interview with Lieutenant General Terry A. Wolff.

5. Michael Eisenstadt and Jeffrey White, "Assessing Iraq's Sunni Arab Insurgency," Policy Focus No. 50, Washington Institute for Near East Policy, December 2005, 2–3.

6. Gordon and Trainor, *Cobra II,* 503.

7. National Security Council, *National Strategy for Victory in Iraq,* 2.

8. Dobbins and Miller, "Overcoming Obstacles to Peace." See also Caplan, *Exit Strategies and State Building,* 316; Kolenda, *The Counterinsurgency Challenge.*

Part VI. Staying the Course in Iraq

1. Gordon and Trainor, *The Endgame,* 158–179.

2. Ibid., 267–311.

3. Peter R. Mansoor, *Surge: My Journey with General David Petraeus and the Remaking of the Iraq War* (New Haven, Conn.: Yale University Press, 2013); Gordon and Trainor, *The Endgame,* 329–350.

4. Ignatius, "How ISIS Spread in the Middle East."

5. Woodward, *State of Denial.*

6. Stanley, *Paths to Peace,* shows that a change in regime is normally needed to alter a "sticky" strategy.

23. Achieving Milestones While Losing the War

1. United Nations Security Council Resolutions 1546, June 8, 2004, and 1723, November 28, 2006; see also US Department of Defense, *Measuring Stability and Security in Iraq,* Report to Congress, October 2005, 6.

2. National Security Council, *National Strategy for Victory in Iraq,* 20.

3. US Department of Defense, *Measuring Stability and Security in Iraq,* October 2005, 2–4.

4. Gareth Porter, "US/IRAQ: General Reveals Rift with Rumsfeld on Insurgents," Inter Press Service, April 15, 2006; National Security Council, *National Strategy for Victory in Iraq,* 6–7.

5. National Security Council, *National Strategy for Victory in Iraq,* 6–7.

6. David Morgan, "U.S. Trying to Understand Iraq Insurgency: Negroponte," Reuters, September 29, 2005.

7. US Department of Defense, *Measuring Stability and Security in Iraq,* Report to Congress, March 2008, 18; events and red trend-lines added. For more on the challenges of assessing insurgencies, see Thomas C. Mayer, *War without Fronts: The American Experience in Vietnam* (Boulder, Colo.: Westview Press, 1985). On analytical measures, see James G. Roche and Barry D. Watts, "Choosing Analytic Measures," *Journal of Strategic Studies* 14, no. 2 (1991): 165–209.

8. National Security Council, *National Strategy for Victory in Iraq,* 13. Security metrics included: "The quantity and quality of Iraqi units; the number of actionable intelligence tips received from Iraqis; the percentage of operations conducted by Iraqis alone or with minor Coalition assistance; the number of car bombs intercepted and defused; offensive operations conducted by Iraqi and Coalition forces; and the number of contacts initiated by Coalition forces, as opposed to the enemy. . . . These indicators have more strategic significance than the metrics that the terrorists and insurgents want the world to use as a measure of progress or failure: number of bombings."

9. See Etienne Vincent, Philip Eles, and Boris Vasiliev, "Opinion Polling in Support of Counterinsurgency," in *The Cornwallis Group XIV: Analysis of Societal Conflict and Counter-Insurgency* (Ottawa, Canada: Canadian Expeditionary Forces Command Operational Research Team Centre for Operational Research and Analysis, Defence Research and Development, 2009).

10. "Nationwide Poll of Iraq," USA Today/CNN/Gallup, March/April 2004.

11. US Department of State, "Opinion Analysis," M-106-04, Office of Research, September 16, 2004, Appendix 6A; Bensahel et al., *After Saddam,* xxvi, note 7.

12. Several polls in 2004 and 2005 showed Sunni support for insurgent attacks ranging from 43 percent to 85 percent. See Michael Eisenstadt, "The Sunni Arab Insurgency: A Spent or Rising Force?" *Policy Watch 1028,* The Washington Institute for Near East Policy, August 26, 2005; "What the Iraqi Public Wants," World Public Opinion.org, January 2–5, 2006; see also "Iraq Index," Brookings Institution, September 27, 2007.

13. Woodward, *State of Denial,* 471.

14. "Iraq Insurgency in 'Last Throes,' Cheney Says," CNN, June 20, 2005. The article refers to a May 30, 2005, interview of Cheney on *Larry King Live.*

15. "Interview with General John Abizaid," *Face the Nation,* CBS, June 26, 2005. In a population of 27 million Iraqis, 0.1 percent would be 27,000 people. If accurate, such a percentage would be historically very low by comparison. See Eisenstadt and White, "Assessing Iraq's Sunni Arab Insurgency," 8–11, who assess a more likely figure of 100,000.

16. "Rumsfeld: Iraq Not Fated to Civil War," CNN, August 23, 2005; Ricks, *Fiasco,* 168–172.

17. US Department of Defense, *Measuring Stability and Security in Iraq,* October 2005, 3.

18. Woodward, *State of Denial,* 475.

19. National Security Council, *National Strategy for Victory in Iraq,* 7. By August 2006, the Department of Defense acknowledged the increasing risk of sectarian

civil war but noted optimism that movement toward civil war could be prevented. US Department of Defense, *Measuring Stability and Security in Iraq,* Report to Congress, August 2006, 34.

20. Interviewee K: Elements in Iraqi society were mobilizing against foreign occupation. See also David Ignatius, "A Shift on Iraq," *Washington Post,* September 26, 2005; Nagl, *Knife Fights,* 161.

21. "Rumsfeld: Don't Call Iraqi Enemy 'Insurgents,'" NBC News, November 29, 2005.

22. Interview with Lieutenant General James Dubik, commander of Multi-National Security Transition Command-Iraq, 2007–2008.

23. Gerry J. Gilmore, "Iraqis, Not Coalition, Must Defeat Insurgents, Rumsfeld Says," American Forces Press Service, March 30, 2005.

24. Gordon and Trainor, *The Endgame,* 137–139.

25. US Department of Defense, *Measuring Stability and Security in Iraq,* August 2006.

26. Woodward, *State of Denial,* 400.

27. Nir Rosen, "If America Left Iraq: The Case for Cutting and Running," *The Atlantic,* December 2005, forecasted, "If the occupation were to end, so, too, would the insurgency. After all, what the resistance movement has been resisting is the occupation."

24. Trapped by Partners in a Losing Strategy

1. Interview with Douglas E. Lute, who emphasized the lack of a coordinated diplomatic-political-military strategy; interviewees A, B, C, D, and O.

2. US Embassy Baghdad cables, "Deputy Secretary Covers Political Process, Economics and Sectarian Violence in May 19 Meeting with Deputy Prime Minister Rowsch Shaways," June 1, 2005; "Charge Advances USG Approach to Constitution and Sunni Inclusion with UNAMI," June 13, 2005; "SCIRI Leader Hakim Offers Charge Assurances of Flexibility on Sunni Inclusion, Constitutional Issues," July 5, 2005; "Iraq Beginning to Focus on Planning for Reconciliation and Consolidating National Unity," June 5, 2006; "The National Reconciliation and Dialogue Project," June 22, 2006; "Prime Minister Ready to Launch National Reconciliation Proposal," June 25, 2006; "Ambassador's August 6 and 7 Meetings with PM Maliki," and "Iraqi MOD—Quelling Violence Top Priority," August 12, 2006; "Iraqi Prime Minister Upbeat to CODEL Frist," October 7, 2006; "PM Raises Sunni Leadership, UNSCR and International Compact with Ambassador," November 13, 2006; "Prime Minister Maliki Discusses Moderate Front and Security with Senators Dodd and Kerry," December 20, 2006.

3. US Embassy Baghdad cable, "Sunni Arab Outreach in Iraq: Mission Plans," September 6, 2005.

4. National Security Council, *National Strategy for Victory in Iraq,* 6–7.

5. US Embassy Baghdad cable, "Iraq Beginning to Focus on Planning for Reconciliation."

6. US Embassy Baghdad cable, "Jafari Pledges More Sunni Dialogue, Says Insurgency Appears Isolated, Desperate," June 1, 2005.

7. US Embassy Baghdad cables, "Iraq Beginning to Focus on Planning for Reconciliation"; "The National Reconciliation and Dialogue Project," June 22, 2006; and "Prime Minister Ready to Launch National Reconciliation Proposal," June 25, 2006.

8. Gordon and Trainor, *The Endgame,* 173. For other examples, see ibid., 26, 35–37, 82–84, 96–97, 131–133, 135–139, 141, 168–175, 216–219, 228–229, 239, 241–263.

9. US Embassy Baghdad cables, "The President's 28 June Speech Provokes Both Applause and Complaints from Iraqi Politicians," July 11, 2005; "July 2 Meeting of Charge MNF-I CG with Iraqi Interior Minister," July 15, 2005; "Sectarian Violence Hampers Sunni Participation in Political Process," August 30, 2005.

10. US Embassy Baghdad cable, "Action Plan to Build Capacity and Sustainability Within Iraqi's Provincial Governments," October 1, 2005.

11. US Embassy Baghdad cables, "MOI Press Conference Misses Human Rights Mark"; July 11, 2005; "Militias and Other Armed Groups in Iraq—Confronting the Sectarian Divide," March 13, 2006; "Security Still Main Concern for Sunni Leaders," March 24, 2006; "Demarche to Iraqi Interior Minister on Site 4," August 7, 2006; and "Ambassador and General Casey Urge PM Maliki to Act Decisively to End Violence," October 16, 2006.

12. US Embassy Baghdad cables, "Deputy Secretary Covers Political Process, Economics and Sectarian Violence," "Jafari Pledges More Sunni Dialogue," "Deputy Speaker of Council of Representatives: Problems with the Speaker and Fear of Baathists," August 6, 2006; and "Senator Brownback Meets with Iraqi PM Maliki," January 12, 2007.

13. US Embassy Baghdad cable, "US Forces in Baghdad Implement Large-Scale Human Rights Initiatives," June 3, 2005.

14. Ibid.; US Embassy Baghdad cables, "Rusafa Prison Conditions Surprisingly Good," July 5, 2005; "Hard-Line Sunni Arab Group Urges U.S. Stay in Iraq," July 7, 2005; "MOI Press Conference Misses Human Rights Mark"; "New Cases of Apparent Abuses Raise More Sunni Arab Complaints against Interior Ministry," July 21, 2005; "Sectarian Violence Hampers Sunni Participation in Political Process"; "MCNS Meets to Discuss Security in Baghdad, Tal Afar, Sectarian Strife," October 6, 2005; "CODEL Shays Calls on Iraqi Minister of Interior," October 10, 2005; "CODEL Hoekstra Meets Iraqi Minister of Interior," and "District Council Members Declare No Trust in Iraqi Police," March 24, 2006; "Fallujah: Army-police Friction and Perceived U.S. 'Mixed Messages,'" April 3, 2006; "Meeting with New Minister of Interior, Jawad Al-Bolani," June 22, 2006; "Demarche to Iraqi Interior Minister on Site 4"; "Allegations of Secret Prisons in Kurdistan"; "SECDEF Meets with Iraqi National Security Team," December 24, 2006.

15. US Embassy Baghdad cable, "Charge Satterfield Meeting with Iraqi Minister of Interior," June 8, 2005.

16. US Embassy Baghdad cable, "July 2 Meeting of Charge MNF-I CG with Iraqi Interior Minister."

17. US Embassy Baghdad cables, "Updated Status of Iraq Detainee Abuse Investigations," December 1, 2005, and "Bunker Investigation Falters, Nationwide Inspections Gather Steam," December 15, 2005.

18. US Embassy Baghdad cable, "Under Secretary of Defense Edelman Meets with Iraqi Minister of Interior," October 20, 2005.

19. Robin Wright and Jim VandeHei, "Unlikely Allies Map Future," *Washington Post,* June 24, 2005.

20. Eric Schmitt, "2,000 More M.P.'s Will Help Train the Iraqi Police," *New York Times,* January 16, 2006.

21. For a description of the sectarian strategy for control of Baghdad, see Gordon and Trainor, *The Endgame,* 298.

22. US Embassy Baghdad cable, "Fallujah: Grass Roots Politics—Leaders Initiate Political and Security Meetings," June 20, 2005.

23. US Embassy Baghdad cable, "Successful Damage Control after Raid on Leading Sunni Party Leader," June 11, 2005.

24. US Embassy Baghdad cable, "Dulame and Rubaie Disucss Iraqi Security with Ambassador," July 27, 2005.

25. Ignatius, "How ISIS Spread in the Middle East"; Knickmeyer and Sebti, "Toll in Iraq's Deadly Surge."

26. US Embassy Baghdad cable, "Fallujah: Army-Police Friction and Perceived U.S. 'Mixed Messages.'"

27. Dexter Filkins, "What We Left Behind in Iraq," *New Yorker,* April 28, 2014.

28. Nussaibah Younis, "The US-Iraq Disconnect over Fighting ISIS," Atlantic Council, December 18, 2015.

29. US Embassy Baghdad cable, "Dawa Hardliner Insists Deep-seated Fear of Former Regime Drives Shi'a Politics," March 20, 2006.

30. US Embassy Baghdad cables, "Prime Minister Maliki Discusses Moderate Front and Security with Senators Dodd and Kerry," December 20, 2006; "Deputy Speaker of Council of Representatives: Problems with the Speaker and Fear of Baathists," August 6, 2006; "Senator Brownback Meets with Iraqi Pm Maliki," January 12, 2007.

31. Gordon and Trainor, *The Endgame,* 227, 239.

32. US Embassy Baghdad cable, "Ambassador's Meeting with Mindef Al-Dulime," September 4, 2005; "PM Ja'afari Plans Major Shakeup of Iraqi Military Leadership; Warned by MNF-I and Embassy," December 4, 2005; "The Disillusioned MINDEF Dulime," March 25, 2006; "MINDEF Dulime Conveys Security Concerns to Ambassador," April 30, 2006.

33. US Embassy Baghdad cable, "Collaboration and Confrontation Between MOI and MOD Forces," March 3, 2006.

34. US Embassy Baghdad cable, "Iraqi MOD—Quelling Violence Top Priority."

35. US Embassy Baghdad cable, "Iraqi Army Officers Discuss Internal Militia Influence, U.S. Troop Levels," January 14, 2007.

36. Filkins, "What We Left Behind in Iraq."

37. US Embassy Baghdad cable, "Meeting with New Minister of Interior, Jawad Al-Bolani."

38. US Embassy Baghdad cable, "Demarche to Iraqi Interior Minister on Site 4."

39. US Embassy Baghdad cable, "Leahy in Iraq," October 9, 2006.

40. US Embassy Baghdad cable, "CODEL Reed Advises MOI That Time Is Running Out," October 13, 2006.

41. Gordon and Trainor, *The Endgame,* 272.

42. US Embassy Baghdad cable, "MOI Bolani Slow to Develop Reform Plan," January 7, 2007.

43. Nouri al-Maliki, "Our Strategy for a Democratic Iraq," *Washington Post,* June 9, 2006; Solomon Moore, "U.S. Diplomat Defends Maliki's Strategy," *Los Angeles Times,* October 1, 2006; Dan Murphy, "Kerry Has Advice for Maliki, but the US Has Few Good Options in Iraq," *Christian Science Monitor,* June 23, 2014.

44. Filkins, "What We Left Behind in Iraq."

45. Nancy Trejos, "U.S. Report Rejected by Iraqi President," *Washington Post,* December 11, 2006.

46. Quoted in Gordon and Trainor, *The Endgame,* 273.

25. Mirror Imaging Civil-Military Relations

1. Huntington, *The Soldier and the State.*

2. Interview with Lieutenant General James Dubik, who recalled that Maliki expressed such concerns to him. Although the prime minister did not use the word coup himself, Dubik said Maliki was clear what was meant. Interviewee A also noted Maliki's fears about a coup.

3. Tim Weiner, *Legacy of Ashes: The History of the CIA* (New York: Anchor Books, 2007).

4. US Embassy Baghdad cables, "Joint Working Group to Assess Iraqi Armed Forces," July 3, 2006; "August 18 Meeting Between Iraqi PM, Ambassador, and MNF-I CG," August 22, 2006; "Ambassador and General Casey Urge PM Maliki to Act Decisively to End Violence"; "Iraqi PM Seeks Greater Authority but Questions Troop Readiness," October 29, 2006; and "PM Raises Sunni Leadership, UNSCR and International Compact with Ambassador," November 13, 2006.

5. US Embassy cable, "Iraqi PM Seeks Greater Authority but Questions Troop Readiness" (emphasis added).

6. Ibid.; US Embassy Baghdad cables, "PM Tells CODEL Mccain Better Weapons and Quicker Transfer Needed, Not More Troops," December 14, 2006; "Iraqi PM Maliki Tells CODEL Pelosi 50,000 U.S. Troops Could Be out in Three to Six

Months, Iraq Seeks Lead on Security," January 30, 2007; and "PM Maliki Frustrated with the Slow Pace of the BSP," February 8, 2007.

7. US Embassy Baghdad cable, "Iraqi Prime Minister Upbeat to CODEL Frist."

8. US Embassy Baghdad cables, "Ambassador's August 6 and 7 Meetings with PM Maliki"; "August 18 Meeting Between Iraqi PM, Ambassador, and MNF-I CG"; "Ambassador and General Casey Urge PM Maliki to Act Decisively to End Violence"; and "PM Maliki Frustrated with the Slow Pace of the BSP."

9. US Embassy Baghdad cable, "Iraqi PM Maliki Urges Iraqi Control over Security, in Meeting with NSA Hadley," November 9, 2006.

10. US Embassy Baghdad cable, "PM Maliki Frustrated with the Slow Pace of the BSP."

11. US Embassy Baghdad cable, "Ambassador and General Casey Urge PM Maliki to Act Decisively to End Violence."

12. Interview with Lieutenant General James Dubik; interviewees A, D, and O.

13. For a discussion of the operation, see Gordon and Trainor, *The Endgame,* 470–504.

26. To Surge or Not to Surge

1. Kahneman, *Thinking, Fast and Slow,* 284.

2. Ibid., 273, 283–286.

3. Daniel Kahneman and Amos Tversky, "Choices, Values, and Frames," *American Psychologist* 39, no. 4 (1984): 341–350.

4. Kahneman, *Thinking, Fast and Slow,* 285.

5. Ibid., 292–299; Daniel Kahneman, Jack L. Knetsch, and Richard H. Thaler, "Experimental Tests of the Endowment Effect and the Coase Theorem," *Journal of Political Economy* 98, no. 6 (1990): 1325–1348; Daniel Kahneman and Amos Tversky, "Prospect Theory: An Analysis of Decision under Risk," *Econometrica* 47, no. 2 (1979): 263–292. For criticism of the endowment effect, see Michael W. Hanemann, "Willingness to Pay and Willingness to Accept: How Much Can They Differ? Reply," *American Economic Review* 81, no. 3 (1991): 635–647.

6. Iklé, *Every War Must End,* 83; Reiter, *How Wars End,* 15–16, notes that belligerents will raise demands after successes and lower them after defeats. Reiter, however, does not account for prospect theory's notion that potential future gains tend to be less important than previous ones.

7. Kahneman, *Thinking, Fast and Slow,* 278–288; Kahneman and Tversky, "Prospect Theory."

8. Downs and Rocke, "Conflict, Agency, and Gambling for Resurrection."

9. Kahneman, *Thinking, Fast and Slow,* 317, 354–345.

10. National Security Council, *National Strategy for Victory in Iraq,* 2–6 (emphasis added).

11. US Department of Defense, *Measuring Stability and Security in Iraq,* Report to Congress, July 2005, October 2005, February 2006, May 2006, and August 2006.

12. Andrew F. Krepinevich Jr., "How to Win in Iraq," *Foreign Affairs* 84, no. 5 (September/October 2005): 87–104.

13. Mansoor, *Surge,* 5–33.

14. Joseph R. Biden Jr. and Leslie H. Gelb, "Unity Through Autonomy in Iraq," *New York Times,* May 1, 2006

15. Bob Woodward, *The War Within: A Secret White House History 2006–2008* (New York: Simon and Schuster, 2008), 4–13.

16. Mansoor, *Surge,* 32–33.

17. James A. Baker III and Lee Hamilton, *The Iraq Study Group Report: The Way Forward: A New Approach* (Washington, D.C.: US Government Printing Office, 2006), 50; Gordon and Trainor, *The Endgame,* 279–280.

18. Quoted in Woodward, *The War Within,* 10.

19. Kimberly Kagan, *The Surge: A Military History* (New York: Encounter Books, 2009), 27–29.

20. Gordon and Trainor, *The Endgame,* 304.

21. Matthew Kaminski, "Why the Surge Worked," *Wall Street Journal,* September 20, 2008; Woodward, *The War Within,* 281.

22. "Text of U.S. Security Adviser's Iraq Memo," *New York Times,* November 29, 2006; Gordon and Trainor, *The Endgame,* 287–291.

23. Bush, *Decision Points,* 93–94; Michael A. Fletcher and Peter Baker, "Bush Ousts Embattled Rumsfeld; Democrats Near Control of Senate," *Washington Post,* November 9, 2006.

24. Woodward, *The War Within,* 232–234.

25. Gordon and Trainor, *The Endgame,* 293–294.

26. US Department of Defense, *Measuring Stability and Security in Iraq,* August 2006.

27. Sabrina Tavernise and John F. Burns, "Promising Troops Where They Aren't Really Wanted," *New York Times,* January 11, 2007; Gordon and Trainor, *The Endgame,* 308.

27. A New Plan on Shaky Foundations

1. George W. Bush, "President Bush Addresses Nation on Iraq War," Congressional Quarterly Transcripts Wire, January 10, 2007.

2. According to a Gallup survey, support for the war continuously decreased from 72 percent on March 22/23, 2003, to 36 percent on January 15–18, 2007, Gallup.com, "In Depth: Topics A to Z: Iraq" (accessed April 30, 2019).

3. Interviewees A, B, C, D, and O; quotation from interviewee O.

4. Interview with Douglas E. Lute.

5. US Department of Defense, *Measuring Stability and Security in Iraq,* Report to Congress, March 2007, 1–2.

6. Biddle et al., "Testing the Surge."

7. Gordon and Trainor, *The Endgame,* 369–388.

8. Filkins, "What We Left Behind in Iraq"; Gordon and Trainor, *The Endgame,* 213–219.

9. Interviewees A and D; Gordon and Trainor, *The Endgame,* 502–503.

10. Interviewees C, K, and B.

11. Interview with Douglas E. Lute; interviewees A, B, and D.

12. Interview with Douglas E. Lute.

13. General (Ret.) James L. Jones, USMC, *The Report of the Independent Commission on the Security Forces of Iraq,* September 6, 2007, 17–20.

14. For a detailed discussion of reform efforts, see Lieutenant General James M. Dubik, "Building Security Forces and Ministerial Capacity: Iraq as a Primer," Institute for the Study of War, 2009.

15. Interview with Lieutenant General James Dubik.

16. Warrick, *Black Flags*; Ignatius, "How ISIS Spread in the Middle East"; McCants, *The ISIS Apocalypse.*

17. Interview with General David H. Petraeus.

18. Interviewee A.

19. Interviewees A, B, D, K, and O.

20. Interviewees A, B, D, and K.

Conclusion to Part VI

1. Clausewitz, *On War,* 100–112.

2. "Public Attitudes Toward the War in Iraq: 2003–2008," Pew Research Center, March 19, 2008.

3. "U.S. Politics and Policy, Section 2: Views of Iraq and Afghanistan," Pew Research Center, September 24, 2008.

4. Barack Obama, "My Plan for Iraq," *New York Times,* July 14, 2008; "Obama's Key Promises," *Washington Post,* January 20, 2010.

5. "Exit Polls: Obama Wins Big Among Young, Minority Voters," CNN Election Center, November 4, 2008.

6. Joseph Logan, "Last U.S. Troops Leave Iraq, Ending War," Reuters, December 18, 2011.

Part VII. Ending the War in Iraq

1. Gordon and Trainor, *The Endgame,* 21–26; 184–189.

2. Ibid., 369–388.

3. Priyanka Boghani, "James Jeffrey: Iraq Was a 'Historic, Dramatic' Failure for Bush and Obama," *PBS Frontline,* July 29, 2014.

4. US CENTCOM commander Admiral Fox Fallon argued in June 2007 that the United States should reduce its presence in Iraq due to failures to advance reconciliation, among other issues. Gates, *Duty,* 68–70.

5. "Iraqi PM backs Obama Troop Exit Plan: Report," Reuters, July 19, 2008.

6. Gordon and Trainor, *The Endgame,* 673–688.

7. Stephen Wicken, "Iraq's Sunnis in Crisis," Institute for the Study of War, May 2013; Jessica Lewis, "The Islamic State of Iraq Returns to Diyala," Institute for the Study of War, April 2014.

8. McCants, *The ISIS Apocalypse*; Warrick, *Black Flags*; Gordon and Trainor, *The Endgame,* 230.

9. Lewis, "The Islamic State of Iraq Returns to Diyala."

10. Michael R. Gordon and Julie Hirschfeld Davis, "In Shift, U.S. Will Send 450 Advisers to Help Iraq Fight ISIS," *New York Times,* June 10, 2015.

28. The Surge Misunderstood

1. Some contend even the military success was illusory, arguing that ethnic cleansing had been so successful that there were simply fewer targets for sectarian rivals to attack. See Nils B. Weidmann and Idean Salehyan, "Violence and Ethnic Segregation: A Computational Model Applied to Baghdad," *International Studies Quarterly* 57, no. 1 (2013): 52–64; Lawrence Korb, Brian Katulis, Sean Duggan, and Peter Juul, *How Does This End? Strategic Failures Overshadow Tactical Gains in Iraq* (Washington, D.C.: Center for American Progress, 2008). For a compelling refutation of this argument, see Biddle et al, "Testing the Surge."

2. Kagan, *The Surge*; John McCain and Joe Lieberman, "The Surge Worked," *Wall Street Journal,* January 10, 2008; Max Boot, "The Truth about Iraq's Casualty Count," *Wall Street Journal,* May 3, 2008; James R. Crider, "A View from Inside the Surge," *Military Review* 89, no. 2 (March/April 2009): 81–88; General David H. Petraeus, "How We Won in Iraq," *Foreign Policy,* October 29, 2013; Linda Robinson, *Tell Me How This Ends: General David Petraeus and the Search for a Way Out of Iraq* (New York: Perseus, 2008); Baker, "Petraeus Parallels Iraq, Afghanistan Strategies."

3. Austin Long, "The Anbar Awakening," *Survival* 50, no. 2 (April/May 2008): 67–94; Steven Simon, "The Price of the Surge," *Foreign Affairs* 87, no. 3 (2008): 57–76; Marc Lynch, "Sunni World," *American Prospect* (September 13, 2007); Jim Michaels, *A Chance in Hell: The Men Who Triumphed Over Iraq's Deadliest City and Turned the Tide of War* (New York: St. Martin's, 2010); Daniel R. Green, "The Fallujah Awakening: A Case Study in Counter-Insurgency," *Small Wars and Insurgencies* 21, no. 4 (December 2010): 591–609.

4. Biddle et al., "Testing the Surge"; Stephen Biddle, "Stabilizing Iraq from the Bottom Up," testimony before the US Senate Foreign Relations Committee, April 2, 2008; Gates, *Duty,* 51; Stephen Biddle, Michael O'Hanlon, and Kenneth Pollack, "How to Leave a Stable Iraq: Building on Progress," *Foreign Affairs* 87, no. 5 (September/October 2008): 40–58; Carter Malkasian, "Did the Coalition Need More Forces in Iraq?," *Joint Force Quarterly* 46, no. 3 (2007): 120–126; Mansoor, *Surge,* 266. Colin H. Kahl, "Walk before Running," *Foreign Affairs* 87, no. 4 (July/August 2008): 151–154, also credits congressional threats of withdrawal.

5. Biddle et al., "Testing the Surge," 9–10, 36–40.

6. Bush, "President Bush Addresses Nation on Iraq War." See also Mansoor, *Surge,* 260–274.

7. Petraeus, "How We Won in Iraq."

8. Ibid.; Zachary Keck, "History's Judgment: The Iraq Surge Failed," *The Diplomat,* June 13, 2014; Peter Beinart, "The Surge Fallacy," *The Atlantic,* September 2015; Mansoor, *Surge,* 269; Emma Sky, *The Unraveling: High Hopes and Missed Opportunities in Iraq* (New York: Perseus, 2015); Amber Phillips, "On Iraq, President Obama Is Getting as Much Blame as George W. Bush," *Washington Post,* June 3, 2015. See also "Who Lost Iraq?" *Politico* (July/August 2015), https://politico.com/magazine/story/2015/06/24/iraq-roundtable; Ignatius, "How ISIS Spread in the Middle East."

29. The Absence of a Political Strategy Erodes US Leverage

1. George W. Bush, "Address to the Nation on the War on Terror from Fort Bragg, North Carolina," June 28, 2005. See also Octavian Manea and John A. Nagl, "COIN Is Not Dead: An Interview with John Nagl," *Small Wars Journal,* February 6, 2012.

2. Biddle et al., "Small Footprint, Small Payoff," 7.

3. Paul et al., *Paths to Victory,* 177–178.

4. Jeanne F. Hull, "Iraq: Strategic Reconciliation, Targeting, and Key Leader Engagement," Strategic Studies Institute, September 2009; Mansoor, *Surge,* 86.

5. Interviewees B and C.

6. Petraeus, "How We Won in Iraq"; Boghani, "James Jeffrey."

7. Gates, *Duty,* 38–57.

8. Hillary Rodham Clinton, "Clinton's Speech on Iraq, March 2008," Council on Foreign Relations, March 17, 2008; Thom Shanker, "Campaign Promises on Ending the War in Iraq Now Muted by Reality," *New York Times,* December 3, 2008.

9. United Nations, "Security Council, Acting on Iraq's Request, Extends 'For Last Time' Mandate of Multinational Force," December 18, 2007, renews UNSCR 1790.

10. Gordon and Trainor, *The Endgame,* 530–539.

11. "Interview with Iraqi Leader Nouri al-Maliki: 'The Tenure of Coalition Troops in Iraq Should Be Limited,'" *Der Spiegel,* July 19, 2008. American forces, said Maliki, should withdraw "As soon as possible, as far as we're concerned. US presidential candidate Barack Obama talks about 16 months. That, we think, would be the right timeframe for a withdrawal, with the possibility of slight changes."

12. Dan Balz, "Obama Makes War Gains," *Washington Post,* July 22, 2008; Chris Weigant, "Maliki's Leverage over Bush," *Huffington Post,* September 10, 2008; "Iraqi Backing of Obama Plan Irks White House," Associated Press, July 21, 2008.

13. US Embassy Baghdad cable, "President Barzani—Nothing Is as It Should Be," November 23, 2008; Beinart, "The Surge Fallacy."

14. Interviewee B.

15. Interviewees A and B; Campbell Robertson and Tariq Maher, "35 Iraq Officials Held in Raids on Key Ministry," *New York Times,* December 17, 2008; Rod Nordland, "Maliki Contests the Result of Iraq Vote," *New York Times,* March 27, 2010; David Ignatius, "Beyond the Coup Rumors, Options for Iraq," *Washington Post,* October 13, 2006.

16. Maliki explored plans in July 2008 for downsizing the ISF, potentially to reduce the influence of opponents. See US Embassy Baghdad cable, "Maliki Calls for Downsizing Iraqi Security Forces," July 24, 2009. Gordon and Trainor, *The Endgame,* 539.

17. US Embassy Baghdad cables, "Maliki Says Neighbors and JAM Are Serious Threats," May 22, 2007, and "The Ambassador's and General Petraeus' April 14 Meeting with PM Maliki," April 17, 2008; Christopher M. Blanchard, "Iraq: Regional Perspectives and U.S. Policy," Congressional Research Service, December 1, 2008; US Embassy Doha cable, "U.S.-Qatar Gulf Security Dialogue (GSD)," December 18, 2007. For the Saudi Arabian government position, as told to US officials, see US Embassy Riyadh cables, "Saudi King Abdullah and Senior Princes on Saudi Policy Toward Iraq," April 20, 2008, and "Saudi Mfa Official on Iraq," August 6, 2008; Robert Kennedy, "Iraqi PM: Saudi Has a 'Culture of Terrorism,'" *al Jazeera,* September 9, 2011); for Kuwaiti government views, see US Embassy Kuwait cable, "For Kuwait, the SOFA a Litmus Test of Iraqi Intentions and Iranian Influence," October 29, 2008.

18. US Embassy Baghdad cable, "Iraqi NSA Rubai on Ahmedi-Nejad's Visit—Corrected Copy," March 25, 2008; Michael Eisenstadt, Michael Knights, and Ahmed Ali, "Iran's Influence in Iraq: Countering Tehran's Whole-of-Government Approach," Washington Institute for Near East Policy, April 2011.

19. Interviewees B and K. For growing tensions with the Kurds, see US Embassy Baghdad cables, "AMB, CG and PM Discuss SOFA, SOI, Ambassador's Trip to Erbil, GOI/KRG Relations and Elections Law," September 21, 2008, "President Barzani—Nothing Is as It Should Be"; "Defense Under Secretary Edelman Meets Leaders of Kirkuk's Ethnic Blocs; Kurds Refuse to Attend," November 1, 2008, and "KRG Officials on Article 140 and Kirkuk," January 11, 2008.

20. US Embassy Baghdad cable, "Maliki Confidante Careful on SOFA and Disputes Importance of Sunni Arab Tribes for Security," August 29, 2008.

21. US Embassy Baghdad cable, "AMB, CG and PM Discuss SOFA, SOI, Ambassador's Trip to Erbil, GOI/KRG Relations and Elections Law."

22. The White House, "Press Briefing by Dana Perino," July 21, 2008; Gordon and Trainor, *The Endgame,* 529–532; Karen DeYoung and Sudarsan Raghavan, "U.S., Iraqi Negotiators Agree on 2011 Withdrawal," *Washington Post,* August 22, 2008; Julian E. Barnes and Paul Richter, "Bush Agrees to 'Horizon' for Pullout," *Los Angeles Times,* July 19, 2008.

23. "Bush Vetoes War-Funding Bill with Withdrawal Timetable," CNN, May 2, 2007.

24. Gates, *Duty,* 51–52.

25. See US Department of State, "Fact Sheet: The New Way Forward in Iraq," January 10, 2007; Petraeus, "How We Won in Iraq"; Gordon and Trainor, *The Endgame,* 585.

26. Gordon and Trainor, *The Endgame,* 532–535.

27. Ibid., 528; US Embassy Baghdad cable, Vice President Hashimi On: the Provincial Elections Law, SFA/SOFA Negotiations," July 10, 2008.

28. Gates, *Duty,* 235–236.

29. Gordon and Trainor, *The Endgame,* 539–540.

30. Interviewee K.

31. Rice, *No Higher Honor,* 694–695. Rice complained that Maliki reneged as soon as Rice returned to Washington.

32. Gordon and Trainor, *The Endgame,* 541; Gates, *Duty,* 236–237. According to Gates, Petraeus reported that an Iranian brigadier general had been arrested for bribing Iraqi officials with $250,000 each to vote against the SOFA.

33. Interviewee K.

34. John R. Crook, "Contemporary Practice of the United States Relating to International Law," *American Journal of International Law* 103, no. 1 (2009): 132–135; Gordon and Trainor, *The Endgame,* 558; Alissa J. Rubin and Campbell Robertson, "Iraq Backs Deal That Sets End of U.S. Role," *New York Times,* November 28, 2008.

35. A revised de-Ba'athification law and a law governing the distribution of oil revenues were among the US congressional benchmarks for Iraq in term of Iraqi national reconciliation (Gates, *Duty,* 60, 231), but the causal linkage between the laws and durable political conclusion are highly suspect. Lionel Beehner and Greg Bruno, "What are Iraq's Benchmarks?" Council on Foreign Relations, March 11, 2008.

36. Nonetheless, Iraq's Vice President Tariq Hashimi, a Sunni Arab, threatened to veto the SOFA, which led to unspecified resolutions for reform in the Council of Representatives (CoR), US Embassy Baghdad cable, "Iraq 201: The Council of Representatives," March 18, 2009.

37. Gordon and Trainor, *The Endgame,* 542–548; interviewees O, A, and D.

38. Gordon and Trainor, *The Endgame,* 569–570.

39. Interviewees K and D.

40. US Embassy Baghdad cable, "PM Maliki: Strengthened Center or Emerging Strongman," February 13, 2009, quoted in Gordon and Trainor, *The Endgame,* 586.

41. Gordon and Trainor, *The Endgame,* 582.

42. Interviewees A, B, C, D, and K; Fred Kaplan, *The Insurgents: David Petraeus and the Plot to Change the American Way of War* (New York: Simon and Schuster, 2013), 263–264, 341.

43. Gordon and Trainor, *The Endgame,* 581–586; Rick Brennan, Charles P. Ries, Larry Hanauer, Ben Connable, Terrence Kelly, Michael J. McNerney, Stephanie Young, Jason H. Campbell, and K. Scott McMahon, *Ending the U.S. War in Iraq: The Final Transition, Operational Maneuver, and Disestablishment of United States Forces-Iraq* (Santa Monica, Calif.: RAND Corporation, 2013), 77–78; Thomas E.

Ricks, "Iraq, the Unraveling (XXIV): U.S. Embassy vs. U.S. Military, Again," *Foreign Policy,* September 28, 2009.

44. US Embassy Baghdad cable, "Scenesetter for Visit of Vice President Biden to Iraq, September 14–17, 2009," September 12, 2009. See also US Embassy Baghdad cable, "Scenesetter: Maliki Heads to the Washington Investment Conference amid Signs of Promise and Risk," October 16, 2009.

45. US Embassy Baghdad cable, "Scenesetter for Iraqi Vice President Hashimi's Visit to Washington," January 27, 2010; Gordon and Trainor, *The Endgame,* 597.

46. US Embassy Baghdad cable, "KRG President Barzani's Visit to Washington," January 20, 2010.

30. New Administration, Similar Challenges

1. A similar strategy review for Afghanistan and Pakistan was ongoing concurrently.

2. US Department of State cable, "U.S. Policy on Political Engagement in Iraq," April 8, 2009.

3. D3 Systems and KA Research Ltd., "Iraq Poll February 2009," survey conducted for ABC News, the BBC and NHK (2009), http://news.bbc.co.uk/1/shared/bsp/hi/pdfs/13_03_09_iraqpollfeb2009.pdf (accessed May 14, 2019), 5ff.

4. Interviewees A, C, and D; Sky, *The Unraveling,* xi.

5. Interviewee C; US Embassy Baghdad cable, "After the Awakening: Tribes as Government in Anbar?" March 19, 2009; Boghani, "James Jeffrey."

6. In an October 16, 2009, cable to Ambassador Susan Rice, Hill discussed reconciliation only in the context of a $225 million program focused on community leaders. US Embassy Baghdad cable, "Scenesetter for Ambassador Rice's Visit to Iraq," October 16, 2009.

7. US Embassy Baghdad cables, "Sunni Sheikhs Feeling Excluded from the Process," August 29, 2008; "Sunni Political Landscape: Sectarian Versus Secular," March 13, 2009; "PM Maliki Salting Intel Agencies with Dawa Loyalists," February 4, 2010.

8. Interviewees A, C, and D; US Embassy Baghdad cable, "Sons of Iraq (SOI) Program Update," August 2, 2008. Maliki began developing "Support Councils" as rival political organizations to SOI, likely to improve grassroots mobilization in advance of the elections. US Embassy Baghdad cables, "Support Councils: What They Are," November 28, 2008, and "Sunni Arab Insider Warns PM Maliki Will Reignite Insurgency," August 29, 2008.

9. Gordon and Trainor, *The Endgame,* 591.

10. US Embassy Baghdad cable, "Arrest of 'Sons of Iraq' Leader Leads to Fighting in Capital, Rising Tension," March 30, 2009.

11. US Embassy Baghdad cable, "Arrest of Sunni Leader Highlights Challenges on Many Fronts," May 18, 2009, quoted in Gordon and Trainor, *The Endgame,* 591.

12. US Embassy Baghdad cable, "Arrest of Sunni Pc Member Stokes Cries of Election Shaping in Diyala," February 20, 2010, and "PRT Diyala: Election Official Points to Frictions with IHEC," August 27, 2009; Gordon and Trainor, *The Endgame,* 592.

13. Ignatius, "How ISIS Spread in the Middle East."

14. Craig Whiteside, "War, Interrupted, Part I: The Roots of the Jihadist Resurgence in Iraq," *War on the Rocks,* November 5, 2014.

15. Ignatius, "How ISIS Spread in the Middle East"; Whiteside, "War, Interrupted, Part I."

16. Political reconciliation was not mentioned at all, according to the cable reporting about the secretary of state's July 24 meeting with Maliki. US Embassy Baghdad cable, "Secretary Clinton's July 24, 2009 Conversation with Iraqi Prime Minister Nuri al-Maliki," July 30, 2009. The issue was not mentioned in a scenesetter to Vice President Biden in advance of his September 2009 trip to Iraq, US Embassy Baghdad cable, "Scenesetter for Visit of Vice President Biden to Iraq, September 14–17, 2009."

17. US Embassy Baghdad cable, "Strategic Framework Agreement with Iraq," May 22, 2009.

18. US Embassy Baghdad cable, "Prime Minister Maliki's Visit: Launching the Strategic Framework Agreement," July 16, 2009.

19. These were representatives of the Strategic Engagement Cell in MNF-I, who facilitated reconciliation efforts between insurgents and the coalition military. Gordon and Trainor, *The Endgame,* 597–598.

20. Gates, *Duty,* 471; US Embassy Baghdad cable, "Uncertainty about Presidency Council, Presidential Veto Authority after Next Election," January 13, 2010. For a good overview of Iraq's Presidency Council and Council of Ministers, see US Embassy Baghdad cable, "Iraq 201: Iraq's Presidency and Cabinet," March 16, 2009.

21. See, for instance, comments by a Shi'a political leader in US Embassy Baghdad cables, "Election Prospects and U.S. Role in Iraq," February 28, 2010, and "GOI Attempts to Link Referendum on Security Agreement to January 2010 Elections," June 11, 2009.

22. US Embassy Baghdad cables, "Iraqi Election Campaign Update: February 22, 2010," February 22, 2010; "Week One of Election Campaign: Political Roundup," February 18, 2010; "Army Intervention in Salah ad-Din 'Governor' Dispute Locks out Provincial Government," February 7, 2010; "Army Withdraws; Salah ad-Din Government Reopens; Governor Dispute Remains," February 8, 2010; "Ayad Allawi Comments on Iraqi Political Environment as Campaign Season Nears," February 11, 2010; "Political Maneuvering and Iraqi Security Forces," February 13, 2010; and "Scenesetter for Visit of Vice President Biden to Iraq, September 14–17, 2009." For a detailed description of Shi'a political parties, see US Embassy Baghdad cable, "Iraq 201: Shi'a Political Landscape Colored by Internal Rivalry," March 18, 2009.

23. US Embassy Baghdad cable, "Iraqi Election Campaign Update: February 22, 2010."

24. US Embassy Baghdad cable, "Delayed Gratification: Election Law Adopted," December 7, 2009. The Obama administration was beginning to establish a pattern of extracting Kurdish concessions, which created animosity among the Kurds that the United States took them for granted. For further discussion on the election law debate, see US Embassy Baghdad cables, "Stage Is Set for Action on Iraq's Election Law," October 12, 2009, and "Impasse on Kirkuk: Leaders Disagree but Offer Few Solutions to Break Election Law Deadlock," October 24, 2009.

25. US Embassy Baghdad cable, "Election Commission Reviews Preparations Ahead of March Elections," January 8, 2010: "After two major elections in 2009, IHEC shows more confidence in asserting itself as Iraq's election authority, and commissioners show a serious commitment to IHEC's obligation to educate parties and the public about the electoral process." According to a later cable, the IHEC was overwhelmingly Shi'a in make-up and its employees' contracts were for only two to three months. They threatened to strike just before the election in hopes of forcing the government to make them full-time civil servants. US Embassy Baghdad cable, "Election Commission Seeks Civil Service Status, Threatens Strike," December 23, 2009. The short-term contracts and sectarian composition likely heightened the risk of the IHEC being influenced by Maliki and the incumbent government.

26. US Embassy Baghdad cables, "Election Commission Reviews Preparations Ahead of March Elections," January 8, 2010; "De-Escalating the De-Baathification Debate: Post Election Vetting Gains Traction," January 18, 2010; "Sunnis Divided in Reaction to Candidate Disqualifications," February 1, 2010.

27. US Embassy Baghdad cables, "Sunnis Divided in Reaction to Candidate Disqualifications," and "PM Maliki Claims De-Ba'athification Controversy under Control; Predicts Appeals Decisions Within Days," January 31, 2010.

28. In the event, Sunnis did not boycott the election. See US Embassy Baghdad cable, "Iraqi Election Campaign Update: February 22, 2010."

29. US Embassy Baghdad cable, "PRT Muthanna: Muthanna Governor Cancels Meeting with Ambassador, Purportedly at PM Maliki's Insistence," February 15, 2010; Maliki reportedly insisted Iraqi officials not meet with Ambassador Hill due to US concerns over disqualification of candidates under de-Ba'athification law.

30. US Embassy Baghdad cable, "The Vice President Discusses Pre- and Post-Election Iraq with Unami," February 6, 2010.

31. US Embassy Baghdad cable, "VPOTUS Meeting with PM Maliki," February 8, 2010. See also US Embassy Baghdad cables, "DPM Shaways Discusses Arab-Kurd, Regional Integration and Presidency Council Issues with A/S Feltman," February 2, 2010, and "Ambassador Meets with PM Maliki, ISCI Chairman; Parliament Session Postponed," February 8, 2010.

32. Gordon and Trainor, *The Endgame,* 607–614.

33. Ibid., 614–615.

34. US Embassy Baghdad cable, "PM Maliki Salting Intel Agencies with Dawa Loyalists." For an example in Anbar, see US Embassy Baghdad cable, "PRT Anbar: Abu Risha Discusses Political Environment, National Elections," February 4, 2010.

35. Quoted in Gordon and Trainor, *The Endgame,* 615.

36. Interviewee C. On February 25, Secretary of State Clinton issued an instruction cable to the US embassy in Iraq that listed talking points on the election, urging nonviolence, inclusion, and rapid formation of a new government. US Department of State cable, "Talking Points on Iraq Government Formation," February 25, 2010.

37. Timothy Williams and Rod Nordland, "Allawi Victory in Iraq Sets Up Period of Uncertainty," *New York Times,* March 26, 2010; Nordland, "Maliki Contests the Result of Iraq Vote."

38. Gordon and Trainor, *The Endgame,* 615.

39. Nordland, "Maliki Contests the Result of Iraq Vote"; Gates, *Duty,* 472.

40. Ned Parker and Caesar Ahmed, "Maliki Seeks Recount in Iraq Elections," *Los Angeles Times,* March 22, 2010; Nordland, "Maliki Contests the Result of Iraq Vote."

41. Sky, *The Unraveling,* 317.

42. For a detailed discussion, see Kenneth Katzman, "Iraq: Politics, Elections, and Benchmarks," Congressional Research Service, July 1, 2010.

43. Gordon and Trainor, *The Endgame,* 617–620.

44. Interviewee C; Ian Black, "Iraq Election Chaos as 52 Candidates Are Disqualified," *The Guardian,* April 26, 2010; Jomana Karadsheh, "Alleged Baath Ties Disqualify Candidates from Iraqi Elections," CNN, April 26, 2010.

45. Richard Spencer, "Iran Attempts to Broker Shi'a Coalition Government in Iraq," *The Telegraph,* March 31, 2010; Rod Nordland," Iran Plays Host to Delegations after Iraq Elections," *New York Times,* April 1, 2010; Gordon and Trainor, *The Endgame,* 639–640.

46. "Iraqi Court Dismisses De-Baathification Cases," Reuters, May 17, 2010.

47. Interviewees A, C, and D; Kenneth M. Pollack, "Middle East Memo, Number 29," Brookings Institution, January 29, 2015, 14–15.

48. Gordon and Trainor, *The Endgame,* 628–640.

49. For Jeffrey's views on the election, see Boghani, "James Jeffrey."

50. Gordon and Trainor, *The Endgame,* 634, 643–644.

51. Eli Lake, "Obama Bid to Pick Iraq Leader Spurned: Talabani Rebuffs Request to Resign," *Washington Times,* November 10, 2010; Gordon and Trainor, *The Endgame,* 628–635.

52. Gordon and Trainor, *The Endgame,* 646–649. See also US Embassy Baghdad cable, "KRG President Barzani's Visit to Washington." Ambassador Hill asked US officials "from the highest levels of government" to dissuade Barzani on a referendum on Kirkuk. US Embassy Baghdad cable, "Secretary of Defense's Meeting with KRG President Masoud Barzani on December 11, 2009," December 13, 2009.

53. Mohammed Tawfeeq, Jomana Karadsheh, and Arwa Damon, "Iraq Leaders' Deal on Power Sharing Appears to Fall Apart," CNN, November 12, 2010; Michael R. Gordon, "In U.S. Exit from Iraq, Failed Efforts and Challenges," *New York Times,* September 22, 2012.

54. Gates, *Duty,* 472.

55. Ignatius, "How ISIS Spread in the Middle East"; Filkins, "What We Left Behind in Iraq"; Brennan et al., *Ending the U.S. War in Iraq,* 108.

56. Peter Beinart, "Obama's Disastrous Iraq Policy: An Autopsy," *The Atlantic,* June 23, 2014.

57. Gates believed the United States "should and would have a residual military presence in Iraq after the end of 2011 . . . even though that would require a follow-on agreement with the Iraqis," but acknowledges that he should have been more realistic about the prospects given the difficult negotiations on the 2008 SOFA. Gates, *Duty,* 238.

58. Ibid., 501–523.

59. Bob Woodward, *Obama's Wars* (New York: Simon and Schuster, 2010), 247–248; 311–313; Ernesto Londoño and Craig Whitlock, "Syria Crisis Reveals Uneasy Relationship Between Obama, Nation's Military Leaders," *Washington Post,* September 18, 2013; Gates, *Duty,* 563, 573–577.

60. Boghani, "James Jeffrey."

61. Gordon and Trainor, *The Endgame,* 654–658; Gates, *Duty,* 552–553.

62. Peter Baker, "Relief over U.S. Exit from Iraq Fades as Reality Overtakes Hope," *New York Times,* June 22, 2014.

63. Gates, *Duty,* 555.

64. The chairman's role as the president's principal uniformed military advisor is enshrined in the 1986 Goldwater-Nichols Act. Donilon was reportedly outraged and called in a tirade to Under Secretary of Defense Michèle Flournoy demanding she keep better control of the military. Gordon and Trainor, *The Endgame,* 658–660.

65. Gates, *Duty,* 554. In an April 2011 meeting with Maliki, for instance, Gates focused on the needs of Iraqi security forces and the strategy to build Iraqi support for the SOFA.

66. Tim Arango and Michael S. Schmidt, "Should U.S. Stay or Go? Views Define Iraqi Factions," *New York Times,* May 10, 2011; Sahar Issa and Roy Gutman, "Iraq's Maliki Signals He May Let U.S. Troops Extend Their Stay," McClatchy, May 12, 2011.

67. Gordon and Trainor, *The Endgame,* 666; Leon Panetta, *Worthy Fights: A Memoir of Leadership in War and Peace* (New York: Penguin Books, 2014), 356.

68. Brennan et al., *Ending the U.S. War in Iraq,* 13.

69. Some US officials believed that Maliki was trying to increase his leverage by delaying any agreement and believed until the end that the Iraqis would ask the United States to keep forces in Iraq. Brennan et al., *Ending the U.S. War in Iraq,* 104.

70. Interviewees B and K.

71. Gates, *Duty,* 555.

72. Gordon and Trainor, *The Endgame,* 672; Panetta, *Worthy Fights,* 392–399.

73. Gordon and Trainor, *The Endgame,* 671–673. After the Islamic State emerged, the United States sent commandos and trainers to Iraq in 2014 under an executive agreement.

74. Baker, "Relief over U.S. Exit from Iraq Fades as Reality Overtakes Hope."

75. Laith Hammoudi, "Iraq's Maliki Lashes out at Sunni Province Seeking Autonomy," McClatchy, October 29, 2011; Waleed Ibrahim, "Provincial Autonomy Risks Sectarian Rift in Iraq," Reuters, November 24, 2011; Gregg Carlstrom, "The Breakup: More Iraqis Bid for Autonomy," *al Jazeera,* December 22, 2011.

76. Kirk H. Sowell, "Iraq's Second Sunni Insurgency," The Hudson Institute, August 9, 2014.

77. The White House, "Remarks by President Obama and Prime Minister al-Maliki of Iraq in a Joint Press Conference," December 12, 2011.

78. Arwa Damon and Mohammed Tawfeeq, "Iraq's Leader Becoming a New 'Dictator,' Deputy Warns," CNN, December 13, 2011.

79. Beinart, "Obama's Disastrous Iraq Policy."

80. Struan Stevenson, "Outside View: A New Dictator Takes Iraq to the Brink," UPI, December 31, 2014.

81. Sowell, "Iraq's Second Sunni Insurgency."

82. Craig Whiteside, "ISIL's Small Ball Warfare: An Effective Way To Get Back into a Ballgame," *War on the Rocks,* April 29, 2015; Malcolm Nance, *The Terrorists of Iraq: Inside the Strategy and Tactics of the Iraq Insurgency 2003–2014* (Boca Raton, Fla.: Taylor and Francis, 2015); McCants, *The ISIS Apocalypse.*

83. David Ignatius, "James Clapper: We Underestimated the Islamic State's 'Will to Fight,'" *Washington Post,* September 18, 2014.

84. Brennan et al., *Ending the U.S. War in Iraq,* 17. The underestimation of US forces' centrality in preventing sectarian violence was noted by interviewees B, C, and O.

Conclusion to Part VII

1. Kenneth M. Pollack, "Iraq Situation Report, Part I: The Military Campaign Against ISIS," Brookings Institution, March 28, 2016.

2. Freedman, *Strategy,* 23–24.

Part VIII. Implications

1. General David H. Petraeus and Michael O'Hanlon, "America's Awesome Military: And How to Make It Even Better," *Foreign Affairs* 95, no. 5 (September /October 2016): 10–17.

2. Chilcot et al., *The Report of the Iraq Inquiry.*

3. See the critical factors framework in chapter 2.

4. Humphrey et al., eds., *Foreign Relations of the United States, 1964–1968,* vol. 3, *Vietnam,* document 40, "Paper by the Under Secretary of State (Ball)."

5. James Dobbins, "Iraq: Winning the Unwinnable War," *Foreign Affairs* 84, no. 1 (January/February 2005): 16–25.

6. Stephen M. Walt, *The Hell of Good Intentions: America's Foreign Policy Elite and the Decline of U.S. Primacy* (New York: Macmillan, 2018).

7. For a discussion of strategic distance in a different context, see Patrick Porter, *The Global Village Myth: Distance, War, and the Limits of Power* (Washington, D.C.: Georgetown Univ. Press, 2015), 2–9.

31. Iraq and Afghanistan Compared

1. Dobbins et al., *America's Role in Nation-Building.*

2. Sunni Arab resistance reemerged quickly after the US withdrawal, with many groups eventually either supporting ISIS or not standing in their way.

32. Implications for US Foreign Policy

1. Interview with Lieutenant General Terry A. Wolff.

2. Interview with General Stanley A. McChrystal.

3. For a Department of Defense dictionary of military terms, see US Department of Defense, *Joint Publication 1-02: Department of Defense Dictionary of Military and Associated Terms,* 2016.

4. Cathal Nolan, *The Allure of Battle: A History of How Wars Have Been Won and Lost* (New York: Oxford Univ. Press, 2019).

5. Clausewitz, *On War,* 81

6. Colin S. Gray, "Defining and Achieving Decisive Victory," Strategic Studies Institute, US Army War College, April 2002, 13; Colin S. Gray, *Modern Strategy* (Oxford: Oxford Univ. Press, 1999), chapter 1; Richard K. Betts, "Is Strategy an Illusion?" *International Security* 25, no. 2 (fall 2000): 5–50.

7. Gray, "Concept Failure?," 22.

8. US Department of Defense, *Joint Publication 1-02,* 227.

9. John Hampden Jackson, *Clemenceau and the Third Republic* (London: Hodder and Stouton, 1946), 228.

10. Waldrop, *Complexity,* 11–13.

11. See also Dennis Blair, Ronald Neumann, and Eric Olsen, "Fixing Fragile States," *The National Interest* (August 27, 2014).

12. Ladwig, "Influencing Clients in Counterinsurgency"; Michael J. McNerney, Angela O'Mahony, Thomas S. Szayna, Derek Eaton, Caroline Baxter, Colin P. Clarke, Emma Cutrufello, Michael McGee, Heather Peterson, Leslie Adrienne Payne, and Calin Trenkov-Wermuth, *Assessing Security Cooperation as a Preventive Tool* (Washington, D.C.: RAND, 2014); Biddle et al., "Small Footprint, Small Payoff."

13. Anne Barnard, "Syrian Opposition Groups Sense U.S. Support Fading," *New York Times,* February 9, 2016; Carol Morello, "Syria Talks in Switzerland Produce Only a Decision to Keep Talking," *Washington Post,* October 15, 2016.

33. Implications for Scholarship

1. Paul et al., *Paths to Victory*; Libicki, "Eighty-Nine Insurgencies."

2. Colonel Gian Gentile, *Wrong Turn: America's Deadly Embrace of Counterinsurgency* (New York: New Press, 2013).

3. Biddle et al., "Small Footprint, Small Payoff," 8; Bruce Stokes, "Which Countries Don't Like America and Which Do?" Pew Research Center, July 15, 2014.

4. Nagl, *Learning to Eat Soup with a Knife*; Kilcullen, *Counterinsurgency*; Galula, *Counterinsurgency Warfare*; Kolenda, *The Counterinsurgency Challenge.*

5. An important effort to define operational art in counterinsurgency is in Lieutenant General James M. Dubik, "Operational Art in Counterinsurgency: A View from the Inside," Institute for the Study of War, May 2012. He views the operational art as a series of geographic and functional transitions (transfers to host nation lead).

6. Allison and Zelikow, *Essence of Decision.*

7. Komer, *Bureaucracy Does Its Thing.*

8. Rasmussen, *The Military's Business.*

9. Ladwig, "Influencing Clients in Counterinsurgency"; Biddle et al., "Small Footprint, Small Payoff."

10. Keefer, ed., *Foreign Relations of the United States, 1961–1963,* vol. 4, *Vietnam,* document 380, "Telegram From the Department of State to the Embassy in Vietnam, Letter from President Johnson to General Minh," Washington, D.C., December 31, 1963, 745–746.

11. Adam Grissom, "Shoulder-to-Shoulder Fighting Different Wars: NATO Advisors and Military Adaptation in the Afghan National Army," in *Military Adaptation in Afghanistan,* ed. Theo Farrell, Frans Osinga, and James A. Russell (Stanford: Stanford Univ. Press, 2013), 263–287, 276.

12. Huntington, *The Soldier and the State.*

13. Eliot A. Cohen, *Supreme Command: Soldiers, Statesmen, and Leadership in Wartime* (New York: Free Press, 2002), 208–224.

14. Feaver, *Armed Servants.*

15. Ibid., location 3893–3902 of 5012.

Acknowledgments

Success has many parents; failure is an orphan. The extent that this book succeeds is due to the extraordinary support of many people.

Professor Theo Farrell, when he chaired the King's College War Studies Department, created a fellowship for a practitioner to teach a course on contemporary war and work toward a doctorate. I was fortunate to have been selected. Theo was a terrific thesis adviser and has become a dear friend. His support, probing questions, and candid feedback were essential to having a successful doctoral thesis and to turning it into a book. Professors David Dunn and Patrick Porter provided me with wonderful advice during my thesis defense, which sharpened the arguments in the book. I am grateful to my sons, Mike, Zach, and Jake, who supported my move to London for the King's College Fellowship and for their love and encouragement during this process.

I have had the good fortune to have worked with many exceptional leaders and mentors who provided me opportunities to serve and make a difference. These wonderful people include John Allen, Stephen Biddle, Paolo Cotta-Ramusino, James Cunningham, Jim and Leslie Cunningham, Jim Dubik, Joseph Dunford, Michèle Flournoy, Ashraf Ghani, Eklil Hakimi, Ibrahim, Jeh Johnson, Fred and Kim Kagan, Graeme Lamb, Peter Lavoy, Doug Lute, Norine MacDonald, Carter Malkasian, Stanley McChrystal, James Miller, Greg Mortenson, Ron Neumann, John "Mick" Nicholson, Mike O'Hanlon, David Petraeus, Chip Preysler, Robin Raphel, Barney Rubin, Greg Ryckman, David Sedney, Michael Semple, Vikram Singh, Masuda Sultan, Wakil, Tony Wayne, Andrew Wilder, Robert Williams, Terry Wolff, Joe Votel, and elders from Naray, Kamdesh, and Ghaziabad, and interlocutors whom I will not name out of concern for their safety. I know that I will look back on this list and be saddened that I missed someone

I should have mentioned. I am grateful for the cavalry paratroopers of 1-91 CAV, 173rd Airborne (Task Force Saber), for your courage and professionalism in showing a better way forward in Afghanistan. Our troopers David Boris, Thomas Bostick, Ryan Fristche, Adrian Hike, Jacob Lowell, and Chris Pfeifer were killed in action fighting for their comrades and country. I hope this book helps America make more worthy policy and strategy.

Chiara Libiseller, a doctoral candidate at King's College, has been a phenomenal research associate. We first met when she was a student in my contemporary war class. She impressed me with her intellectual curiosity, attention to detail, and analytical rigor. I needed all of that help to turn a clunky doctoral thesis into a readable volume, so I asked her if she could support. She agreed, thankfully, and the quality of this book would not have been possible without her candor, acumen, and professional excellence. The errors and shortcomings in this book are all mine. Chiara reduced them significantly. I could not have asked for a better associate. I also want to thank Emily Ashbridge, Caroline Bechtel, and Helene Olsen for their superb assistance. I am indebted to Drs. Stephen Biddle and Carter Malkasian for helping me focus the arguments in the introduction and conclusion.

The University Press of Kentucky has been a wonderful publisher every step of the way. Natalie O'Neal has guided me carefully through this process and has been fun to work with. Ila McEntire, the supervising editor, and Brooke Raby, Ashley Runyon, and Jewell Boyd in marketing have been fabulous. Derik Shelor has had the unenviable copyediting task, which he has performed masterfully.

Finally, I want to thank my wife, Nicole Kauss, for her understanding and support through the Ph.D. process and the work to turn it into a book. She is a former intelligence community official who has deployed to combat zones several times and spent her career providing vital information to commanders and policymakers. Our debates and discussions about the issues have sharpened my arguments. She's been my champion through the entire process and has made working on it a joy.

Index

1st Marine Division, 185

Abdullah, Foreign Minister Abdullah, 62, 77, 101, 152, 160, 162
Abizaid, General John P., 75, 85, 199
Accountability and Justice Commission, 235–238
adverse selection, 92. *See also* patron-client relationship
Afghan civil war, 52–53
Afghan Development Zone (ADZ), 90
Afghan Interim Transitional Administration, 59
Afghanistan-Pakistan Annual Review (APAR), 115–116, 134
Afghan National Army (ANA), 71–72, 91, 96, 268–269
Afghan Transitional Administration (ATA), 39, 59, 63, 75
Agha, Mohammad Tayyab, 40, 54, 134, 143–145, 149–151, 156
Akhund, Mullah Obaidullah, 40
al-Bolani, Jawad, 205
al-Dulaimi, Minister of Defense Saadoun, 204–205
al-Hakim, Abd al-Aziz, 206
al-Hashimi, Vice President Tariq, 235, 241
al-Jaafari, Prime Minister Ibrahim, 188, 203–204, 206
Allawi, Prime Minister Ayad, 181, 184, 187, 204, 236–238
al-Maliki, Nouri Kamal, 190, 203–206, 208–209, 216, 221, 226–230, 234–241, 243
al Qaeda, 41, 43, 47, 53–55, 62, 64–65, 95, 103, 143–145, 164. *See also* bin Laden, Osama
al Qaeda in Iraq [AQI], 58, 215–216, 221–222, 225–226, 234. *See also* al-Zarqawi, Abu Musab; Tawhid wal-Jihad
al-Sadr, Muqtada, 182–183, 185–186, 216
al-Zarqawi, Abu Musab, 185, 197
American Revolution, 26, 31
Anbar Awakening, 216, 226, 234
Anbar province, Iraq, 185, 187–188, 190, 216, 241
Arab Spring, 120, 239
Austin, General Lloyd, 238–240

Ba'ath Party, 180–181, 184. *See also* de-Ba'athification
Baghdad, Iraq, 172, 174, 177, 181, 183, 185, 189, 204–205, 213, 234–235, 238
Bagram Air Base, 41, 74, 77, 157
Ball, Under Secretary of State George, 246
Baluchistan province, Pakistan, 50, 65
Baradar, Mullah Abdul Ghani, 15, 146–147
bargaining asymmetries, 128–129, 250, 262, 269; in Afghanistan, 165; in Iraq, 227
Barzani, Massoud, 179, 235, 238
Benson, Colonel Kevin, 172–173
Bergdahl, Sergeant Bowe, 149–150, 161
Biddle, Stephen, 92–94, 112, 223
Biden, Vice President Joe, 101, 175, 212, 236

bin Laden, Osama, 52–53, 55, 64, 119. *See also* al Qaeda
Blackwater (private US security company), 185
Blackwill, Robert, 152, 184
Blair, Tony, 173
Bolger, Lieutenant General Daniel F., 2–3, 44
Bonn Agreement, 63
Bonn Conference, 43, 59–62
Bonn II Conference, 152
Bonn process, 62–66, 77, 78, 142
Brahimi, Lakhdar, 76–77
Bremer, L. Paul, 180–181, 184–185
bureaucratic friction, 6–7, 10, 69, 148–149, 249, 259–261, 267; in Afghanistan, 69–70, 78, 83–84, 100, 104–05, 117; in Iraq, 200; in Vietnam, 20
Bush, President George W., 45, 47, 70, 94–95, 171, 187, 209, 212–213, 215, 223, 227–229

Casey, General George, 187, 191, 200, 203–204, 212–213
Central Command, US (CENTCOM), 85, 100, 139, 172, 174, 177, 199
Chalabi, Ahmed, 180–1
Charge of the Knights operation, 15, 209, 216
Chayes, Sarah, 44, 86, 96
Cheney, Vice President Richard, 46, 193, 199
Chilcot Report, 173–174, 245
Clinton, Secretary of State Hillary, 134–136, 226, 231–232. *See also* US Department of State
Coalition Provisional Authority (CPA), 167–168, 180–181, 183–185, 187, 193, 260
Combined Forces Land Component Command (CFLCC), 172–177
Combined Security Training Command—Afghanistan (CSTC-A), 96
complexity (in war), 57–58
confirmation bias, 83, 196, 258; in Afghanistan, 85, 95, 115, 122, 143; in Iraq, 200
counterinsurgency, 18, 32–36, 73, 92, 168, 213, 264–267, 268
Cowper-Coles, Sir Sherard, 130
CPA Order Number 1. *See* de-Ba'athification
CPA Order Number 2. *See* Iraqi Army
Crocker, Ambassador Ryan, 195, 215–216, 229–230
Cuban Missile Crisis, 18–19, 148

de-Ba'athification, 167–168, 180–183, 188–189, 193, 238
Defense Intelligence Agency, 90, 190
Democratic Republic of Vietnam (DRV), 18–21
Desert Crossing exercise, 177–178
Dobbins, Ambassador James "Jim," 76, 129, 159, 168–169, 176, 181–182, 246
Doha, Qatar, 125, 145–147, 149–150, 156–157, 159, 161, 164
Dubik, Lieutenant General James, 96, 200, 216–217
Durand Line, 50, 55

Eikenberry, Lieutenant General Karl, 90, 101, 105
Electoral Complaints Commission, 101

Fahim, Mohammad Qassim, 59–60, 62, 71–74, 86
Fallujah, Iraq, 182, 185, 241
Flournoy, Under Secretary of Defense for Policy Michèle A., 102, 115
Franks, General Tommy, 64, 172–177, 193
Future of Iraq Project, 178

Gates, Secretary of Defense Robert "Bob," 99–102, 107, 115, 213, 226–227, 238, 240. *See also* US Department of Defense
Ghani, President Ashraf, 11, 160, 162–164
Global War on Terrorism, 13, 43, 47
Good Friday Agreement, 26, 128
Government of National Unity, 11, 162–163, 206

Grossman, Marc, 136–137, 139–140, 150–151, 153, 156
Group of Eight (G8), 70–71
guerrilla warfare, 18, 31, 68, 127, 246
Gulf War (1991), 5, 17, 72

Hadley, National Security Advisor Stephen, 208–209, 213–214, 227–228
Hagel, Secretary of Defense Chuck, 161, 175, 201
Hamas, 47, 161
Haqqani network, 123, 146, 161–162
Harvey, Derek, 184–185, 195
Hekmatyar, Gulbuddin, 41, 51–54, 61
Helmand province, Afghanistan, 39, 54, 108, 121, 129
High Peace Council (HPC), 142, 152, 157–158
Hill, Ambassador Chris, 230, 233, 235–237, 243
Hizb-i-Islami Gulbuddin (HiG), 12, 51–52, 61–62, 64–65
Holbrooke, Richard, 99, 101, 105–106, 110, 115, 135–136, 140
Huntington, Samuel, 93, 207, 269–270
Hussein, Saddam, 167, 171–175, 184

Ibrahim, Haji, 40
Ignatius, David, 182, 185, 234
Interim Iraqi Government (IIG), 184, 187
Internal Look war game, 174–175
International Security Assistance Forces—Afghanistan (ISAF), 74, 99–100, 111, 113, 119–121, 123, 139, 150
Inter-Services Intelligence (ISI) (Pakistan), 145–146
Iraqi Army, 175, 199, 205, 231; disbanding of, 168, 181–182, 193
Iraqi Governing Council (IGC), 180–181, 183–184
Iraqi National Alliance (INA), 237
Iraqi Security Forces (ISF), 185, 187, 190, 203–204, 208–209, 216–217, 227, 234, 251
Iraqiya Party, 236–238
Iraq Study Group, 212–213
Islamabad Accords (1993), 52
Islamic Emirate of Afghanistan, 54, 144, 150, 155, 158–159
Islamic State of Afghanistan (ISA), 40, 52–55, 60, 62, 64–65, 66, 142
Islamic State of Iraq and Syria (ISIS), 11, 217, 221–222, 251, 265

Jabr, Bayan, 188, 203
Jaish al-Mahdi (JAM), 182–183, 186, 190–191, 204
Jamiat-e-Islami Afghanistan Party, 51, 53, 142

Kabul Bank, 114–115
Kandahar province, Afghanistan, 39, 46, 53–54, 89, 99, 108
Karzai, Hamid, 39–41, 43, 61–63, 75, 80, 85–86, 89–90, 101, 119, 123, 141–142, 149–150, 152, 155–159, 160, 252
Keane, General Jack, 212–213
Khalilzad, Ambassador Zalmay, 61, 84, 164, 189–190
Khan, Mohammad Daoud, 51
kleptocracy, 86–87, 96–97, 124, 251, 352–353
Korean War, 26–27

loss aversion, 196, 210–211, 259

Malkasian, Carter, 76, 129, 162
Maples, Lieutenant General Michael D., 90
Marjah, Afghanistan, 108
Massoud, Ahmad Shah, 52–53, 55, 60–61
Mattis, Secretary of Defense Jim, 185
McChrystal, General Stanley A., 58, 108–110, 255; assessment, 99–101
McKiernan, General David, 99, 174
McNamara, Defense Secretary Robert S., 18–21
mujahideen, 39–41, 51–52
Mullen, Chairman of the Joint Chiefs of Staff Admiral Michael, 117, 119, 240
Multi-National Force—Iraq (MNF-I), 187, 226, 232

Nasr, Vali, 106, 129
National Security Council (NSC), 84–85, 102, 120, 134–135, 148, 213, 215, 227, 237, 258–259
nation-building, 45, 70
Nixon, Richard, 21–22
Northern Alliance, 39, 46, 54–55, 60–62, 63–66, 76

Obama, Barack, 81–82, 99, 101–103, 105–107, 120, 122–125, 135, 157–159, 221–222, 226, 235, 238–241, 251
Odierno, General Ray, 229, 235, 237–238
Office of Reconstruction and Humanitarian Assistance (ORHA), 179
Omar, Mullah Mohammad, 39–41, 54, 144, 151, 155
Operation Cobra II, 174
Operation Enduring Freedom, 43, 46, 164
Operation Hamkari (Afghanistan), 108
Operation Moshtarak (Afghanistan), 108
Operation Vigilant Resolve, 185

patron-client relationship(s), 92–94, 225, 261, 267–269; in Afghanistan, 96–97, 118, 123, 251; in Iraq, 208, 226, 242–243; in Vietnam, 21, 141. *See also* kleptocracy
Pentagon. *See* US Department of Defense
Peshawar Accords of 1992, 52–53
Peshawar Seven, 51
Petraeus, General David H., 111–112, 135–136, 181, 215, 216–217, 223, 226–227
Phase IV planning (Operation Enduring Freedom), 173, 175–177
planning fallacy, 49, 192
Powell, Secretary of State Colin L., 70, 76–77, 192–193. *See also* US Department of State
principal-agent theory. *See* patron-client relationship(s)
Provincial Reconstruction Teams (PRTs), 84–85

Rabbani, President Burhanuddin, 51–52, 152
Raphel, Ambassador Robin L., 53, 164
Reidel, Bruce, 99
Rice, Secretary of State Condoleezza, 213, 228. *See also* US Department of State
Rohrabacher, Congressman Dana, 75
Rumsfeld, Secretary of Defense Donald, 46, 70, 73, 75, 77, 80, 84, 171–172, 176–178, 180, 191, 199. *See also* US Department of Defense

Sánchez, Lieutenant General Ricardo, 185
Sayyaf, Abdul Rasul, 40, 51, 53
Scowcroft, Brent, 178
Shah, King Zahir, 51, 61
Shura-e-Nazar (Supreme Council of the North). *See* Massoud, Ahmad Shah; Northern Alliance
Sistani, Ayatollah Ali, 183, 188
Sons of Iraq (SoI), 225, 234
Soviet-Afghan War, 51–52
Spanta, Afghan National Security Advisor Dadfar, 158
Special Representative for Afghanistan and Pakistan (SRAP), 99, 136, 139–140. *See also* Holbrooke, Richard
Status of Forces Agreement (SOFA), 226–229, 239–240
Strategic Framework Agreement (SFA), 226, 228, 235
Supreme Council for the Islamic Revolution in Iraq (SCIRI), 180, 206

Talabani, Jalal, 206, 228, 237
Tawhid wal-Jihad, 183–184, 185
Tehrik-e-Taliban Pakistan (TTP), 146
Third Army (US), 172–177

UN Assistance Mission for Iraq (UNAMI), 231–232
United Iraqi Alliance, 188
US Central Command (CENTCOM), 172, 174, 177
US Department of Defense, 4–5, 67–69, 115–116, 118–119, 139–140, 148, 153–154, 169, 174–175, 178–179, 212,

215. *See also* Gates, Secretary of Defense Robert "Bob"; Rumsfeld, Secretary of Defense Donald
US Department of State, 68–69, 105, 148, 178. *See also* Clinton, Secretary of State Hillary; Powell, Secretary of State Colin L.; Rice, Secretary of State Condoleezza

von Clausewitz, Carl, 3, 26, 28–29, 257

weapons of mass destruction (WMD), 47, 167, 171
Wolfowitz, Deputy Secretary of Defense Paul D., 175, 177

Zinni, Anthony, 177–178